Diagnosing Musculoskeletal Conditions using Artificial Intelligence and Machine Learning to Aid Interpretation of Clinical Imaging

Diagnosing Musculoskeletal Conditions using Artificial Intelligence and Machine Learning to Aid Interpretation of Clinical Imaging

Edited by

Rakesh Kumar

Department of Computer Science and Engineering, Chandigarh University, Gharuan, Punjab, India

Meenu Gupta

Department of Computer Science and Engineering, Chandigarh University, Gharuan, Punjab, India

Ajith Abraham

School of Artificial Intelligence, Bennett University, Greater Noida, Uttar Pradesh, India

Paul Antony

Department of Orthopedics, Apollo Super Speciality Hospital, Muscat, Oman

ELSEVIER

ACADEMIC PRESS

An imprint of Elsevier

Academic Press is an imprint of Elsevier
125 London Wall, London EC2Y 5AS, United Kingdom
525 B Street, Suite 1650, San Diego, CA 92101, United States
50 Hampshire Street, 5th Floor, Cambridge, MA 02139, United States

Notices

ISBN: 978-0-443-32892-3

For information on all Academic Press publications visit our website at https://www.elsevier.com/books-and-journals

Publisher: Stacy Masucci
Acquisitions Editor: Elizabeth Brown
Editorial Project Manager: Sara Pianavilla
Production Project Manager: Omer Mukthar
Cover Designer: Mark Rogers

Typeset by TNQ Technologies
Transferred to Digital Printing 2024

Contents

Contributors

Ketema Adere
Computer Science and Engineering, Adama Science and Technology University, Adama, Ethiopia

Paul Antony
Department of Orthopedics, Apollo Super Speciality Hospital, Muscat, Oman

Ritesh Bhat
Department of Mechatronics Engineering, Rajalakshmi Engineering College, Thandalam, Tamil Nadu, India

John Blake
University of Aizu, Aizuwakamatsu, Japan

Mahsa Borhani Peikani
University of Nevada Reno, Physiology and Cell Biology, Reno, NV, United States

Seyed Buhari
Universiti Teknologi Brunei, School of Business, Bandar Seri Begawan, Brunei

Eshita Dey
Department of Artificial Intelligence & Machine Learning, Techno International New Town, Kolkata, West Bengal, India

Ricardo Fitas
Department of Mechanical Engineering, Technical University of Darmstadt, Hesse, Germany

Krishnendu Ghosh
Department of Computer Science & Engineering, Indian Institute of Information Technology Dharwad, Dharwad, Karnataka, India

Shilpa Gite
Symbiosis Institute of Technology, Symbiosis International (Deemed University), Pune, Maharashtra, India; Symbiosis Centre for Applied AI, Symbiosis International (Deemed University), Pune, Maharashtra, India

Shubham Gupta
Model Institute of Engineering and Technology, Jammu, Jammu and Kashmir, India

Meenu Gupta
Department of Computer Science and Engineering, Chandigarh University, Gharuan, Punjab, India

K.S. Hemanth
Department of Computer Science, Christ University, Bangalore, Karnataka, India

Sailesh Suryanarayan Iyer
Rai School of Engineering, Rai University, Department of Computer Science and Engineering, Ahmedabad, Gujarat, India

Anupriya Jain
School of Computer Applications MRIIRS, Faridabad, Haryana, India

Balaram Yadav Kasula
University of the Cumberlands, Williamsburg, KY, United States

Harashleen Kour
Model Institute of Engineering and Technology, Jammu, Jammu and Kashmir, India

Rakesh Kumar
Department of Computer Science and Engineering, Chandigarh University, Gharuan, Punjab, India

Vishal Kumar
Department of Orthopaedics, Postgraduate Institute of Medical Education and Research, Chandigarh, India

Vidya Lakshmi
REVA University, School of Computer Science and Applications, Bengaluru, Karnataka, India

Chandrakant Mahobiya
Anjaneya University, Raipur, Chhattisgarh, India

Md Mehedi Hassan
Computer Science and Engineering Discipline, Khulna University, Khulna, Bangladesh

Nithesh Naik
Department of Mechanical and Industrial Engineering, Manipal Institute of Technology, Manipal Academy of Higher Education, Manipal, Karnataka, India

Munmun Patra
Department of Artificial Intelligence & Machine Learning, Techno International New Town, Kolkata, West Bengal, India

Chander Prabha
Chitkara University Institute of Engineering and Technology, Chitkara University, Punjab, India

Rajesh Sharma R
Computer Science and Engineering, Alliance University, Bengaluru, Karnataka, India

Chinnaiyan R
Lingaya's Vidyapeeth University, Faridabad, Haryana, India

Seyed Mahmoud Sajjadi Mohammadabadi
University of Nevada Reno, Computer Science and Engineering, Reno, NV, United States

Vajjiram Santhanam
Department of Mechatronics Engineering, Rajalakshmi Engineering College, Thandalam, Tamil Nadu, India

Ketan Sarvakar
Ganpat University, Information Technology, Mehsana, Gujarat, India

Karuppannan Sekar
Department of Artificial Intelligence and Machine Learning, Rajalakshmi Engineering College, Thandalam, Tamil Nadu, India

Palvi Sharma
Department of Computer Science & Engineering, Model Institute of Engineering and Technology, Jammu, Jammu and Kashmir, India

Seema Sharma
School of Computer Applications MRIIRS, Faridabad, Haryana, India

Bahiru Shifaw
Computer Science and Engineering, Adama Science and Technology University, Adama, Ethiopia

Swetza Singh
School of Pharmacy, BBD University, Lucknow, Uttar Pradesh, India

Akey Sungheetha
Computer Science and Engineering, Alliance University, Bengaluru, Karnataka, India

Ali Talyshinskii
Department of Urology, Astana Medical University, Astana, Kazakhstan

Vamakshi Thaker
School of Pharmacy, BBD University, Lucknow, Uttar Pradesh, India

Deepika Tigulla
REVA University, School of Computer Science and Applications, Bengaluru, Karnataka, India

Shivam Verma
School of Pharmacy, BBD University, Lucknow, Uttar Pradesh, India

Pawan Whig
Vivekananda Institute of Professional Studies-TC, Delhi, New Delhi, India

Nikhitha Yathiraju
University of the Cumberlands, Williamsburg, KY, United States

Preface

Bone deformations can lead to many musculoskeletal disorders, significantly affecting an individual's quality of life. These conditions often result in pain, reduced mobility, and decreased functionality, influencing daily activities and overall well-being. Early and accurate detection of bone deformations is crucial for effective medical intervention, because timely diagnosis can lead to better treatment outcomes and improved patient prognosis. This book proposes a comprehensive approach that integrates artificial intelligence (AI) and machine learning (ML) techniques to identify bone deformations promptly and accurately. By leveraging advanced image analysis and pattern recognition, this approach aims to revolutionize the field of orthopedic diagnostics. AI and ML offer powerful tools for analyzing complex medical images, detecting subtle abnormalities, and predicting the progression of musculoskeletal conditions. The integration of these technologies promises to enhance the precision and efficiency of diagnostic processes, ultimately leading to better patient care.

In this edited volume, we aim to cover the myriad challenges, cutting-edge technologies, diverse applications, and future trends in the use of AI and ML for bone deformation analysis. The book is structured to provide a holistic view of the field, encompassing theoretical foundations, innovative techniques, and practical implementations. Our goal is to present a comprehensive resource that will benefit researchers, clinicians, and practitioners by offering insights into the latest advances and practical solutions for early prediction and management of bone deformations. We begin with an exploration of the fundamental challenges in detecting and diagnosing bone deformations, highlighting the limitations of traditional methods and the need for advanced technological interventions. Subsequent chapters delve into state-of-the-art AI and ML frameworks, showcasing how these technologies can be harnessed to overcome existing barriers. We present detailed discussions on various methodologies from deep learning models for image analysis to sophisticated algorithms for pattern recognition and anomaly detection.

Practical applications are a focal point of this book, with case studies and real-world examples demonstrating the implementation of AI and ML in clinical settings. These chapters provide valuable insights into how theoretical concepts are translated into practice, offering a road map for integrating these technologies into everyday diagnostic workflows. This book represents a collaborative effort by leading AI, ML, and orthopedic research experts. We are grateful to the contributors who have shared their knowledge and expertise, making this volume a valuable addition to the existing literature. We hope this book will serve as a comprehensive guide and inspire further research and innovation in bone deformation diagnostics. We invite readers to explore the advances presented in this book and consider how AI and ML can be applied to improve the diagnosis and treatment of musculoskeletal disorders. By fostering a deeper understanding of these technologies and their potential, we aim to contribute to the ongoing evolution of orthopedic care, ultimately enhancing the quality of life for individuals affected by bone deformations.

The book is composed of 14 chapters contributed from diverse researchers across the world. Chapter 1, "**An Introductory approach for Bone deformities, Osteoarthrosis, Osteoporosis, and Spondylosis of Spine using Machine Learning**," provides a systematic review for the identification of musculoskeletal problems using Preferred Reporting Items for Systematic Review and Meta-Analysis guidelines. This chapter considered three databases: IEEE Xplore, Scopus, and Web of Science. The search query is applied to database using keywords such as "bone deformation," "osteoporosis," "osteoarthrosis," "spondylosis," and "machine learning." A total of 970 articles were extracted from the database and 40 articles were selected after rigorous analysis. In addition, important investigations are highlighted based on the literature review, offering insights for future research directions. Possible solutions to these investigations are also provided in this chapter. Chapter 2, "**Identification and Classification of Musculoskeletal Conditions Using Artificial Intelligence and Machine Learning**," discusses the latest advances in AI-driven diagnostics, noting that these algorithms could enable more accurate medical imaging analysis than before, supporting the early detection and accurate diagnosis of arthritis and fractures. The chapter also discusses the advancement of personalized treatment plans after the deep data analysis that AI affords, emphasizing the realization of better patient outcomes. Chapter 3, "**Early-stage detection of osteoarthritis of the joints (hip and knee) using Machine Learning**," examines to what extent ML can be used to bring about radical changes. It fuses numerous algorithms and uses large datasets to place importance on the selection of relevant features and data preparation that will increase model accuracies. The use of ML has been linked to many ethical issues, including safeguarding data privacy and ensuring the accountability of employed models. Chapter 4, "**Bone cancer classification and detection using Machine Learning technique**," delves into the classification of bone cancer, focusing on leveraging AI techniques for precisely and effectively identifying various bone disease types using clinical imaging and medical data. It explores a range of supervised and unsupervised AI algorithms, addressing critical challenges such as data imbalance, feature selection, and model interpretability. Chapter 5, "**Identification of the risk of osteoporosis in older Vietnamese women using Artificial Intelligence and Machine Learning**," develops and evaluates several ML models for predicting osteoporosis risk using data collected from Hanoi Medical University Hospital in Vietnam. The dataset includes information on patient demographics, lifestyle factors, medical history, and specific blood test parameters. The primary objective is to leverage data from the Osteoporosis Self-assessment Tool to create tools to assess the likelihood of osteoporosis in women aged 50 years and greater in Northern Vietnam. A secondary objective is to validate the accuracy of the Ho-Pham et al. and the Osteoporosis Self-Assessment Tool (OSTA) models by incorporating additional factors such as height, area, and various blood test results. Chapter 6, "**Identification of Paget's Disease in the Human Body Using Artificial Intelligence and Machine Learning**," discusses the potential uses of AI and ML, including treatment planning, quantitative evaluation, improved visualization, early detection, and research developments. Medical image analysis using AI algorithms can detect unique characteristics of bone remodeling, predict fracture risk, assess treatment outcomes, and provide novel insights into disease pathophysiology. Chapter 7, "**Identification and Classification of Rheumatoid Arthritis using Artificial Intelligence and Machine Learning**," explores the identification and classification of rheumatoid arthritis (RA) using advanced AI and ML techniques. This chapter examines the field of AI, particularly ML applications for RA classification, showing its potential to affect clinical practices significantly and enhance patient health. Chapter 8, "**Early detection approach for Analysis of Osteoarthritis using Artificial Intelligence and Machine Learning**," focuses on methods used to detect osteoarthritis (OA) early using ML and AI approaches. Datasets from imaging data, patient medical histories, biomarkers, and so on are employed by ML and AI models to detect early symptoms of OA before severe symptoms emerge. The methodology uses emerging ML algorithms to predict and draw insights from intricate behavior present in the data, leading to prompt diagnosis and intervention. This chapter presents the importance of merging AI and ML in health care, offering a path for further research in the early identification of OA. Chapter 9, "**Pre-analysis of ankylosing spondylitis using Machine Learning**," explores the use of ML for the preanalysis and early detection of ankylosing spondylitis (AS). This chapter offers information on its application at various phases of AS diagnosis and therapy and describes the various methods employed, datasets, contributions, and limits of previous work. Chapter 10, "**Analysis and identification of gout flares using Machine Learning**," addresses the underlying mechanisms of frequent gout flares compared to infrequent ones using metabolomics. The development of the early detection of acute gout is also covered in this chapter. This is important because it allows medical professionals to provide targeted pain relief interventions quickly and stop the disease from progressing, which guarantees proven long-term joint function. The potential for prompt detection of gout flares based on nurses' primary observations of patient complaints in the emergency room was thoroughly investigated in this study. It also explores the creation of prediction models for the risk of hospitalization associated with flare-ups in gout patients. Chapter 11, "**Impact of Amyloidosis on Bones and its relation with dementia**," illustrates common manifestations of osteoporosis, fractures, joint pains, amyloid arthropathy, and spinal involvement and sheds light on rarer symptoms. It also investigates the link between amyloidosis and dementia, emphasizing the role of amyloid-beta proteins in developing neurodegenerative disorders. Chapter 12, "**Real-world application of Machine learning in Bone Deformities identification**," concludes with an in-depth exploration of how ML is adopted to identify bone defects in real-world scenarios. It highlights the integration of medical informatics with the diagnosis of bone deformities, demonstrating the transformative potential of ML to enhance clinicians' assessments of skeletal abnormalities. By promoting further research, this chapter underscores the significant impact of these technologies on improving diagnostic accuracy and patient outcomes in orthopedics. Chapter 13, "**Artificial Intelligence and Machine Learning for Foot and**

Ankle Disorders," helps to improve the efficiency of health care delivery and patient outcome by enabling providers to offer truly patient-centered care tailored to individual needs. Chapter 14, "**Conclusion: A future perspective of Diagnosing Musculoskeletal Conditions Using Artificial Intelligence And Machine Learning**," offers a forward-looking perspective on diagnosing musculoskeletal conditions using AI and ML. It discusses the transformative potential of these technologies in enhancing diagnostic accuracy, enabling early detection, and personalizing treatment plans.

We hope this book empowers both healthcare professionals and patients on the journey of improved diagnosis and management of musculoskeletal conditions through AI-driven insights. We invite you to delve deeper and explore the transformative possibilities that AI-based interventions offer for the treatment of musculoskeletal conditions and overall patient well-being.

Rakesh Kumar

Meenu Gupta

Ajith Abraham

Paul Antony

CHAPTER

An introductory approach to bone deformities, osteoarthrosis, osteoporosis, and spondylosis of spine using machine learning

1

Palvi Sharma[1], Rakesh Kumar[2], Meenu Gupta[2], Vishal Kumar[3]

[1]*Department of Computer Science & Engineering, Model Institute of Engineering and Technology, Jammu, Jammu and Kashmir, India;* [2]*Department of Computer Science and Engineering, Chandigarh University, Gharuan, Punjab, India;* [3]*Department of Orthopaedics, Postgraduate Institute of Medical Education and Research, Chandigarh, India*

1.1 Introduction

The human body's bones serve as its structural framework, enabling mobility, support, and protection. The size, shape, and density of bones vary because they are made of a matrix of minerals and collagen fibers. Long bones such as the femur serve as support and provide mobility whereas flat bones such as the cranium shield internal organs (Aslam Khan et al., 2023). However, disturbances in bone structure that can arise from birth defects, injuries, infections, or metabolic imbalances may disturb the balance and affect any segment of the skeletal system, such as the long bones, spine, pelvis, or joints (Yadahalli et al., 2021). Furthermore, osteoporosis, spondylosis, osteoarthrosis (OA), and bone deformities are important musculoskeletal conditions that affect the quality of life of millions of people globally, as shown in Fig. 1.1. Despite variations in each condition's presentation, they are all related by their impact on the health of bones and structural integrity. Developing successful diagnostic and treatment strategies requires an understanding of these connections. Bone deformity refers to any abnormality in the position, structure, or shape of bones. These abnormalities are frequently the consequence of birth defects, trauma, infections, or developmental disorders. In contrast, osteoporosis is characterized by a reduction in bone density and increased stiffness, which increases the risk of fractures, particularly in elderly people (Sasirekha et al., 2023). Osteoarthrosis is a degenerative disease of the joints that mostly affects cartilage, causing pain, stiffness, and reduced mobility, especially in the knees and hips, which bear weight (Brahim et al., 2019). During movement, cartilage functions as a cushion, enabling bones

Diagnosing Musculoskeletal Conditions using Artificial Intelligence and Machine Learning to Aid Interpretation of Clinical Imaging
https://doi.org/10.1016/B978-0-443-32892-3.00001-4

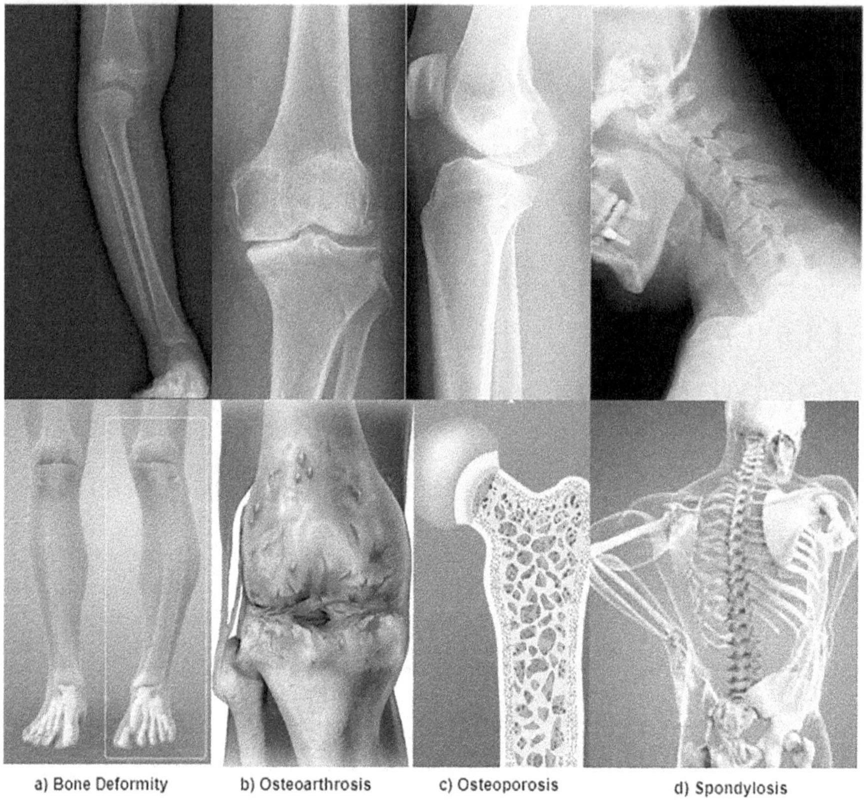

FIGURE 1.1

Samples of bone deformity, osteoarthrosis, osteoporosis, and spondylosis.

to move easily over one another. This cartilage progressively decreases and wears away in OA, causing bone-on-bone contact. This results in inflammation of the afflicted joints, causing discomfort, stiffness, swelling, and decreased range of motion (Bowes et al., 2021). Multiple variables lead to the emergence of osteoarthrosis. Age is a major variable because cartilage degradation can result from joints that have been subjected to wear and tear over time. Genetics also have a role because certain people inherit genes that make them more prone to developing OA. Joint trauma or traumas such as fractures or torn ligaments may accelerate the development of osteoarthrosis in afflicted joints. Furthermore, obesity increases the chance of developing osteoarthrosis by placing additional strain on weight-bearing joints such as the knees and hips (Bonakdari et al., 2021; Schiratti et al., 2021). Furthermore, the bone condition known as osteoporosis is characterized by a loss in bone quality and density that raises the risk of fractures. It happens when the body produces too little bone, loses too much bone, or does both, leaving the bones brittle and weak (Kerketta and Ghosh, 2021). Osteoporosis usually develops silently, showing no signs until a fracture happens, usually to the wrist, hip, or spine. Increasing age, heredity, hormonal shifts (such as women experiencing menopause), specific medical disorders (such as rheumatoid arthritis or hyperthyroidism), inadequate calcium or vitamin D intake, a sedentary lifestyle, smoking, excessive alcohol intake, and long-term use of certain medications (such as glucocorticoids) are risk factors for osteoporosis. Bone density tests, such as dual-energy x-ray absorptiometry (DXA), are commonly used to make the diagnosis (Bui et al., 2022; Albuquerque et al., 2023). Spondylosis is characterized by degenerative changes in the spine, including bone spur formation and spinal canal compression, which result in pain, stiffness, and neurologic problems (Wang et al., 2018). The main factor contributing to spondylosis is normal aging, which causes each component of the spine to become weaker. Intervertebral discs become thinner and less effective in supporting the vertebrae as people age because the discs lose elasticity and water content (Sreeraj et al., 2022). Osteophytes can produce pain and stiffness by irritating and inflaming tissues around them (Hopkins et al., 2019). Despite their different etiologies, these disorders share common risk factors, including aging, genetic predisposition, hormone imbalances, and lifestyle variables such as poor nutrition and sedentary behavior. Furthermore, the existence of one disorder can frequently worsen symptoms or signs of another, highlighting the importance of comprehensive treatment approaches (Joo et al., 2020; Lee et al., 2022).

In recent years, the diagnosis and classification of musculoskeletal problems such as bone deformities, osteoporosis, OA, and spondylosis have been made possible by the development of machine learning (ML) techniques. ML algorithms are capable of extracting significant patterns and features from vast amounts of medical imaging data that may not be visible to the human eye. To help with the early detection and diagnosis of various disorders, these algorithms are able to distinguish between healthy and sick tissues. In addition, ML models can help medical professionals make quicker and more accurate clinical choices. Personalized risk assessments and treatment suggestions can be generated by ML algorithms by using patient-specific data such as imaging findings, laboratory results, and medical history. This enhances the precision of diagnoses while optimizing treatment plans, improving patient outcomes, and lowering medical expenses. In the real world, ML technologies are enabling more accurate and efficient diagnosis and treatment planning, which is transforming musculoskeletal care. Health care professionals may provide patients who have bone deformities, osteoporosis, OA, and spondylosis with proactive, individualized care by leveraging data-driven insights, which will eventually enhance their well-being and quality of life.

The chapter is organized as follows: Section 1.2 presents the review work used to extract the relevant articles according to Preferred Reporting Items for Systematic Reviews and Meta-Analyses (PRISMA) guidelines. Section 1.3 discusses previous work done by researchers on bone deformity, osteoarthrosis, osteoporosis, and spondylosis. In Section 1.4 different investigations are extracted from previous work. Section 1.5 introduces the methodology used to identify these four disease. Furthermore, proposed solutions of investigations are discussed in Section 1.6. Finally, the work is concluded in Section 1.7.

1.2 Review process

In this section, the PRISMA guidelines were followed during the review process. Research articles relevant to the problem statement were collected from three databases: IEEE Xplore, Scopus, and Web of Science. A total of 970 publications with publication dates ranging from 2016 onward were collected. Of those, 170 articles were removed by manual screening because they contained only review-based information, had duplicate data, were linguistically problematic, and were not relevant to the problem statement (970−170 = 800). Subsequently, the remaining 800 articles underwent abstract and title review, resulting in the exclusion of 551 articles not aligned with the problem statement (800−551 = 249). In the eligibility phase, 249 articles were accessed and 209 articles were removed that did not contain primary data or were not relevant to the scope. Finally, only 40 publications (249−209) were considered for the identification of osteoporosis, osteoarthrosis, and malformations of the bones.

1.3 Literature survey

This section discusses the research analysis done by different researchers for the identification of bone disorders with deep learning (DL) technologies.

1.3.1 Identification of bone deformity

Shen et al. (2021), proposed employing CNN (convolutional neural network), residual network (ResNet), and transfer learning for the detection of bone deformities. The authors used high-resolution synchrotron radiation microcomputed tomography images obtained during uniaxial continuous compression to classify and predict mechanical states of cortical and trabecular bone tissue. Furthermore, optimized CNN architectures were able to categorize new images between pristine and failed states for cortical bone with over 98% accuracy, and trabecular bone with over 90% accuracy. El-Saadawy et al. (2021) discussed a VGG (Visual Geometry Group) network for x-ray classification and abnormality detection. An x-ray of the bone is first classified into seven categories and then is sent to seven classifiers that have been trained to identify abnormalities in the bone based on the type of image. The MURA (musculoskeletal radiographs) dataset was used to test the model. The dataset consisted of a set of bone x-ray scans. In the first phase, the ideal average sensitivity and specificity were 95.86% and 99.63%, respectively. The maximum average sensitivity and specificity attained in the following phase were 92.50% and 92.12%, respectively. Liao et al. (2024) presented a lesion-guided adaptive graph network for abnormality detection from a musculoskeletal radiograph. Specifically, the Adaptive GraphSAGE Attention module was developed to perform bone abnormality detection on the located musculoskeletal abnormality regions and the Lesion-Guided Recurrent Feature Sampling module was designed to localize corresponding musculoskeletal abnormality regions. Experiments were performed on the MURA dataset and the proposed model achieved an accuracy of 87.81%.

Saif et al. (2019) introduced the Capsule Network for abnormality detection in musculoskeletal radiographs. A MURA dataset consisted of 14,656 images and was used to train and test the model. Different types of image sizes were used to train the model, including 64 × 64, 128 × 128, and 224 × 224. The model achieved 95.5 accuracy. Hajianfar (2021) suggested ResNet 50, Inception V-03, and Inception ResNet 50–V02 to detect bone abnormalities. After removing cases with missing or low-quality pictures, a dataset of 2253 patients (762 men and 1491 women), mean age 50.31 ± 18.1 years, was collected. After training, Inception ResNet 50–V02 with spatial pyramid pooling had the best performance for normal–abnormal discrimination, with an area under the curve (AUC) of 0.68, sensitivity of 0.66, and specificity of 0.71.

1.3.2 Identification of osteoarthrosis

Lazzarini et al. (2017) proposed an ML-based analytics pipeline to identify knee osteoarthrosis in overweight women. The pipeline used many clinical and structural outcome indicators for osteoarthrosis, including questionnaires, imaging-based data, biochemical markers, and clinical factors. The dataset used in this study was collected from a preventive randomized controlled trial called PROOF, which included 407 middle-aged women without clinical knee OA (KOA) at the start of the study and a body mass index of 27 kg/m^2 or higher. After training the dataset, the model achieved an AUC of >0.7. Nelson et al. (2019) discussed a K-means–based ML approach to identify phenotypes that may respond differentially to therapies by characterizing various knee osteoarthrosis progression markers. The authors used publicly available data from the Foundation for the National Institute of Health, which consisted of 597 patients, including 59% women, mean age 62 years. Progressors and nonprogressors were distinguished from one another using clustering techniques, direction-projection-permutation testing, and distance weighted discrimination. Variables based on magnetic resonance imaging (MRI) showed greater significance in identifying progressors. Whereas bone marrow lesions, osteophytes, and other parameters were linked to progression, baseline characteristics such as WOMAC (Western Ontario and McMaster Universities Arthritis Index) discomfort, meniscal extrusion, and serum markers were associated with nonprogression. Bayramoglu et al. (2022) introduced the gradient boosting machine (GBM) and CNN to predict patellofemoral OA. The authors collected the dataset from the Multicenter Osteoarthritis Study (MOST). The dataset consisted of 3026 individuals aged 50 to 79 years who were at high risk of developing radiographic KOA or already had the condition at baseline. After the experimental work, this model achieved a 93.7 AUC.

Kwon et al. (2020) studied Inception-ResNet-v2 for the automatic classification of KOA severity using gait data and radiographic images. This study focused on the Kellgren–Lawrence grading system. The model's F1 score, sensitivity, and precision were 0.71, 0.76, and 0.70 respectively. Du et al. (2018) proposed artificial neural network (ANN), support vector machine (SVM), random forest (RF), and naive Bayes (NB) to predict KOA progression on MRI using ML techniques. The feature set was split into medial and lateral compartments and trials were conducted independently for each. After training, the proposed model achieved 79.6, 81.1, 81.7, and 84.5 F1 scores for ANN, SVM, RF, and NB, respectively. Antony et al. (2017) proposed a CNN technique for the automatic detection of knee joints and quantification of KOA. Two phases are involved in automatically assessing the severity of KOA: first, automatically localizing the knee joints, and then automatically classifying images of the localized knee joints. For the experimental work the authors used two datasets: Osteoarthritis Initiative (OAI) and MOST. The authors achieved 60.3% accuracy. Leung et al. (2020) discussed a multitask DL model with transfer learning and used ResNet 34 with 34 layers to predict knee replacement and the diagnosis of OA. To train the model the authors used an OAI dataset consisting of 4796 individuals. The proposed model achieved an 87% AUC.

1.3.3 Detection of osteoporosis

Cuaya-Simbro et al. (2021) proposed NB, SVM, AdaBoost, RF, and K-nearest algorithm for fall risk detection in elderly people with osteoporosis from balanced data. The authors examined the complex problem of falls among the elderly population, focusing on those who have osteoporosis, because they are more vulnerable to these types of accidents. Through an examination of balance metrics taken from posturographic research done with and without eyes, as well as prospective fall monitoring, they were able to develop predictive computer models that can be used to detect people who are at danger of falling. The authors used a dataset of 296 elderly people and obtained a 0.75 AUC. Scanlan et al. (2018) discussed ANN for the detection of osteoporosis from percussion responses using an electronic stethoscope. The technique applies stimuli to the tibia with a reflex hammer and records impulse responses with an electronic stethoscope. The authors used a limited dataset of 12 patients and achieved 60% sensitivity for the proposed model. Kwon et al. (2022) suggested RF, AdaBoost, and GBM for prescreening osteoporosis in postmenopausal Korean women. Data from 1431 postmenopausal women aged of 40 to 69 years were analyzed and 20 traits related to osteoporosis were selected using recursive feature elimination and feature significance. After training the model the authors achieved an accuracy of 83.2%, 84.9%, and 82.9% for RF, AdaBoost, and GBM, respectively. Yang et al. (2021) introduced ANN, SVM, RF, k-nearest neighbors (KNN), and logistic regression (LR) to predict the presence of osteoporosis. The authors built the prediction models using ML algorithms that could be used as osteoporosis screening tools for persons aged over 50 years. The authors also compared the effectiveness of recently created models with that of conventional prediction models. Data collected from the health checkup programs at a northern Taiwanese medical institution yielded 5982 eligible participants after 12,647 were disqualified owing to age or incomplete DXA scans. Training and testing datasets were divided between groups of men and women (80:20). The results of the ANN, SVM, RF, KNN, and LR (linear regression) models demonstrated AUROC (Area under the Receiver Operating Characteristic Curve) values of 0.837, 0.840, 0.843, 0.821, and 0.827 for males and 0.781, 0.807, 0.811, 0.767, and 0.772 for women.

Liu et al. (2022) presented a hierarchical model that covers the lumbar vertebrae and was analyzed by ML models based on clinical data and computed tomography (CT) images. The authors employed six ML algorithms including LR, SVM with radial-basis function kernel, ANN, GB (gradient boosting), and stacking. The authors collected the dataset from clinical trials consisting of 2210 people, in which 246 had CT images. For the three layers of the hierarchical model, LR was able to discriminate osteoporosis from normal bone mineral density with AUROCs of 0.818, 0.838, and 0.962. Tanphiriyakun et al. (2021) proposed seven ML algorithms: RF, GB, LR, SVM, NB, neural network (multilayer perceptron), and KNN. This study employed a dataset of 13,562 osteoporosis therapy profiles collected between January 2011 and December 2019. From the seven models, the model with the highest performance was RF, with an accuracy of 0.69, precision of 0.70, recall of 0.89, and area under the receiver operating curve of 0.70. Jang et al. (2021) discussed the VGG16 model for the prediction of osteoporosis from simple hip radiography. The authors collected 1001 datasets from female patients aged 55 years and older that included proximal femur DXA scans matched with corresponding same-side cropped simple hip bone radiography images. Finally, after the execution of the model the authors achieved 81.2% accuracy, 91.1% sensitivity, and 68.9% specificity.

1.3.4 Identification of spondylosis

Joo et al. (2020), suggested an ML-based generalized linear model (GLM) and SVM model for the prediction of radiographic progression in patients with axial spondylarthritis (axSpA). In multidimensional data, ML was used to predict outcomes and characterize risk. There were few data available to predict how axSpA may proceed radiographically. Therefore, the authors investigated whether radiographic progression in axSpA can be predicted using supervised ML techniques. The authors collected the dataset from the clinic and laboratory. The dataset was divided into training (253) and testing (173) information. After the experimental analysis the SVM outperformed GLM with values of 0.56 and 0.51 for precision and recall, respectively. Lee et al. (2022) discussed CNN for detecting lateral cervical spine radiograph to detect cervical spondylotic myelopathy (CSM). The authors collected the data of 270 patients from the spine center of a university hospital, 96 of whom had CSM and 111 who did not have CSM. To train the model, 145 images were used; 62 images were used for testing. After the analysis, the CNN model achieved 87.1% accuracy. Ji et al. (2023) proposed five traditional classifiers, SVM, LR, KNN, DT (decision tree), and RF, to classify gait patterns between CSM patients and healthy controls (HCs). The authors measured nine spatiotemporal characteristics for each participant as well as two kinematic and four nonlinear parameters of the unilateral hip, knee, and ankle angles in the sagittal, coronal, and transverse planes. The authors collected the data of 45 age-matched asymptomatic HCs and recorded locomotion data using a three-dimensional motion capture system. From the five discussed models, the SVM achieved the highest accuracy of 95.56%. Castro-Zunti et al. (2020) discussed KNN, RF, and the Inception V3 model for the early detection of ankylosing spondylitis (AS) using texture features. The authors also discussed the textural feature-based classifier that distinguishes the sacroiliac joints of young and elderly control patients and those afflicted by erosion, to identify erosion in both age groups of patients. The dataset consisted of 681 gray-scale JPG images representing a single sacroiliac joint. These particular regions of interest were extracted from 53 patients' labeled DICOM frames from CT scans of the abdomen and pelvis. All images of the right joints were flipped horizontally to guarantee orientation uniformity. After the model was trained, the RF outperformed KNN and achieved 96.0%, 92.9%, and 0.97% accuracy, recall, and AUC, respectively.

Zhu et al. (2022) introduced RF, least absolute shrinkage and selection operator (LASSO) regression, and SVM to predict AS. Based on the results of patients' blood tests, the authors used ML techniques to create diagnostic models for AS. At the First Affiliated Hospital of Guangxi Medical University, the authors examined 348 AS patients. They then randomly assigned patients to training and validation groups, with 258 AS patients and 247 non-AS patients in the former group and 90 AS patients and 113 non-AS patients in the latter. The nomogram in the training group showed an AUC value of 0.8779462 and a C-value of 0.878. The nomogram in the validation group showed an AUC value of 0.8232055 and a C-value of 0.823. Huo et al. (2022) proposed an ANN model for the analysis of the treatment effect of acupuncture on cervical spondylosis. The authors evaluated, through DL techniques, the effectiveness of acupuncture in treating cervical spondylosis and neck discomfort. The authors collected the data of 270 patients and divided them into acupuncture needle (group A), acupoint latent acupuncture (group B), and puncture bloodletting (group C). In the experimental analysis the model achieved 82.64% accuracy. Sun et al. (2023) introduced K means clustering, LASSO, and RF for identifying AS. Based on the clinical classification, they created a novel subtype predictive model for AS. Subsequently, the authors examined variations in the immunologic microenvironment to determine the pathophysiology of AS. The authors collected the data of 196 patients from the Gene Expression Omnibus database. Using the three parameters of absolute neutrophil count, absolute monocyte count, and C-reactive protein, a predictive model was created for the AS, and the AUC was 0.983.

1.4 Proposed investigations

This study explores the identification of bone disorders, osteoarthrosis, osteoporosis, and spondylosis. The following investigations are relevant to the evaluation of the study:

Investigation 1: What genetic and environmental factors contribute to bone deformities, osteoarthrosis, osteoporosis, and spondylosis, and how can ML help identify these factors?
Investigation 2: How do the symptoms of osteoarthrosis, osteoporosis, and spondylosis differ, and what are the key diagnostic criteria for each condition?
Investigation 3: How do MRI and DXA scanning advances aid in early detection, monitoring, and treatment planning for bone deformities, osteoarthrosis, osteoporosis, and spondylosis?
Investigation 4: What are the socioeconomic implications of bone deformities, osteoarthrosis, osteoporosis, and spondylosis in terms of health care costs, disability rates, and quality of life?
Investigation 5: What are the challenges in applying ML for identifying bone disorders, and how can these challenges be addressed to improve the reliability and effectiveness of ML-based diagnostic tools in clinical practice?

1.5 Methodology

There are several important processes involved in training an ML model for the diagnosis of spondylosis, osteoarthrosis, osteoporosis, and bone deformities, as shown in Fig. 1.2. First, a large dataset composed of various imaging investigations, patient data, and medical records is collected. This data undergoes preprocessing and cleaning to handle missing values, normalize numerical features, encode categorical variables, and select relevant features. Then, meaningful information is extracted from the raw data to create relevant features. After the selection of appropriate classification algorithms such as LR, SVMs, or DL models, the dataset is split into training, validation, and test sets for model training and evaluation. The performance of the model is evaluated using a variety of evaluation criteria such as accuracy, F1 score, recall, and precision. After validation, the completed model is used in clinical or research environments, ensuring compliance with ethical and regulatory requirements. To keep the model relevant and useful for identifying bone problems while fostering interpretability and explainability for medical professionals, it is important to prioritize regular evaluation and continuous enhancement.

1.6 Discussion

This section provides a comprehensive overview of the various investigations undertaken in the study.

Investigation 1: What genetic and environmental factors contribute to bone deformities, osteoarthrosis, osteoporosis, and spondylosis, and how can ML help identify these factors?

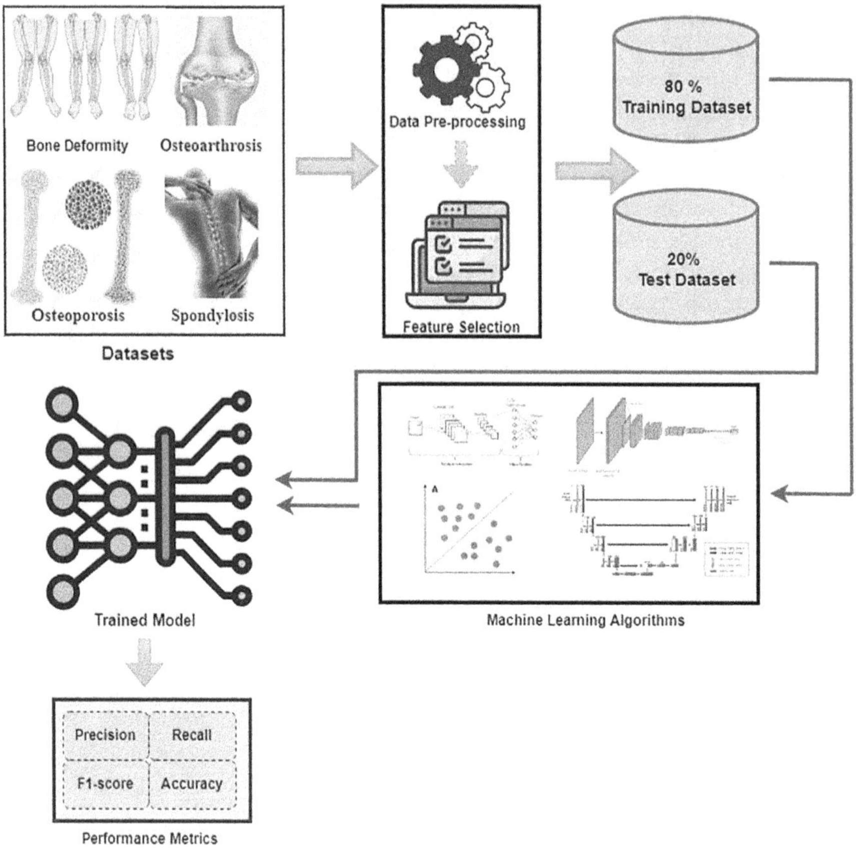

FIGURE 1.2

Proposed methodology.

P.S: Osteoarthrosis, osteoporosis, spondylosis, and other bone abnormalities are caused by a variety of hereditary and environmental variables. Inherited characteristics that affect metabolism, bone structure, and the likelihood of acquiring certain disorders are examples of genetic factors, as shown in Fig. 1.3. Environmental factors include lifestyle decisions such as nutrition, the exercise regimen, chemical exposure, and socioeconomic status. Through the analysis of large-scale datasets containing genetic, medical, lifestyle, and environmental exposure data, ML can assist in identifying these aspects. Patterns, correlations, and interactions among different genetic characteristics, environmental exposures, and clinical outcomes can be found with the aid of these algorithms. Through the application of ML, researchers are able to decode new information about the fundamental causes of bone diseases and pinpoint possible areas for treatment or avoidance.

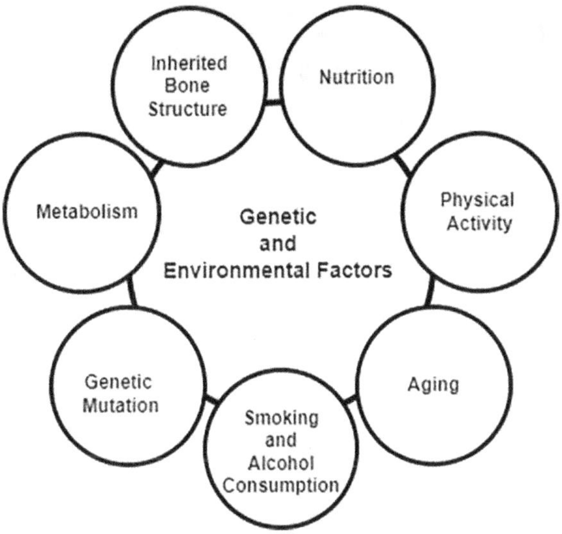

FIGURE 1.3

Genetic and environmental factors.

Investigation 2: How do the symptoms of osteoarthrosis, osteoporosis, and spondylosis differ, and what are the key diagnostic criteria for each condition?

P.S: Different musculoskeletal disorders such as osteoarthrosis, osteoporosis, and spondylosis have different clinical presentations and diagnostic standards, as shown in Fig. 1.4.

- **Osteoarthrosis**

The usual symptoms of osteoarthrosis, also referred to as OA, include joint pain, stiffness, swelling, and decreased range of motion. Osteophyte bony enlargements or crepitus may be present in the afflicted joints. Diagnostic criteria for OA include clinical evaluation, radiographic findings such as joint space narrowing and osteophyte formation, and exclusion of other causes of joint pain.

- **Osteoporosis**

Weakened bones and a higher chance of fractures, particularly in the hip, spine, and wrist, are the main symptoms of osteoporosis. Clinical manifestations can include low-force fractures, vertical loss of height, and back pain from compression fractures of the vertebrae. Using methods such as dual-energy x-ray absorptiometry (DEXA) scanning, osteoporosis diagnostic criteria can be determined. Low bone density is indicated by a T-score of−2.5 or less.

- **Spondylosis**

Spinal OA, or spondylosis, is a condition that affects the spine and can cause stiffness, pain in the neck or back, and neurologic symptoms such as limb paralysis

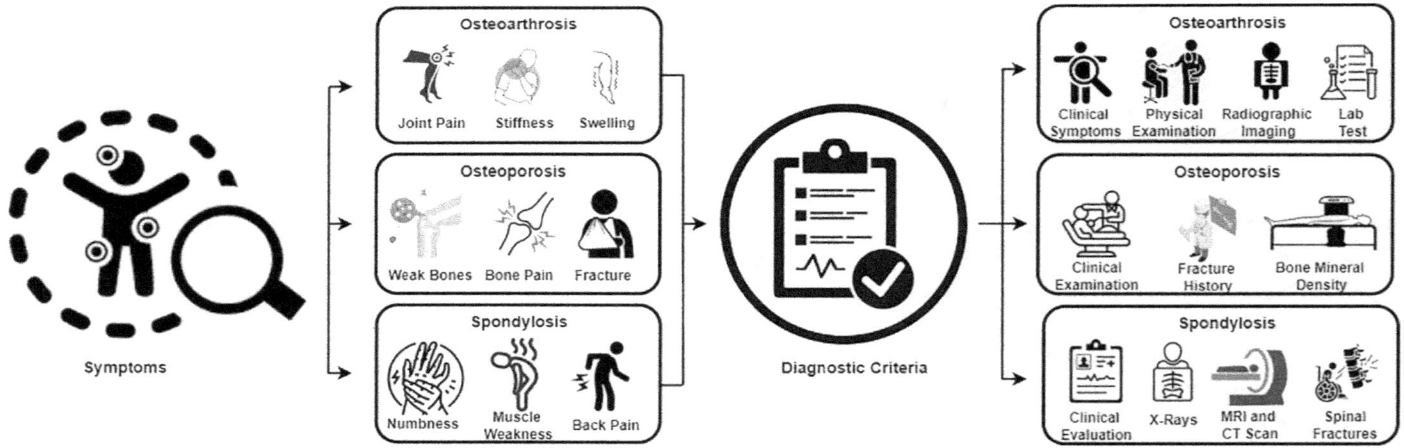

FIGURE 1.4

Symptoms and diagnostic criteria. *CT*, computed tomography; *MRI*, magnetic resonance imaging.

or numbness. Diagnostic procedures such as x-rays and MRI, which show degenerative changes in the spine such as disc degeneration, osteophyte growth, and facet joint arthritis, are commonly used to diagnose spondylosis.

Spondylosis, osteoporosis, and osteoarthrosis all have different clinical presentations and diagnostic standards despite having some joint and spinal involvement. Osteoporosis causes weakening bones and an increased risk of fractures, whereas osteoarthrosis mostly affects joints and manifests as joint pain and stiffness. In addition to causing neurologic symptoms and back or neck pain, spondylosis affects the spine. Diagnostic approaches to each condition involve a combination of clinical evaluation and imaging studies tailored to their specific manifestations.

Investigation 3: How do MRI and DXA scanning advancements aid in early detection, monitoring, and treatment planning for bone deformities, osteoarthrosis, osteoporosis, and spondylosis?

P.S: The early detection, monitoring, and treatment planning of bone abnormalities, osteoarthrosis, osteoporosis, and spondylosis are greatly aided by advances in MRI and DXA scanning, as shown in Fig. 1.5. When soft tissue anomalies such as

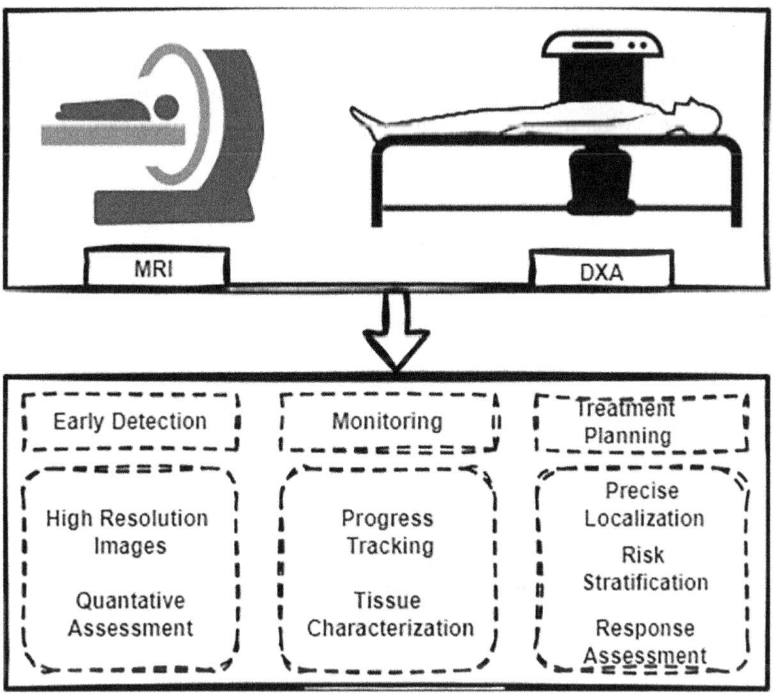

FIGURE 1.5

Magnetic resonance imaging (MRI) and dual-energy x-ray absorptiometry (DXA) scanning advancements.

cartilage degradation and joint inflammation are detected early in the imaging process, MRI allows for a thorough visualization of the soft tissues and prompt intervention. In addition, by giving accurate anatomical information for therapies such as corticosteroid injections or surgical operations, MRI facilitates therapy planning and enables long-term monitoring of disease development. On the other hand, DXA scanning collects data on bone mineral density, which is useful for tracking the progression of the disease and the effectiveness of treatment over time. It also helps detect osteoporosis early. Measurements obtained from DXA scans are used to guide treatment decisions, including whether to start pharmaceutical medication and take precautions to lower the risk of fracture. When combined, MRI and DXA screening are essential for improving patient care, treatment results, and quality of life for people with disorders relating to the bones.

Investigation 4: What are the socioeconomic implications of bone deformities, osteoarthrosis, osteoporosis, and spondylosis in terms of health care costs, disability rates, and quality of life?

P.S: Osteoarthrosis, osteoporosis, spondylosis, and other bone disorders have significant socioeconomic implications that affect medical costs, disability rates, and quality of life in a number of ways, as shown in Fig. 1.6:

- **Health care Expenses:** Because of the high expenses of diagnosis, treatment, and management, illnesses related to the bones place a considerable financial strain on health care systems. Medical consultations, diagnostic imaging, lab testing, prescription of drugs, surgery, recovery, and long-term care are among

FIGURE 1.6

Socioeconomic implications.

the costs. Rising health care costs are a result of the combined costs of managing spondylosis, osteoarthrosis, osteoporosis, and bone abnormalities. These costs affect both patients and health care providers.

- **Disability Rates**: Bone diseases can cause functional impairment, chronic pain, and disability, which makes it harder for people to go about their everyday lives and take part in social, occupational, and recreational activities. Under severe circumstances, a person may become permanently disabled and require assistance with daily living activities, personal care, and mobility. Therefore, disorders relating to the bones influence worker productivity and economic involvement by raising rates of disability and absenteeism from work.
- **Quality of Life:** Bone abnormalities, osteoporosis, osteoarthrosis, and spondylosis can severely impair a person's quality of life. Physical and mental well-being are hampered by chronic pain, restricted movement, and functional limitations, which can result in diminished independence, social isolation, and psychological discomfort. Decreased life quality poses significant obstacles for people and their families because it affects many areas of life, such as social interactions, emotional well-being, physical function, and general life satisfaction.

Addressing these socioeconomic challenges requires comprehensive strategies focused on preventive measures, early intervention, access to affordable health care services, rehabilitation programs, and social support systems to mitigate the impact of bone disorders on individuals and society as a whole.

Investigation 5: What are the challenges in applying ML to identify bone disorders, and how can these challenges be addressed to improve the reliability and effectiveness of ML-based diagnostic tools in clinical practice?

P.S: Several challenges and limitations (as shown in Fig. 1.7) in applying ML for identifying bone disorders are data availability and quality, the interpretability and explainability of ML models, generalization across different populations, and ethical and legal issues. To improve the reliability and effectiveness of ML-based diagnostic tools in clinical practice, efforts are under way to address these challenges. This includes standardizing medical imaging protocols, enhancing data sharing across institutions, developing explainable artificial intelligence techniques to improve model transparency, and exploring methods such as transfer learning and domain adaptation to enhance model generalization. Additionally, regulations are being established to ensure the safe and legal use of machine learning-based diagnostic tools in healthcare. By addressing these challenges, machine learning (ML) has the potential to significantly improve bone disease identification and treatment, leading to better patient outcomes and care.

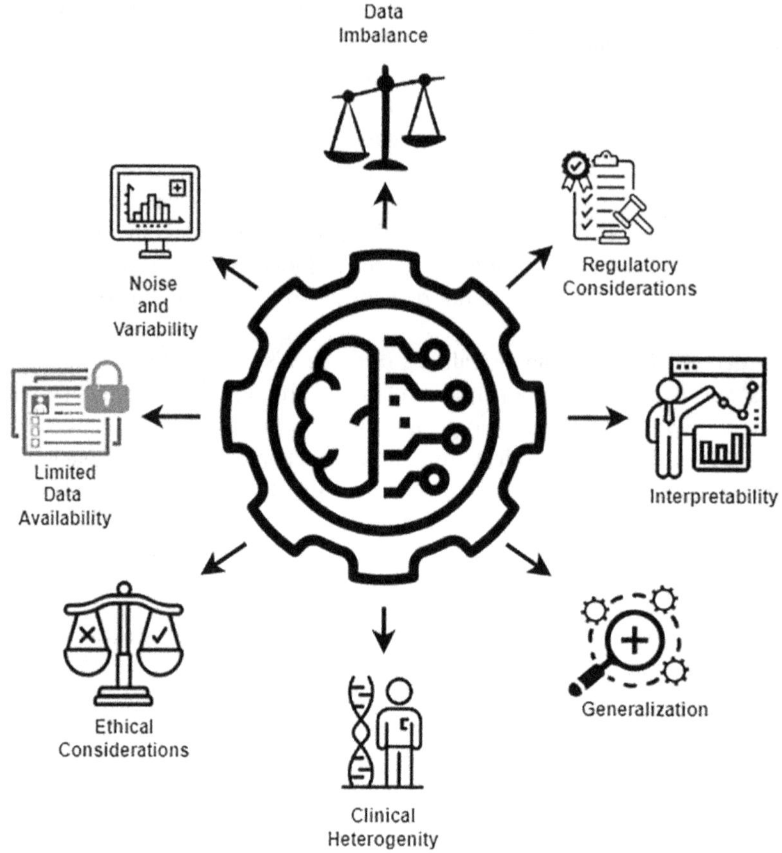

FIGURE 1.7

Challenges in identification of musculoskeletal problems.

1.7 Conclusion

ML technologies offer unique opportunities for the diagnosis and treatment of musculoskeletal conditions, including spondylosis, osteoporosis, OA, and bone deformities. Using advanced computing algorithms and analyzing vast amounts of medical data, medical professionals can improve the precision of diagnosis, predict the development of a disease, and customize treatment plans to each patient's specific needs. With additional refinements and developments, these technologies hold great promise for improving patient outcomes and expanding musculoskeletal health care globally. By adopting these innovative approaches, health care systems handle the difficulties presented by musculoskeletal disorders may more successfully, which will ultimately result in better patient care and lower costs.

References

Albuquerque, G.A., Carvalho, D.D.A., Cruz, A.S., Santos, J.P.Q., Machado, G.M., Gendriz, I.S., Fernandes, F.R.S., Barbalho, I.M.P., Santos, M.M., Teixeira, C.A.D., Henriques, J.M.O., Gil, P., Neto, A.D.D., Campos, A.L.P.S., Lima, J.G., Paiva, J.C., Morais, A.H.F., Lima, T.S., Valentim, R.A.M., 2023. Osteoporosis screening using machine learning and electromagnetic waves. Sci. Rep. 13 (1). https://doi.org/10.1038/s41598-023-40104-w.

Antony, J., McGuinness, K., Moran, K., O'Connor, N.E., 2017. Automatic detection of knee joints and quantification of knee osteoarthritis severity using convolutional neural networks. Lect. Notes Comput. Sci. 10358, 376–390. https://doi.org/10.1007/978-3-319-62416-7_27.

Aslam Khan, M., Yaseen Ahmed, M., Safadar Hussain, S., Abid, K., 2023. Machine learning for the identification of bone deformities. Int. J. Innovat. Res. Eng. Multidiscip. Phy. Sci. 11 (1). https://doi.org/10.37082/ijirmps.v11.i1.230311.

Bayramoglu, N., Nieminen, M.T., Saarakkala, S., 2022. Machine learning based texture analysis of patella from X-rays for detecting patellofemoral osteoarthritis. Int. J. Med. Inf. 157. https://doi.org/10.1016/j.ijmedinf.2021.104627.

Bonakdari, H., Jamshidi, A., Pelletier, J.P., Abram, F., Tardif, G., Martel-Pelletier, J., 2021. A warning machine learning algorithm for early knee osteoarthritis structural progressor patient screening. Ther. Adv. Musculoskel. Dis. 13. https://doi.org/10.1177/1759720X21993254.

Bowes, M.A., Kacena, K., Alabas, O.A., Brett, A.D., Dube, B., Bodick, N., Conaghan, P.G., 2021. Machine-learning, MRI bone shape and important clinical outcomes in osteoarthritis: data from the osteoarthritis initiative. Ann. Rheum. Dis. 80 (4), 502–508. https://doi.org/10.1136/annrheumdis-2020-217160.

Brahim, A., Jennane, R., Riad, R., Janvier, T., Khedher, L., Toumi, H., Lespessailles, E., 2019. A decision support tool for early detection of knee osteoarthritis using X-ray imaging and machine learning: data from the osteoarthritis initiative. Comput. Med. Imag. Graph. 73, 11–18. https://doi.org/10.1016/j.compmedimag.2019.01.007.

Bui, H.M., Ha, M.H., Pham, H.G., Dao, T.P., Nguyen, T.T.T., Nguyen, M.L., Vuong, N.T., Hoang, X.H.T., Do, L.T., Dao, T.X., Le, C.Q., 2022. Predicting the risk of osteoporosis in older vietnamese women using machine learning approaches. Sci. Rep. 12 (1). https://doi.org/10.1038/s41598-022-24181-x.

Castro-Zunti, R., Park, E.H., Choi, Y., Jin, G.Y., Ko, S.B., 2020. Early detection of ankylosing spondylitis using texture features and statistical machine learning, and deep learning, with some patient age analysis. Comput. Med. Imag. Graph. 82. https://doi.org/10.1016/j.compmedimag.2020.101718.

Cuaya-Simbro, G., Perez-Sanpablo, A.I., Morales, E.F., Quiñones Uriostegui, I., Nuñez-Carrera, L., 2021. Comparing machine learning methods to improve fall risk detection in elderly with osteoporosis from balance data. J. Healthcare Eng. 2021. https://doi.org/10.1155/2021/8697805.

Du, Y., Almajalid, R., Shan, J., Zhang, M., 2018. A novel method to predict knee osteoarthritis progression on MRI using machine learning methods. IEEE Trans. NanoBiosci. 17 (3), 228–236. https://doi.org/10.1109/TNB.2018.2840082.

El-Saadawy, H., Tantawi, M., Shedeed, H.A., Tolba, M.F., 2021. A hybrid two-stage GNG-modified VGG method for bone X-rays classification and abnormality detection. IEEE Access 9, 76649–76661. https://doi.org/10.1109/ACCESS.2021.3081915.

Hajianfar, G., 2021. Dual input scintigraphy image-based fused deep neural networks for bone abnormalities detection and differentiation. 2021 IEEE Nucl. Sci. Symp. Med. Imag. Conf. 1–3.

Hopkins, B.S., Weber, K.A., Kesavabhotla, K., Paliwal, M., Cantrell, D.R., Smith, Z.A., 2019. Machine learning for the prediction of cervical spondylotic myelopathy: a post hoc pilot study of 28 participants. World Neurosurg. 127, e436–e442. https://doi.org/10.1016/j.wneu.2019.03.165.

Huo, H., Chang, Y., Tang, Y., 2022. Analysis of treatment effect of acupuncture on cervical spondylosis and neck pain with the data mining technology under deep learning. J. Supercomput. 78 (4), 5547–5564. https://doi.org/10.1007/s11227-021-03959-2.

Jang, R., Choi, J.H., Kim, N., Chang, J.S., Yoon, P.W., Kim, C.H., 2021. Prediction of osteoporosis from simple hip radiography using deep learning algorithm. Sci. Rep. 11 (1). https://doi.org/10.1038/s41598-021-99549-6.

Ji, X., Zeng, W., Dai, Q., Zhang, Y., Du, S., Ji, B., 2023. Machine learning-based detection of cervical spondylotic myelopathy using multiple gait parameters. Biomimetic Int. Robot. 3 (2). https://doi.org/10.1016/j.birob.2023.100103.

Joo, Y.B., Baek, I.W., Park, Y.J., Park, K.S., Kim, K.J., 2020. Machine learning–based prediction of radiographic progression in patients with axial spondyloarthritis. Clin. Rheumatol. 39 (4), 983–991. https://doi.org/10.1007/s10067-019-04803-y.

Kerketta, S.R., Ghosh, D., 2021. Detection of onset and progression of osteoporosis using machine learning. Mach. Lear. Healthcare Appl. 137–149. https://doi.org/10.1002/9781119792611.ch9.

Kwon, S.B., Han, H.S., Lee, M.C., Kim, H.C., Ku, Y., Ro, D.H., 2020. Machine learning-based automatic classification of knee osteoarthritis severity using gait data and radiographic images. IEEE Access 8, 120597–120603. https://doi.org/10.1109/ACCESS.2020.3006335.

Kwon, Y., Lee, J., Park, J.H., Kim, Y.M., Kim, S.H., Won, Y.J., Kim, H.Y., 2022. Osteoporosis pre-screening using ensemble machine learning in postmenopausal Korean women. Healthcare (Switzerland) 10 (6). https://doi.org/10.3390/healthcare10061107.

Lazzarini, N., Runhaar, J., Bay-Jensen, A.C., Thudium, C.S., Bierma-Zeinstra, S.M.A., Henrotin, Y., Bacardit, J., 2017. A machine learning approach for the identification of new biomarkers for knee osteoarthritis development in overweight and obese women. Osteoarthritis Cartilage 25 (12), 2014–2021. https://doi.org/10.1016/j.joca.2017.09.001.

Lee, G.W., Shin, H., Chang, M.C., 2022. Deep learning algorithm to evaluate cervical spondylotic myelopathy using lateral cervical spine radiograph. BMC Neurol. 22 (1). https://doi.org/10.1186/s12883-022-02670-w.

Leung, K., Zhang, B., Tan, J., Shen, Y., Geras, K.J., Babb, J.S., Cho, K., Chang, G., Deniz, C.M., 2020. Prediction of total knee replacement and diagnosis of osteoarthritis by using deep learning on knee radiographs: data from the osteoarthritis initiative. Radiology 296 (3), 584–593. https://doi.org/10.1148/radiol.2020192091.

Liao, Y., Li, X., Peng, C., 2024. LGAG-net: lesion-guided adaptive graph network for bone abnormality detection from musculoskeletal radiograph. IEEE Access 12, 26710–26718. https://doi.org/10.1109/ACCESS.2023.3345896.

Liu, L., Si, M., Ma, H., Cong, M., Xu, Q., Sun, Q., Wu, W., Wang, C., Fagan, M.J., Mur, L.A.J., Yang, Q., Ji, B., 2022. A hierarchical opportunistic screening model for osteoporosis using machine learning applied to clinical data and CT images. BMC Bioinf. 23 (1). https://doi.org/10.1186/s12859-022-04596-z.

Nelson, A.E., Fang, F., Arbeeva, L., Cleveland, R.J., Schwartz, T.A., Callahan, L.F., Marron, J.S., Loeser, R.F., 2019. A machine learning approach to knee osteoarthritis phenotyping: data from the FNIH biomarkers consortium. Osteoarthritis Cartilage 27 (7), 994–1001. https://doi.org/10.1016/j.joca.2018.12.027.

Saif, A.F.M., Shahnaz, C., Zhu, W.P., Ahmad, M.O., 2019. Abnormality detection in musculoskeletal radiographs using capsule network. IEEE Access 7, 81494–81503. https://doi.org/10.1109/ACCESS.2019.2923008.

Sasirekha, K., Sivaraman, M., Sumitha, J., 2023. Identification and classification of leukemia using machine learning approaches. Int. Conf. Self Sust. Artif. Int. Syst., ICSSAS 2023 - Proc. 282–288. https://doi.org/10.1109/ICSSAS57918.2023.10331767.

Scanlan, J., Li, F.F., Umnova, O., Rakoczy, G., Lövey, N., Scanlan, P., 2018. Detection of osteoporosis from percussion responses using an electronic stethoscope and machine learning. Bioengineering 5 (4). https://doi.org/10.3390/BIOENGINEERING5040107.

Schiratti, J.B., Dubois, R., Herent, P., Cahané, D., Dachary, J., Clozel, T., Wainrib, G., Keime-Guibert, F., Lalande, A., Pueyo, M., Guillier, R., Gabarroca, C., Moingeon, P., 2021. A deep learning method for predicting knee osteoarthritis radiographic progression from MRI. Arthritis Res. Ther. 23 (1). https://doi.org/10.1186/s13075-021-02634-4.

Shen, S.C.y., Peña Fernández, M., Tozzi, G., Buehler, M.J., 2021. Deep learning approach to assess damage mechanics of bone tissue. J. Mech. Behav. Biomed. Mater. 123. https://doi.org/10.1016/j.jmbbm.2021.104761.

Sreeraj, M., Joy, J., Jose, M., Varghese, M., Rejoice, T.J., 2022. Comparative analysis of machine learning approaches for early stage cervical spondylosis detection. J. King Saud Uni. Comp. Inf. Sci. 34 (6), 3301–3309. https://doi.org/10.1016/j.jksuci.2020.08.010.

Sun, X., Zhou, C., Zhu, J., Wu, S., Liang, T., Jiang, J., Chen, J., Chen, T., Huang, S.S., Chen, L., Ye, Z., Guo, H., Zhan, X., Liu, C., 2023. Identification of clinical heterogeneity and construction of a novel subtype predictive model in patients with ankylosing spondylitis: an unsupervised machine learning study. Int. Immunopharm. 117. https://doi.org/10.1016/j.intimp.2023.109879.

Tanphiriyakun, T., Rojanasthien, S., Khumrin, P., 2021. Bone mineral density response prediction following osteoporosis treatment using machine learning to aid personalized therapy. Sci. Rep. 11 (1). https://doi.org/10.1038/s41598-021-93152-5.

Wang, N., Cui, L., Huang, X., Xiang, Y., Xiao, J., 2018. EasiCSDeep: a deep learning model for cervical spondylosis identification using surface electromyography signal. arXiv. https://doi.org/10.48550/arxiv.1812.04912.

Yadahalli, S., Parmar, A., Zambare, P., Sawant, R., 2021. Bone deformity identification using machine learning. Proc. 5th Int. Conf. Int. Comp. Con. Syst. ICICCS 2021 1355–1358. https://doi.org/10.1109/ICICCS51141.2021.9432272.

Yang, W.Y.O., Lai, C.C., Tsou, M.T., Hwang, L.C., 2021. Development of machine learning models for prediction of osteoporosis from clinical health examination data. Int. J. Environ. Res. Publ. Health 18 (14). https://doi.org/10.3390/ijerph18147635.

Zhu, J., Lu, Q., Liang, T., JieJiang, Li, H., Zhou, C., Wu, S., Chen, T., Chen, J., Deng, G., Yao, Y., Liao, S., Yu, C., Huang, S., Sun, X., Chen, L., Chen, W., Ye, Z., Guo, H., Chen, W., Jiang, W., Fan, B., Tao, X., Zhan, X., Liu, C., 2022. Development and validation of a machine learning-based nomogram for prediction of ankylosing spondylitis. Rheumatol. Ther. 9 (5), 1377–1397. https://doi.org/10.1007/s40744-022-00481-6.

Further reading

Jamshidi, A., et al., 2020. Identification of the most important features of knee osteoarthritis structural progressors using machine learning methods. Ther. Adv. Musculoskelet. Dis. 12, 1–12. https://doi.org/10.1177/1759720X20933468.

CHAPTER

2

Identification and classification of musculoskeletal conditions using artificial intelligence and machine learning

Ritesh Bhat[1], Vajjiram Santhanam[1], Karuppannan Sekar[2], Shilpa Gite[3,4], Nithesh Naik[5], Ali Talyshinskii[6]

[1]*Department of Mechatronics Engineering, Rajalakshmi Engineering College, Thandalam, Tamil Nadu, India;* [2]*Department of Artificial Intelligence and Machine Learning, Rajalakshmi Engineering College, Thandalam, Tamil Nadu, India;* [3]*Symbiosis Institute of Technology, Symbiosis International (Deemed University), Pune, Maharashtra, India;* [4]*Symbiosis Centre for Applied AI, Symbiosis International (Deemed University), Pune, Maharashtra, India;* [5]*Department of Mechanical and Industrial Engineering, Manipal Institute of Technology, Manipal Academy of Higher Education, Manipal, Karnataka, India;* [6]*Department of Urology, Astana Medical University, Astana, Kazakhstan*

2.1 Introduction

Musculoskeletal disorders (MSDs) contribute to a major portion of global health issues, causing disability (Briggs et al., 2020). Despite traditional techniques such as physical examinations and imaging, the need for innovation in diagnostic and therapeutic methods has become important (Conaghan et al., 2011). The advent of the use of artificial intelligence (AI) and machine learning (ML) in health care has undoubtedly given rise to a new era, promising the better and more economic management of MSDs (Ahmed et al., 2020). Fig. 2.1 represents the mind map of AI and ML in health care, drawing the critical elements. The application of AI and ML facilitates a better patient management experience and also speeds the overall medical process (Shajari et al., 2023). However, the issues of data security and ethical considerations always exist (Hirschmann et al., 2019). This chapter thus focuses on discussing the integration and challenges of AI and ML in musculoskeletal care.

Figure 2.1 diagrammatically explains the relational mapping of all core elements of AI and ML in MSD treatment along with their significant roles.

Diagnosing Musculoskeletal Conditions using Artificial Intelligence and Machine Learning to Aid Interpretation of Clinical Imaging
https://doi.org/10.1016/B978-0-443-32892-3.00002-6

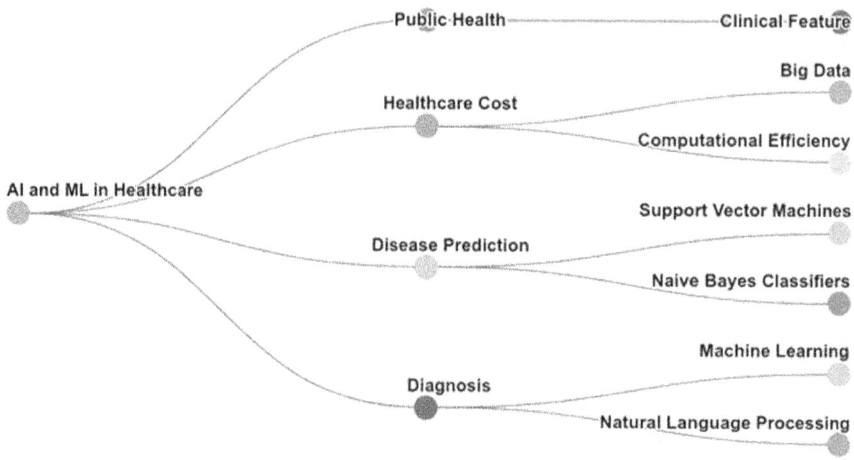

FIGURE 2.1

Relational mapping of core elements and their significance. *AI*, artificial intelligence; *ML*, machine learning.

Created using SCOPUS AI.

2.2 Artificial intelligence and machine learning in health care

2.2.1 Basics of artificial intelligence and machine learning

AI and ML have been changing almost every sector including health care, especially diagnostic testing and personalized medicine. Table 2.1 lists a few critical studies based on the application of AI in different sectors. AI is a tool with the ability to replicate certain important human intelligence traits such as continuous learning and decision-making. Turing's basic concept of machine intelligence is the foundation of the explained trait (Turing, 1950). Most decision-making by AI is accomplished by its ability to recognize patterns. This ability is a fundamental revolution in health care, providing more accurate analytics and results that can be improved by complex analysis (Shrestha et al., 2021) The applications of AI and ML in health care are expanding from improving diagnostic imaging to optimizing treatment protocols. Deep learning, a sophisticated ML technique, has greatly advanced medical image interpretation, contributing to diagnostic accuracy (Lecun et al., 2015), but the adoption of AI poses ethical challenges (Taddeo et al., 2019), such as ensuring business justice for the responsible use and privacy protection of this technology. The subsequent sections highlight the important yet challenging role of AI and ML in contemporary health care, highlighting their potential in addressing musculoskeletal diseases while considering ethics regarding their translation among widespread users.

Table 2.1 Synthesized insights from a few critically reviewed articles.

Study	Year	Significant insights
AI advancements in pandemic management (Syrowatka et al., 2021)	2020	ML's role in enhancing public health responses and clinical applications detailed.
Ethical frameworks in AI/ML technologies (Piano, 2020)	2020	Discusses the ethical landscapes of AI, emphasizing the importance of algorithmic accountability.
Cardiology innovations through AI in Food and Drug Administration–approved devices (Benjamens et al., 2020)	2020	Reviews advances in cardiology via AI, focusing on regulatory approvals.
AI's potential for societal benefit (Tomašev et al., 2020)	2020	Explores AI's application in social welfare projects, with an emphasis on realistic and impactful initiatives.
AI's role in supporting Sustainable Development Goals (Vinuesa et al., 2020)	2020	Analyzes the influence of AI on achieving economic sustainability objectives.
Comparing AI model and physician accuracy in disease diagnostics (Park et al., 2021)	2021	Demonstrates AI's superior predictive accuracy in diagnosing conditions based on lab tests.
The confluence of AI with precision medicine for personalized health care (Johnson et al., 2021)	2021	Highlights the integration of AI in customizing patient care through precision medicine.
Enhancing breast cancer diagnosis and prognosis with AI (Li and Giger, 2021)	2021	Investigates AI's impact on breast cancer diagnosis, enhancing detection and treatment strategies.
Promoting ethical considerations in AI research (Srikumar et al., 2022)	2022	Stresses the need for ethical diligence among AI researchers and practitioners.
Deep learning in materials science for electron microscopy (Jacobs, 2022)	2022	Discusses the application and challenges of deep learning in analyzing microscopy images.
AI's application in climate change adaptation strategies (Leal Filho et al., 2022)	2022	Reviews how AI aids in climate adaptation, particularly in managing water resources.
ML-based forecasting of AI development trends (Krenn et al., 2023)	2023	Uses ML for predictive analyses of AI's evolution, focusing on the significance of temporal data.

AI, *artificial intelligence;* ML, *machine learning.*

2.2.2 Artificial intelligence and machine learning in medical imaging and diagnostics

As discussed earlier, AI and ML have greatly improved medical imaging and diagnostics, making them more precise and efficient. These technologies rely on sophisticated algorithms that analyze complex medical images. The algorithms are designed to detect patterns and anomalies indicating various health conditions (Najjar, 2023). Thus, the applications of AI and ML in diagnostics can be used to detect crucial health problems such as cancers and cardiovascular irregularities while integrating them with techniques including echocardiography and advanced imaging methods. The use of AI and ML reduces the time needed for accurate diagnoses and speeds decision-making for patient management. Moreover, these technologies enhance image quality, automate repetitive tasks, and provide predictive insights, helping to develop customized treatment plans and enhance patient outcomes (Panayides et al., 2020).

Table 2.2 lists a few studies demonstrating how AI and ML are used in medical imaging and diagnostics considering various medical cases. The potential for significant enhancement in the quality of patient care is immense. The literature supports the fact that integrating AI and ML into clinical practice is not a theoretical concept but a tangible and transformative reality. This innovation promises to revolutionize the field of medical imaging and diagnostics, offering a remarkable degree of precision and effectiveness in disease diagnosis and treatment planning. The papers referenced in Table 2.2 provide concrete illustrations of the practical applications of AI and ML in real-life scenarios, showcasing recent uses and technological breakthroughs. As discussed earlier, the application of the AI and ML in the health care sector poses challenges, but the studies listed in the table suggest that these could be handled effectively. The studies also suggest that the focus must not be solely on the application of AI and ML techniques but also on ensuring that the outcome has a positive effect on the medical practices and personalized treatments.

2.3 Artificial intelligence–driven diagnostics in musculoskeletal condition treatment

Fig. 2.2 represents a relational mind-map of AI-driven diagnostics and MSD treatment.

Mind-map of AI-driven diagnostics and MSD treatment.

Continual improvements in AI and ML techniques have enhanced the preliminary diagnostics and treatment of MSDs. For example, AI's role in augmenting magnetic resonance imaging (MRI) analyses enables the superior diagnosis of complex conditions such as septic arthritis and spinal osteomyelitis, offering deeper insights than previously possible (Palestro et al., 2006). In addition, innovations such as fluorodeoxyglucose positron-emission tomography (FDG-PET) are emerging as valuable tools in spinal infection diagnosis, representing a significant leap forward in imaging techniques (Blemker et al., 2007). The use of whole-body MRI in specialties such as oncology and rheumatology demonstrates the broad applicability of AI-enhanced imaging, assisting in early disease detection and thorough disease management (Lecouvet, 2016). Moreover, AI-driven quantitative image analysis, including radiomics and texture analysis, has become instrumental in extracting

Table 2.2 Refined overview of progress in AI and ML for medical imaging and diagnostics.

Study focus	Year	Core insights
Evaluating convolutional neural network approaches in medical imaging: comprehensive training versus fine-tuning (Tajbakhsh et al., 2016)	2016	Explores the effectiveness of different convolutional neural network training methods for medical image analysis.
Pathological insights via AI: enhancing breast cancer metastasis detection (Liu et al., 2019)	2019	Advances nodal metastasis detection in breast cancer using AI technologies.
CheXNeXt: benchmarking deep learning in pneumonia diagnosis against radiologists (Rajpurkar et al., 2018)	2020	Demonstrates AI's ability to match or surpass radiologists in diagnosing pneumonia from chest radiographs.
Deep learning in diagnosing diabetic retinopathy: a fundus image analysis (Sarki et al., 2020)	2020	Reviews deep learning models for early detection of diabetic retinopathy and related eye conditions.
Exploring predictive analytics in health care: a machine learning perspective (Lamba et al., 2021)	2021	Surveys the landscape of ML in predictive health care analytics.
Forecasting Alzheimer's disease: ML models for progression from MCI (Grueso and Viejo-Sobera, 2021)	2021	Assesses ML techniques for predicting Alzheimer's disease progression from mild cognitive impairment.
Deep learning in magnetic resonance imaging for multiple sclerosis: automated detection and quality enhancement (Shoeibi et al., 2021)	2021	Focuses on improving magnetic resonance imaging–based multiple sclerosis detection using deep learning.
AI enhanced ultrasound in fatty liver disease diagnosis: a systematic exploration (Nduma et al., 2023)	2023	Discusses AI's role in enhancing ultrasound diagnostics for challenging liver conditions.
AI's emerging role in colorectal cancer screening via colonoscopy (Young et al., 2023)	2023	Evaluates AI's effectiveness in colorectal lesion detection and characterization.
Advancing precision pathology through explainable AI (Klauschen et al., 2024)	2024	Highlights the significance of explainability in AI for improving diagnostic precision in pathology.

AI, *artificial intelligence;* ML, *machine learning.*

nuanced details from images, leading to more precise diagnoses and tailored treatment approaches (Piórkowski et al., 2023). Table 2.3 lists a selection of studies demonstrating AI's diverse applications within medical imaging. The studies emphasize the use of various convolutional neural network models to improve image analysis, encompassing rapid MRI techniques and superresolution imaging, automating and refining diagnostic procedures for better accuracy and personalized care (Cha et al., 2022).

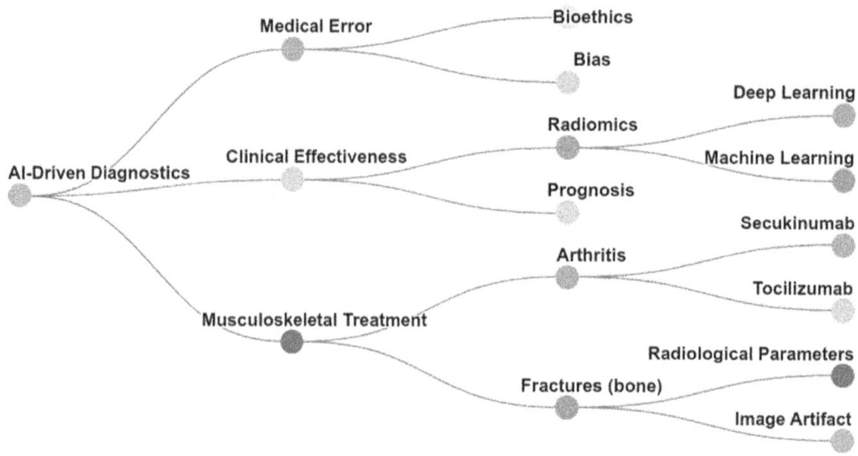

FIGURE 2.2

A mind map showcasing artificial intelligence (AI)-driven diagnostics in treating musculoskeletal disorders.

Created using SCOPUS AI.

AI's transformative impact extends to automating routine imaging tasks, enabling predictive modeling, and supporting personalized treatment plans while treating musculoskeletal conditions. Thus, AI can facilitate quicker and more informed clinical decision-making by improving diagnostic accuracy and reducing interpretation times. AI's contribution to early disease detection and progression monitoring is crucial, opening a new era for medical imaging in MSD treatment. The deployment of AI in clinical settings is set to address the shortage of subspecialized radiologists. AI enhances radiologists' workflows by prioritizing exams, assisting in image interpretation, and expediting report generation, promising to enhance patient care and operational efficiency in musculoskeletal radiology (Guermazi et al., 2024). Furthermore, the incorporation of digital twin models in customized health care has the ability to lead to notable progression in MSD treatment, because these models provide intricate simulations that inform treatment options (Gyftopoulos and Subhas, 2020).

2.4 Personalized musculoskeletal disorder treatment plans through artificial intelligence

Preparing a traditional personalized MSD treatment plan is a challenging task because no single cause exists and the response to the treated conditions differs among various individuals (El-Tallawy et al., 2021). However, the advent of AI and ML provides a ray of hope to address this problem and offers tailored treatment plans for each MSD patient. Fig. 2.3 is a mind map showcasing the possible

Table 2.3 Artificial intelligence and machine learning applications in imaging studies (Cha et al., 2022).

Study	Artificial intelligence name	CNN architecture	ROI or important region labeling
Adams et al. (2019)	DCNN	AlexNet/ GoogLeNet	–
Urakawa et al. (2019)	CNN	VGG16	–
Cheng et al. (2019)	DCNN	DenseNet-121	Assessed label accuracy and image quality.
Krogue et al. (2020)	Deep learning	DenseNet-169	Used single-shot detector with Resnet-50 FPN architecture
Yu et al. (2020)	CNN	Inception-V3	Local expert approved or revised each ROI
Mutasa et al. (2020)	CNN	Custom ResNet	Employed Grad-CAM for ROI identification
Beyaz et al. (2020)	CNN	GA blocks	Manually cropped images of fractured and nonfractured femoral necks
Mawatari et al. (2020)	DCNN	GoogLeNet	Radiographs checked and annotated by radiologists with computed tomography and magnetic resonance imaging for ROI
Yamada et al. (2020)	CNN	Xception	Orthopedic surgeon preprocessed images by cropping relevant femoral regions
Cheng et al. (2020)	DCNN	DenseNet-121	Used Grad-CAM for important region highlighting
Yoon et al. (2020)	Faster R–CNN	VGG-16	Applied Grad-CAM for ROI identification
Sato et al. (2021)	CNN	EfficientNet-B4	Used Grad-CAM for ROI marking
Bae et al. (2021)	CNN	CBAM++	ResNet18 and highlighted regions with Grad-CAM
Murphy et al. (2022)	CNN	GoogLeNet	Labeled images using MATLAB training image Labeler application

CNN, *convolutional neural network;* ROI, *region of interest.*

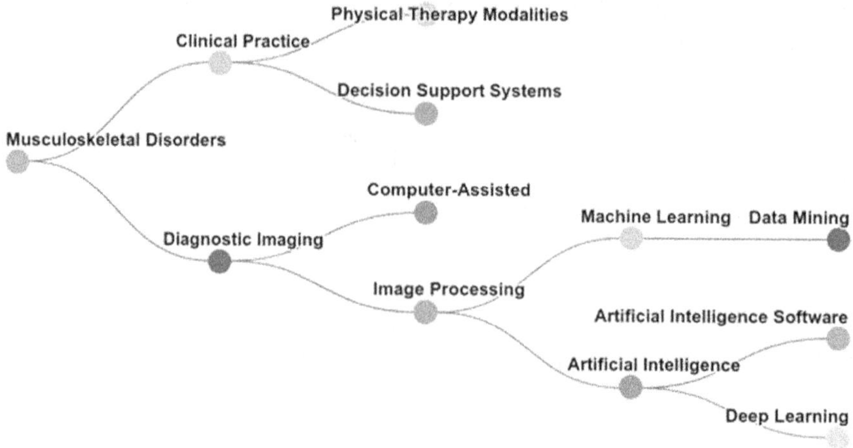

FIGURE 2.3

Detailed mind map showcasing the convergence of musculoskeletal disorder treatment, machine learning, and clinical practices.

Created using SCOPUS AI.

integration of AI and ML at various stages that facilitates an effective treatment plan for individuals. The success of this integration is based on the abilities of AI and ML to identify patterns and accurately predict the best effective treatment plans for individual cases. Such an integration would optimize the management of personalized treatment plans for individuals and also be useful in managing more complex health conditions that were undiscovered while being traditionally diagnosed.

2.4.1 Data analysis for personalized patient care

Incorporating AI and ML in the preparation of personalized patient care (PPC) for individuals with MSDs tends to use extensive datasets to analyze various variables such as genetic information, preliminary diagnostic information, and patient history (Padhi et al., 2023). These types of data analytics have a significant role in preparing the PPC. Descriptive analysis may facilitate an understanding of historical trends whereas predictive analysis may help in forecasting. Similarly, prescriptive analysis may be used to discover innovative treatment plans. By effectively applying data analytics principles in medical sciences, the simple act of storing medical records can be transformed into the discovery of new medications. Data science and analytics are crucial tools in health care because they enable more informed decision-making and the identification and elimination of inefficiencies. The most basic part is the pattern recognition ability of AI and ML while using these datasets that improve the forecasting precision. One such exemplary model was developed and discussed in a study to forecast the progression of osteoarthritis (Lee et al.,

2022). Forecasting techniques are the ones that in reality help tailor the treatment procedures for a selected individual for a particular MSD. Thus, by employing accurate and precise data analytics integrated with the traditional techniques of prevention, health care practitioners can significantly augment the efficacy of interventions, heighten patient contentment, and boost overall results.

2.4.2 Predictive modeling for treatment strategies

Incorporating predictive modeling with AI and ML is revolutionizing the treatment strategies for MSDs. This innovative approach empowers clinicians to anticipate disease progression accurately, patient responses to treatments, and the likelihood of adverse effects with unparalleled precision (Haider, 2023; Rasool et al., 2023). The complexities of predictive modeling concerning treatment strategies lies in investigating the extensive and varied datasets in-depth, including genetic information, electronic health records, environmental exposures, and lifestyle choices (Prosperi et al., 2018; Krittanawong et al., 2018). AI algorithms facilitates analyzing these datasets, recognizing patterns, and generating insights that otherwise might not have been immediately apparent. Thus, predictive models offer a much more precise approach to managing prolonged MSD conditions compared with the traditional trial-and-error method. The technique promises to facilitate the process of identifying the best possible treatments for a specific patient group by analyzing disease patterns across large populations. It significantly enhances the precision of clinical outcomes by identifying the right individuals who are more likely to develop MSDs (Tiwari et al., 2023).

Furthermore, predictive modeling could be of great help in shaping health policies and guidelines. Thus, in simple words, it can be said that by using predictive modelling technique, the healthcare professionals can analyze data at the population level, gain valuable insights, allocate resources more effectively and implement targeted public health interventions for MSD patients.

2.5 Surgical applications of artificial intelligence in orthopedics

Incorporating AI into orthopedic surgery has led to a significant shift toward patient-centered care, focusing on personalized surgical planning and execution. AI has been proven to enhance surgical precision, improve surgical outcomes, and promote the development of personalized interventions, especially for MSD conditions in which accuracy is crucial. AI-driven technologies, particularly ML algorithms and computer vision, have become an integral part of the modern surgical process, from preoperative planning to postoperative care.

During the preoperative phase, AI systems can help to analyze patient data, including imaging and biomechanical information, which helps a surgeon to develop personalized surgical strategy. This kind of tailored planning undoubtedly has a significant role in orthopedics, where unique anatomical differences and specific intricacies of conditions require customized approaches. The emergence of robot-assisted surgery, driven by ML and predictive analytics, represents a major step forward in the surgical applications of AI. The involvement of robots along with advanced navigation systems powered by AI have begun to revolutionize orthopedic surgery by enhancing the precision of complex procedures such as bone cutting and joint replacement. These systems minimize errors and optimize implant positioning and alignment, significantly improving the overall surgical outcomes and long-term effectiveness.

Considering the postoperative phase, AI facilitates monitoring recovery trajectories, predicting complications, and providing the rehabilitation protocols for individual needs. Wearable AI devices along with predictive models are used to analyze patient data to provide customized recovery plans, ensuring that postsurgical care is as personalized and effective as the surgical intervention itself (Singam, 2023). Thus, the continuous development of AI technologies promises significant improvements in surgical accuracy, patient well-being, and overall health care delivery. Table 2.4 lists important studies demonstrating how AI has improved surgical techniques and patient care protocols. Embracing these advances, the future of AI-driven orthopedic surgery offers exciting possibilities for enhancing patient care, inspiring health care professionals and researchers.

Table 2.4 Studies that incorporated AI in surgical techniques and patient care.

Study	Year	Central discoveries
Evaluation of robot-assisted spinal surgery effectiveness (Li et al., 2020)	2020	Enhanced the precision of bone resection and prosthetic placement.
Wearable AI devices in post-surgical monitoring (Liu et al., 2020)	2020	Improved the precision of postoperative monitoring.
AI-driven preoperative protocols for knee arthroplasty (Lambrechts et al., 2022)	2022	Improved implant sizing and positioning precision.
AI for tailored orthopedic implant design (Ren and Yi, 2022)	2022	Facilitated the achievement of customized implant designs concerning each patients' anatomical needs.
AI in predicting outcomes of rotator cuff surgery (Potty et al., 2023)	2023	Helped in early identification of patients at elevated risk.
Computer-assisted image registration in surgery (Taleb et al., 2023)	2023	Contributed to improvements in surgical navigation accuracy while reducing errors during operations.

AI, *artificial intelligence.*

2.6 Challenges and ethical considerations

Identifying and classifying MSDs with the help of AI and ML, like any their other applications, presents significant challenges and raises important ethical considerations (Kumar et al., 2024). Various studies on AI/ML implementation in health care have highlighted concerns regarding data privacy and the safeguarding of patient information. For instance, research has drawn attention to the vulnerability of health care data to breaches, highlighting the need for strong cybersecurity measures that should adhere to legal frameworks such as the Health Insurance Portability and Accountability Act of 1996. Algorithmic bias is another critical issue, because the biased training datasets are known to affect treatment outcomes significantly and intensify health care inequalities (Anikwe et al., 2022; Chen et al., 2023). Also, the integration of AI into clinical practice requires substantial investments in infrastructure and training. Nevertheless, researchers are exploring the various difficulties of incorporating AI technologies into existing health care systems and trying to find solutions to overcome these hurdles. Ethical concerns include ensuring informed consent is given and maintaining the relationship between health care professionals and patients. The lack of transparency in a few AI algorithms tends to complicate the process of obtaining informed consent and thus potentially affects patient autonomy (Durán and Jongsma, 2021).

2.7 Conclusion

This chapter explored the significant impacts of AI and ML applications in MSD treatment. It discussed the improvements while offering the potential for improved accuracy in diagnosing and treating MSDs. The advances represent a significant shift toward a health care model that may focus on personalization and placing patients at the center, highlighting the importance of being able to anticipate and meet the specific needs of every individual.

The integration of AI/ML technologies is forecasted to have the potential of greatly enhancing the effectiveness, efficiency, and inclusivity of patient care. AI and ML advances exhibit the potential for completely transforming traditional trial-and-error treatment methods while improving outcomes. However, there are significant challenges that need to be addressed to use the potential of AI/ML integration fully in MSD treatment, particularly addressing data privacy and potential biases in algorithms. To address such difficulties effectively while providing the best patient care with precise treatment, it is important for clinicians, technologists, researchers, and the policymakers to engage in collaborative efforts and have a cohesive and unified approach to harness the potential of AI and ML fully in treating MSD.

AI disclosure

During the preparation of this work the author(s) used ChatGPT 4 and SCOPUS AI to improve the writing language. The SCOPUS AI tool was used to create the unique mind map for the description provided in the chapter in a few places. However, I wish to declare that no programs were used to generate the content. After using this tool/service, the author(s) reviewed and edited the content as needed and take(s) full responsibility for the content of the publication.

References

Adams, M., Chen, W., Holcdorf, D., McCusker, M.W., Howe, P.D.L., Gaillard, F., 2019. Computer vs human: deep learning versus perceptual training for the detection of neck of femur fractures. J. Med. Imag. Radiat. Oncol. 63 (1), 27–32. https://doi.org/10.1111/1754-9485.12828.

Ahmed, Z., Mohamed, K., Zeeshan, S., Dong, X.Q., 2020. Artificial intelligence with multi-functional machine learning platform development for better healthcare and precision medicine. Database 2020. https://doi.org/10.1093/database/baaa010.

Anikwe, C.V., Nweke, H.F., Ikegwu, A.C., Egwuonwu, C.A., Onu, F.U., Alo, U.R., Teh, Y.W., 2022. Mobile and wearable sensors for data-driven health monitoring system: state-of-the-art and future prospect. Expert Syst. Appl. 202.

Bae, J., Yu, S., Oh, J., Kim, T.H., Chung, J.H., Byun, H., Yoon, M.S., Ahn, C., Lee, D.K., 2021. External validation of deep learning algorithm for detecting and visualizing femoral neck fracture including displaced and non-displaced fracture on plain X-ray. J. Digit. Imag. 34 (5), 1099–1109. https://doi.org/10.1007/s10278-021-00499-2.

Benjamens, S., Dhunnoo, P., Meskó, B., 2020. The state of artificial intelligence-based FDA-approved medical devices and algorithms: an online database. Npj Digit. Med. 3 (1). https://doi.org/10.1038/s41746-020-00324-0.

Beyaz, S., Açici, K., Sümer, E., 2020. Femoral neck fracture detection in X-ray images using deep learning and genetic algorithm approaches. Joint Dis. Relat. Surg. 31 (2), 175–183. https://doi.org/10.5606/EHC.2020.72163.

Blemker, S.S., Asakawa, D.S., Gold, G.E., Delp, S.L., 2007. Image-based musculoskeletal modeling: applications, advances, and future opportunities. J. Magn. Reson. Imag. 25 (2), 441–451. https://doi.org/10.1002/jmri.20805.

Briggs, A.M., Shiffman, J., Shawar, Y.R., Åkesson, K., Ali, N., Woolf, A.D., 2020. Global health policy in the 21st century: challenges and opportunities to arrest the global disability burden from musculoskeletal health conditions. Best Pract. Res. Clin. Rheumatol. 34 (5). https://doi.org/10.1016/j.berh.2020.101549.

Cha, Y., Kim, J.T., Park, C.H., Kim, J.W., Lee, S.Y., Yoo, J.I., 2022. Artificial intelligence and machine learning on diagnosis and classification of hip fracture: systematic review. J. Orthop. Surg. Res. 17 (1). https://doi.org/10.1186/s13018-022-03408-7.

Chen, R.J., Wang, J.J., Williamson, D.F.K., Chen, T.Y., Lipkova, J., Lu, M.Y., Sahai, S., Mahmood, F., 2023. Algorithmic fairness in artificial intelligence for medicine and healthcare. Nat. Biomed. Eng. 7 (6), 719–742. https://doi.org/10.1038/s41551-023-01056-8.

Cheng, C.T., Ho, T.Y., Lee, T.Y., Chang, C.C., Chou, C.C., Chen, C.C., Chung, I.F., Liao, C.H., 2019. Application of a deep learning algorithm for detection and visualization of hip fractures on plain pelvic radiographs. Eur. Radiol. 29 (10), 5469–5477. https://doi.org/10.1007/s00330-019-06167-y.

Cheng, C.-T., Chen, C.-C., Cheng, F.-J., Chen, H.-W., Su, Y.-S., Yeh, C.-N., Chung, I.-F., Liao, C.-H., 2020. A human- algorithm integration system for hip fracture detection on plain radiography: system development and validation study. JMIR Med. Inf. 8.

Conaghan, P.G., O'Connor, P., Isenberg, D.A., 2011. Musculoskeletal Imaging. Oxford University PressOxford. https://doi.org/10.1093/med/9780199235773.001.1, 9780191726118.

Durán, J.M., Jongsma, K.R., 2021. Who is afraid of black box algorithms? On the epistemological and ethical basis of trust in medical AI. J. Med. Ethics 47 (5), 329–335. https://doi.org/10.1136/medethics-2020-106820.

El-Tallawy, S.N., Nalamasu, R., Salem, G.I., LeQuang, J.A.K., Pergolizzi, J.V., Christo, P.J., 2021. Management of musculoskeletal pain: an update with emphasis on chronic musculoskeletal pain. Pain Ther. 10 (1), 181–209. https://doi.org/10.1007/s40122-021-00235-2.

Grueso, S., Viejo-Sobera, R., 2021. Machine learning methods for predicting progression from mild cognitive impairment to Alzheimer's disease dementia: a systematic review. Alzheimer's Res. Ther. 13 (1). https://doi.org/10.1186/s13195-021-00900-w.

Guermazi, A., Omoumi, P., Tordjman, M., Fritz, J., Kijowski, R., Regnard, N.E., Carrino, J., Kahn, C.E., Knoll, F., Rueckert, D., Roemer, F.W., Hayashi, D., 2024. How AI may transform musculoskeletal imaging. Radiology 310 (1). https://doi.org/10.1148/radiol.230764.

Gyftopoulos, S., Subhas, N., 2020. Musculoskeletal imaging applications of artificial intelligence. Semin. Muscoskel. Radiol. 24 (1). https://doi.org/10.1055/s-0039-3400511.

Haider, R., 2023. The future of pharmaceuticals industry 2024. J. Pharm. Pharmacol. Res. 6.

Hirschmann, A., Cyriac, J., Stieltjes, B., Kober, T., Richiardi, J., Omoumi, P., 2019. Artificial intelligence in musculoskeletal imaging: review of current literature, challenges, and trends. Semin. Muscoskel. Radiol. 23 (3), 304–311. https://doi.org/10.1055/s-0039-1684024.

Jacobs, R., 2022. Deep learning object detection in materials science: current state and future directions. Comput. Mater. Sci. 211. https://doi.org/10.1016/j.commatsci.2022.111527.

Johnson, K.B., Wei, W.Q., Weeraratne, D., Frisse, M.E., Misulis, K., Rhee, K., Zhao, J., Snowdon, J.L., 2021. Precision medicine, AI, and the future of personalized health care. Clin. Translat. Sci. 14 (1), 86–93. https://doi.org/10.1111/cts.12884.

Klauschen, F., Dippel, J., Keyl, P., Jurmeister, P., Bockmayr, M., Mock, A., Buchstab, O., Alber, M., Ruff, L., Montavon, G., Müller, K.R., 2024. Toward explainable artificial intelligence for precision pathology. Annu. Rev. Pathol. 19, 541–570. https://doi.org/10.1146/annurev-pathmechdis-051222-113147.

Krenn, M., Buffoni, L., Coutinho, B., Eppel, S., Foster, J.G., Gritsevskiy, A., Lee, H., Lu, Y., Moutinho, J.P., Sanjabi, N., Sonthalia, R., Tran, N.M., Valente, F., Xie, Y., Yu, R., Kopp, M., 2023. Forecasting the future of artificial intelligence with machine learning-based link prediction in an exponentially growing knowledge network. Nat. Mach. Intell. 5 (11), 1326–1335. https://doi.org/10.1038/s42256-023-00735-0.

Krittanawong, C., Bomback, A.S., Baber, U., Bangalore, S., Messerli, F.H., Wilson Tang, W.H., 2018. Future direction for using artificial intelligence to predict and manage hypertension. Curr. Hypertens. Rep. 20 (9). https://doi.org/10.1007/s11906-018-0875-x.

Krogue, J.D., Cheng, K.V., Hwang, K.M., Toogood, P., Meinberg, E.G., Geiger, E.J., Zaid, M., McGill, K.C., Patel, R., Sohn, J.H., Wright, A., Darger, B.F., Padrez, K.A., Ozhinsky, E.,

Majumdar, S., Pedoia, V., 2020. Automatic hip fracture identification and functional subclassification with deep learning. Radiol.: Artif. Intell. 2 (2). https://doi.org/10.1148/RYAI.2020190023.

Kumar, R., Gupta, M., Abraham, A., 2024. A critical analysis on vertebra identification and cobb angle estimation using deep learning for scoliosis detection. IEEE Access 12, 11170–11184. https://doi.org/10.1109/ACCESS.2024.3353794.

Lamba, D., Hsu, W.H., Alsadhan, M., 2021. Predictive analytics and machine learning for medical informatics: a survey of tasks and techniques. In: Machine Learning, Big Data, and IoT for Medical Informatics. Elsevier, United States, pp. 1–35. https://doi.org/10.1016/B978-0-12-821777-1.00023-9, 9780128217771.

Lambrechts, A., Wirix-Speetjens, R., Maes, F., Van Huffel, S., 2022. Artificial intelligence based patient-specific preoperative planning algorithm for total knee arthroplasty. Front. Robot. AI 9. https://doi.org/10.3389/frobt.2022.840282.

Leal Filho, W., Wall, T., Rui Mucova, S.A., Nagy, G.J., Balogun, A.L., Luetz, J.M., Ng, A.W., Kovaleva, M., Safiul Azam, F.M., Alves, F., Guevara, Z., Matandirotya, N.R., Skouloudis, A., Tzachor, A., Malakar, K., Gandhi, O., 2022. Deploying artificial intelligence for climate change adaptation. Technol. Forecast. Soc. Change 180. https://doi.org/10.1016/j.techfore.2022.121662.

Lecouvet, F.E., 2016. Whole-body MR imaging: musculoskeletal applications. Radiology 279 (2), 345–365. https://doi.org/10.1148/radiol.2016142084.

Lecun, Y., Bengio, Y., Hinton, G., 2015. Deep learning. Nature 521 (7553), 436–444. https://doi.org/10.1038/nature14539.

Lee, L.S., Chan, P.K., Wen, C., Fung, W.C., Cheung, A., Chan, V.W.K., Cheung, M.H., Fu, H., Yan, C.H., Chiu, K.Y., 2022. Artificial intelligence in diagnosis of knee osteoarthritis and prediction of arthroplasty outcomes: a review. Arthroplasty 4 (1). https://doi.org/10.1186/s42836-022-00118-7.

Li, H., Giger, M.L., 2021. Artificial Intelligence and Interpretations in Breast Cancer Imaging. Elsevier BV, pp. 291–308. https://doi.org/10.1016/b978-0-12-821259-2.00015-6, 9780128212592.

Li, J., Fang, Y., Jin, Z., Wang, Y., Yu, M., 2020. The impact of robot-assisted spine surgeries on clinical outcomes: a systemic review and meta-analysis. Int. J. Med. Robot. Comput. Assist. Surg. 16 (6), 1–14. https://doi.org/10.1002/rcs.2143.

Liu, Y., Kohlberger, T., Norouzi, M., Dahl, G.E., Smith, J.L., Mohtashamian, A., Olson, N., Peng, L.H., Hipp, J.D., Stumpe, M.C., 2019. Artificial intelligence–based breast cancer nodal metastasis detection insights into the black box for pathologists. Arch. Pathol. Lab Med. 143 (7), 859–868. https://doi.org/10.5858/arpa.2018-0147-OA.

Liu, Y., Liu, C., Gao, M., Wang, Y., Bai, Y., Xu, R., Gong, R., 2020. Evaluation of a wearable wireless device with artificial intelligence, iThermonitor WT705, for continuous temperature monitoring for patients in surgical wards: a prospective comparative study. BMJ Open 10 (11). https://doi.org/10.1136/bmjopen-2020-039474.

Mawatari, T., Hayashida, Y., Katsuragawa, S., Yoshimatsu, Y., Hamamura, T., Anai, K., Ueno, M., Yamaga, S., Ueda, I., Terasawa, T., Fujisaki, A., Chihara, C., Miyagi, T., Aoki, T., Korogi, Y., 2020. The effect of deep convolutional neural networks on radiologists' performance in the detection of hip fractures on digital pelvic radiographs. Eur. J. Radiol. 130. https://doi.org/10.1016/j.ejrad.2020.109188.

Murphy, E.A., Ehrhardt, B., Gregson, C.L., von Arx, O.A., Hartley, A., Whitehouse, M.R., Thomas, M.S., Stenhouse, G., Chesser, T.J.S., Budd, C.J., Gill, H.S., 2022. Machine learning outperforms clinical experts in classification of hip fractures. Sci. Rep. 12 (1). https://doi.org/10.1038/s41598-022-06018-9.

Mutasa, S., Varada, S., Goel, A., Wong, T.T., Rasiej, M.J., 2020. Advanced deep learning techniques applied to automated femoral neck fracture detection and classification. J. Digit. Imag. 33 (5), 1209–1217. https://doi.org/10.1007/s10278-020-00364-8.

Najjar, R., 2023. Redefining radiology: a review of artificial intelligence integration in medical imaging. Diagnostics 13 (17). https://doi.org/10.3390/diagnostics13172760.

Nduma, B.N., Al-Ajlouni, Y.A., Njei, B., 2023. The application of artificial intelligence (AI)-based ultrasound for the diagnosis of fatty liver disease: a systematic review. Cureus. https://doi.org/10.7759/cureus.50601.

Padhi, A., Agarwal, A., Saxena, S.K., Katoch, C.D.S., 2023. Transforming clinical virology with AI, machine learning and deep learning: a comprehensive review and outlook. Virus-Disease 34 (3), 345–355. https://doi.org/10.1007/s13337-023-00841-y.

Palestro, C.J., Love, C., Miller, T.T., 2006. Imaging of musculoskeletal infections. Best Pract. Res. Clin. Rheumatol. 20 (6), 1197–1218. https://doi.org/10.1016/j.berh.2006.08.009.

Panayides, A.S., Amini, A., Filipovic, N.D., Sharma, A., Tsaftaris, S.A., Young, A., Foran, D., Do, N., Golemati, S., Kurc, T., Huang, K., Nikita, K.S., Veasey, B.P., Zervakis, M., Saltz, J.H., Pattichis, C.S., 2020. AI in medical imaging informatics: current challenges and future directions. IEEE J. Biomed. Health Inform. 24 (7), 1837–1857. https://doi.org/10.1109/JBHI.2020.2991043.

Park, D.J., Park, M.W., Lee, H., Kim, Y.J., Kim, Y., Park, Y.H., 2021. Development of machine learning model for diagnostic disease prediction based on laboratory tests. Sci. Rep. 11 (1). https://doi.org/10.1038/s41598-021-87171-5.

Piano, S.L., 2020. Ethical principles in machine learning and artificial intelligence: cases from the field and possible ways forward. Humanit. Soc. Sci. Commun. 7, 1–7.

Piórkowski, A., Obuchowicz, R., Urbanik, A., Strzelecki, M., 2023. Advances in musculoskeletal imaging and their applications. J. Clin. Med. 12 (20). https://doi.org/10.3390/jcm12206585.

Potty, A.G., Potty, A.S.R., Maffulli, N., Blumenschein, L.A., Ganta, D., Mistovich, R.J., Fuentes, M., Denard, P.J., Sethi, P.M., Shah, A.A., Gupta, A., 2023. Approaching artificial intelligence in orthopaedics: predictive analytics and machine learning to prognosticate arthroscopic rotator cuff surgical outcomes. J. Clin. Med. 12 (6). https://doi.org/10.3390/jcm12062369.

Prosperi, M., Min, J.S., Bian, J., Modave, F., 2018. Big data hurdles in precision medicine and precision public health. BMC Med. Inf. Decis. Making 18 (1). https://doi.org/10.1186/s12911-018-0719-2.

Rajpurkar, P., Irvin, J., Ball, R.L., Zhu, K., Yang, B., Mehta, H., Duan, T., Ding, D., Bagul, A., Langlotz, C.P., Patel, B.N., Yeom, K.W., Shpanskaya, K., Blankenberg, F.G., Seekins, J., Amrhein, T.J., Mong, D.A., Halabi, S.S., Zucker, E.J., Ng, A.Y., Lungren, M.P., 2018. Deep learning for chest radiograph diagnosis: a retrospective comparison of the CheXNeXt algorithm to practicing radiologists. PLoS Med. 15 (11). https://doi.org/10.1371/journal.pmed.1002686.

Rasool, S., Husnain, A., Saeed, A., Gill, A.Y., Hussain, H.K., 2023. Harnessing predictive power: exploring the crucial role of machine learning in early disease detection. J. Inovasi dan Hum. 1, 302–315.

Ren, M., Yi, P.H., 2022. Artificial intelligence in orthopedic implant model classification: a systematic review. Skeletal Radiol. 51 (2), 407–416. https://doi.org/10.1007/s00256-021-03884-8.

Sarki, R., Ahmed, K., Wang, H., Zhang, Y., 2020. Automatic detection of diabetic eye disease through deep learning using fundus images: a survey. IEEE Access 8, 151133–151149. https://doi.org/10.1109/ACCESS.2020.3015258.

Sato, Y., Takegami, Y., Asamoto, T., Ono, Y., Hidetoshi, T., Goto, R., Kitamura, A., Honda, S., 2021. Artificial intelligence improves the accuracy of residents in the diagnosis of hip fractures: a multicenter study. BMC Muscoskel. Disord. 22 (1). https://doi.org/10.1186/s12891-021-04260-2.

Shajari, S., Kuruvinashetti, K., Komeili, A., Sundararaj, U., 2023. The emergence of AI-based wearable sensors for digital health technology: a review. Sensors 23 (23). https://doi.org/10.3390/s23239498.

Shoeibi, A., Khodatars, M., Jafari, M., Moridian, P., Rezaei, M., Alizadehsani, R., Khozeimeh, F., Gorriz, J.M., Heras, J., Panahiazar, M., Nahavandi, S., Acharya, U.R., 2021. Applications of deep learning techniques for automated multiple sclerosis detection using magnetic resonance imaging: a review. Comput. Biol. Med. 136. https://doi.org/10.1016/j.compbiomed.2021.104697.

Shrestha, Y.R., He, V.F., Puranam, P., von Krogh, G., 2021. Algorithm supported induction for building theory: how can we use prediction models to theorize? Organ. Sci. 32 (3), 856–880. https://doi.org/10.1287/orsc.2020.1382.

Singam, A., 2023. Revolutionizing patient care: a comprehensive review of artificial intelligence applications in anesthesia. Cureus. https://doi.org/10.7759/cureus.49887.

Srikumar, M., Finlay, R., Abuhamad, G., Ashurst, C., Campbell, R., Campbell-Ratcliffe, E., Hongo, H., Jordan, S.R., Lindley, J., Ovadya, A., Pineau, J., 2022. Advancing ethics review practices in AI research. Nat. Mach. Intell. 4 (12), 1061–1064. https://doi.org/10.1038/s42256-022-00585-2.

Syrowatka, A., Kuznetsova, M., Alsubai, A., Beckman, A.L., Bain, P.A., Craig, K.J.T., Hu, J., Jackson, G.P., Rhee, K., Bates, D.W., 2021. Leveraging artificial intelligence for pandemic preparedness and response: a scoping review to identify key use cases. Npj Digit. Med. 4 (1). https://doi.org/10.1038/s41746-021-00459-8.

Taddeo, M., McCutcheon, T., Floridi, L., 2019. Trusting artificial intelligence in cybersecurity is a double-edged sword. Nat. Mach. Intell. 1 (12), 557–560. https://doi.org/10.1038/s42256-019-0109-1.

Tajbakhsh, N., Shin, J.Y., Gurudu, S.R., Hurst, R.T., Kendall, C.B., Gotway, M.B., Liang, J., 2016. Convolutional neural networks for medical image analysis: full training or fine tuning? IEEE Trans. Med. Imag. 35 (5), 1299–1312. https://doi.org/10.1109/TMI.2016.2535302.

Taleb, A., Guigou, C., Leclerc, S., Lalande, A., Bozorg Grayeli, A., 2023. Image-to-Patient registration in computer-assisted surgery of head and neck: state-of-the-art, perspectives, and challenges. J. Clin. Med. 12 (16). https://doi.org/10.3390/jcm12165398.

Tiwari, P.C., Pal, R., Chaudhary, M.J., Nath, R., 2023. Artificial intelligence revolutionizing drug development: exploring opportunities and challenges. Drug Dev. Res. 84 (8), 1652–1663. https://doi.org/10.1002/ddr.22115.

Tomašev, N., Cornebise, J., Hutter, F., Mohamed, S., Picciariello, A., Connelly, B., Belgrave, D.C.M., Ezer, D., Haert, F.C.v.d., Mugisha, F., Abila, G., Arai, H., Almiraat, H., Proskurnia, J., Snyder, K., Otake-Matsuura, M., Othman, M., Glasmachers, T., Wever, W.D., Teh, Y.W., Khan, M.E., Winne, R.D., Schaul, T., Clopath, C., 2020. AI for social good: unlocking the opportunity for positive impact. Nat. Commun. 11 (1). https://doi.org/10.1038/s41467-020-15871-z.

Turing, A.M., 1950. Computing Machinery and Intelligence.

Urakawa, T., Tanaka, Y., Goto, S., Matsuzawa, H., Watanabe, K., Endo, N., 2019. Detecting intertrochanteric hip fractures with orthopedist-level accuracy using a deep convolutional neural network. Skeletal Radiol. 48 (2), 239–244. https://doi.org/10.1007/s00256-018-3016-3.

Vinuesa, R., Azizpour, H., Leite, I., Balaam, M., Dignum, V., Domisch, S., Felländer, A., Langhans, S.D., Tegmark, M., Fuso Nerini, F., 2020. The role of artificial intelligence in achieving the sustainable development goals. Nat. Commun. 11 (1). https://doi.org/10.1038/s41467-019-14108-y.

Yamada, Y., Maki, S., Kishida, S., Nagai, H., Arima, J., Yamakawa, N., Iijima, Y., Shiko, Y., Kawasaki, Y., Kotani, T., Shiga, Y., Inage, K., Orita, S., Eguchi, Y., Takahashi, H., Yamashita, T., Minami, S., Ohtori, S., 2020. Automated classification of hip fractures using deep convolutional neural networks with orthopedic surgeon-level accuracy: ensemble decision-making with antero-posterior and lateral radiographs. Acta Orthop. 91 (6), 699–704. https://doi.org/10.1080/17453674.2020.1803664.

Yoon, S.J., Kim, T.H., Joo, S.B., Oh, S.E., 2020. Automatic multi-class intertrochanteric femur fracture detection from ct images based on ao/ota classification using faster r-cnn-bo method. J. Appl. Biomed. 18 (4), 97–105. https://doi.org/10.32725/jab.2020.013.

Young, E., Edwards, L., Singh, R., 2023. The role of artificial intelligence in colorectal cancer screening: lesion detection and lesion characterization. Cancers 15 (21). https://doi.org/10.3390/cancers15215126.

Yu, J.S., Yu, S.M., Erdal, B.S., Demirer, M., Gupta, V., Bigelow, M., Salvador, A., Rink, T., Lenobel, S.S., Prevedello, L.M., White, R.D., 2020. Detection and localisation of hip fractures on anteroposterior radiographs with artificial intelligence: proof of concept. Clin. Radiol. 75 (3), 237.e1–237.e9. https://doi.org/10.1016/j.crad.2019.10.022.

Further reading

Rhon, D.I., Teyhen, D.S., Shaffer, S.W., Goffar, S.L., Kiesel, K., Plisky, P.P., 2016. Developing predictive models for return to work using the military power, performance and prevention (mp3) musculoskeletal injury risk algorithm: a study protocol for an injury risk assessment programme. Inj. Prev. 24 (1).

CHAPTER

Early stage detection of osteoarthritis of the joints (hip and knee) using machine learning

3

K.S. Hemanth[3], Deepika Tigulla[1], Vidya Lakshmi[1], Seyed Buhari[2]

[1]*REVA University, School of Computer Science and Applications, Bengaluru, Karnataka, India;* [2]*Universiti Teknologi Brunei, School of Business, Bandar Seri Begawan, Brunei;* [3]*Department of Computer Science, Christ University, Bangalore, Karnataka, India*

3.1 Introduction

3.1.1 Bridging the gap: Health care meets technology

Technology and health care have always been seen as separate domains, but they are now working together to form a sustainable alliance. The integration of technology and health care has become essential for the future of health care in a world where innovation is driven by technology and the latter seeks to improve patient outcomes. Technology has revolutionized diagnostics and treatment options, leading to improved accuracy and outcomes. Health care providers may now detect illnesses early, personalize treatments, and keep a closer eye on patients' progress with the help of modern technologies such as genetic testing, wearable sensors, and medical imaging equipment. These developments might lead to better quality of life and perhaps save lives. What was once a requirement during the peak of COVID guidelines has opened doors for patients across the world to receive more inclusive and easily accessible care. Health care practitioners may now contact patients in faraway regions and deliver virtual treatment with the use of video consultations, remote monitoring equipment, and mobile health applications. Patients now have far easier access to health care services, lower expenses, and more convenience thanks to this (Embracing the Future: The Impactful Intersection between Technology and Healthcare, 2023). Telehealth use is 38 times higher than before the pandemic. Vaccine development time is cut short to months instead of decades, and wearable tech such as Fitbits and smart contact lenses are turning people into personal health monitors (Council post, 2023).

Health care is undergoing a technological revolution. Every day, the health care sector produces enormous volumes of data. Technology has therefore made it possible to harness its power. Health care professionals may uncover patterns, derive useful insights, and arrive at well-informed judgments by using data analytics and predictive modeling techniques. This data-driven strategy promotes community health management, facilitates early intervention, and improves patient care (Fig. 3.1).

Diagnosing Musculoskeletal Conditions using Artificial Intelligence and Machine Learning to Aid Interpretation of Clinical Imaging
https://doi.org/10.1016/B978-0-443-32892-3.00003-8

3.2 The crippling cost: Osteoarthritis and global health

Among the musculoskeletal disorders, osteoarthritis (OA) is one of the major contributors to years lived with disability. OA is a degenerative joint condition that restricts one's ability to function properly by causing pain, swelling, and stiffness. Tissues around the joint are also affected by OA. The knees, hips, back, and hands are the most typical places for it. OA can arise for a variety of reasons, including a history of overuse or joint injuries, advanced age, and obesity. Women are more affected than are males. OA is more common in elderly people (about 70% of those over age 55). Hence, as the population ages, it is anticipated that its prevalence will rise globally. OA usually develops in the late forties to midfifties, although it can occasionally occur in younger people, especially athletes and those who have had joint damage or injury (Knee Osteoarthristies, 2024).

A systematic review and metaanalysis were conducted to estimate the epidemiologic characteristics of knee OA in the 21st century. The global majority of knee OA was 16.0% in individuals aged 15 and over. The incidence increased with age at 70–79 years for incidence and were higher in females than in males. The majority varied across continents, in which Asia had the highest amount of knee OA and South America the lowest. Radiographic knee OA was more prevalent than symptomatic knee OA. The incidence of knee OA was lower in smokers than in nonsmokers (Cui et al., 2020).

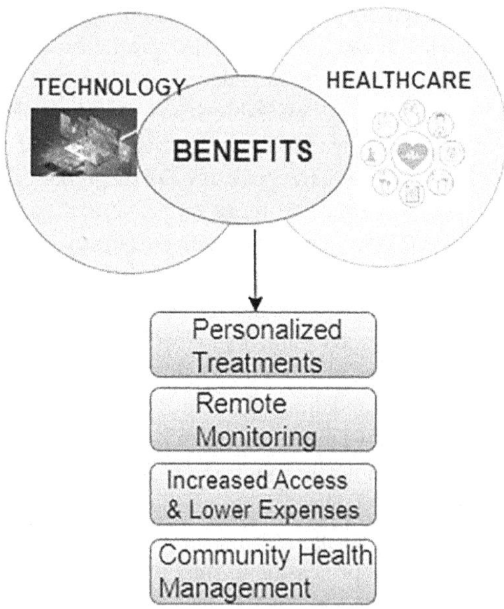

FIGURE 3.1

Integration of technology and health care.

A review of 34 cross-sectional studies from the past few decades found that OA is a significant public health problem in lower middle- and low-income countries, affecting about one in six persons. The increasing prevalence of OA may be due to factors such as rising life expectancy, obesity, occupational factors, and poverty, and has profound implications for individuals and society, particularly in settings with limited health care resources (Yahaya et al., 2021).

3.2.1 Diagnosing osteoarthritis: The achilles' heel of traditional methods

3.2.1.1 Unreliable reliance on subjective assessments

OA is diagnosed when persistent pain and functional impairments significantly affect everyday activities. Early stage OA can be difficult to diagnose and classify because of its shifting symptoms (Knee Osteoarthritis, 2024). Traditional classification criteria for OA, such as the ACR (American College of Rheumatology) criteria or radiographic Kallgren and Lawrence grading, indicate that the disease is advanced and disease-modifying treatments have been unsuccessful. To improve OA management and limit its impact on patients and society, it is recommended to prioritize early stage diagnosis and treatment. Radiography can help diagnose OA by identifying structural joint abnormalities including joint space narrowing and osteophyte production. This is universally supported. At an early stage, pain severity and transition to chronic pain may be averted. Functional limits are not yet severe enough to hinder a healthy lifestyle, and anatomical abnormalities may progress. To develop novel management techniques for OA, the focus is on diagnosing and treating early stage cases. It is critical to separate diagnosis from classification of early stage OA. Diagnosis occurs in clinical practice, whereas classification is the process of classifying people with OA in clinical research (Oei and Runhaar, 2023) (Fig. 3.2).

The figure depicts the percentage of types of OA obseved from the individuals suffering from the OA pain. A total of 69% of individuals in the world have knee OA, 9% hip OA, and 22% other OA (Oei et al., 2022).

Traditional diagnosis of OA is limited by the subjectivity of clinical evaluations, mainly because:

- There are no standard criteria against which such conditions should be judged, which makes it hard for doctors to provide an accurate indication when diagnosing the same condition. Hence, there is variability in diagnosis among health care providers. This can lead to unstable diagnostic reliability.
- In most cases, patients are predominant in helping to diagnose OA on the strength of the symptoms they report, which is often about pain, inability to move, and stiffness. However, such symptoms are subjective and can differ severely in

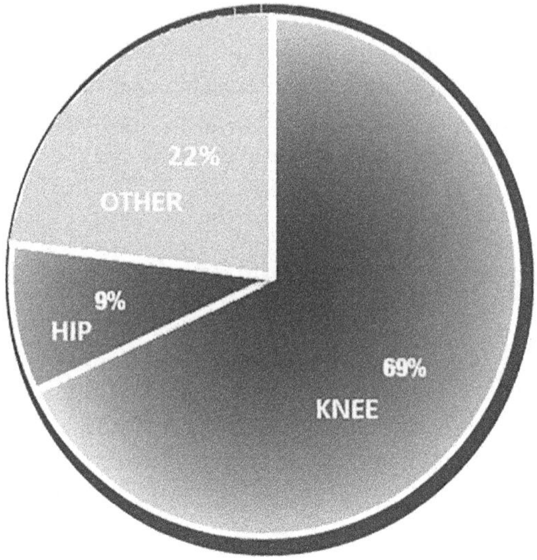

FIGURE 3.2

People with osteoarthritis in knee and hip.

their strength and duration. Thus, it is difficult to make an objective assessment of the extent to which the condition has reached the threshold of a particular health outcome.

- It is hard to make a clear diagnosis of OA from other conditions because it presents the same symptoms as other diseases affecting musculoskeletal system, such as rheumatoid arthritis or gout. Distinguishing between these situations may be difficult particularly at the beginning of the illness course, which makes it easy for people to mistake one for the other. This is a limitation that may result in a wrong diagnosis or the late diagnosis of OA.
- Traditional clinical tests fall short in detecting early OA changes, resulting in poor sensitivity and specificity and rendering them unreliable. Physical examination and range of motion tests may not be sensitive or specific enough to detect these changes. Thus, there is a potential for both overestimation and underestimation (Wang et al., 2021).

3.2.1.2 The lag in detection with standard imaging

Image analysis is widely employed for early OA detection and characterization as well as conventional imaging methods for diagnosing OA results in time delays. Clinical studies continue to enroll participants based only on radiographic imaging. Advanced imaging modalities and imaging analytic approaches may be possible for detecting persons with early stage knee OA, because additional time and financial resources are available for patient screening.

The sensitivity and specificity of diagnostic criteria are critical. Magnetic resonance imaging (MRI) might be a useful imaging modality to meet these requirements.

Unlike radiography, MRI may detect structural deterioration in OA-affected joints and surrounding tissues. Knee OA appears to affect several tissues, including articular cartilage, bone marrow, meniscus, ligaments, synovium, fat pads, and periarticular musculature, especially in the early stages of the illness. MRI can detect changes in soft tissues around the hip joint, including structures such as the labrum and ligamentum teres, which are thought to have a role in OA. Ultrasound and computed tomography (CT) are other methods for assessing OA characteristics. MRI methods can improve sensitivity even further. Quantitative MRI methods have largely been used on the knee, followed by the hip joint.

Various radiotracers can be employed for MRI. Owing to the high cost and restricted availability, its application in extensive clinical investigations is impractical. On the other hand, MRI has the capacity to unlock several underlying causes as well as subtypes of OA at their initial stages, based on its applications (Ehmig et al., 2023) (Fig. 3.3).

OA has been widely investigated using usual imaging techniques such as MRI, ultrasound, and conventional radiography, which cannot capture all features of the illness simultaneously. In addition, the processes are time-consuming.

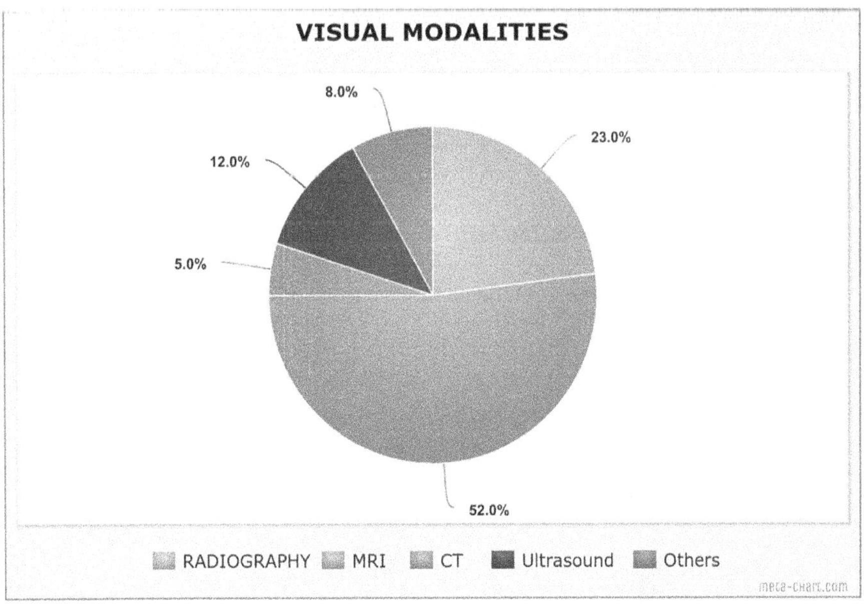

FIGURE 3.3

Identifying osteoarthritis using different imaging modalities.
CT, computed tomography; *MRI*, magnetic resonance imaging.

- **Identification of Early changes:** Standard imaging techniques might not always show early variations that are associated with OA. For example, synovitis or early cartilage damage might not be clearly depicted by usual imaging procedures such as radiographs. In its early stages, the disease may be difficult to diagnose until it becomes severe, and thus treatment would not yield positive results.
- **Soft Tissue Assessment**: Although people are increasingly recognizing that changes in fat and tissue inflammation greatly contribute to OA problems, ordinary imaging techniques may be unable to detect these kinds of soft tissue alterations accurately. X-rays are used for imaging purposes, but MRI and ultrasound procedures are better at detecting these kinds of soft tissue abnormalities. Nevertheless, they still may fail to detect even small ones, leading to delays in clinical assessment and treatment.
- **Quantification of Inflammation**: An attempt to quantify swelling has many difficulties because MRI and ultrasonography remain the best tools to assess synovitis. However, standard MRI procedures are not always the most suitable for contrast to assess swelling, and may lead to a delay in the accurate measurement of synovitis as well as other inflammatory transformations.
- **Limited Resolution**: The constraints posed on standard imaging methods may be low, especially in terms of detecting small morphologic variations of bone and cartilage. Hence, it will take longer to reveal preclinical OA-associated pathologic abnormalities such as osteonodules with cartilage damage or focal subchondral bone osteopenia.
- **Interpretation Time:** Interpreting a usual image study can be difficult, especially if it involves huge datasets or complicated cases. This may lead to deferred treatment interventions or medication for patients with AO because interpretation takes longer than necessary.

3.3 Potential of machine learning for early osteoarthritis detection

OA is the most common joint disease and is a widespread disability that carries a significant health burden around the world. Timely diagnosis and aggressive therapeutic intervention are important ameliorating the outcome for patients with OA and preventing its worsening. Machine learning (ML) technology is becoming an essential way for the better and early detection of OA. ML algorithms equipped with enormous computing power and the capacity to comprehend large and heterogeneous datasets can have unbelievable accuracy. Medical information made of images and biomechanical data is a good example. ML can discover hidden patterns, helping to diagnose OA early and improving its treatment. Thanks to this innovative ability, predictive modeling, individualized treatment protocols, and remote sensing are possible that eventually lead to the proactive and more progressive management of OA. Apart from the development of a fast diagnosis, research on using ML for the early detection of OA points to the beginning of innovations likely to improve precision medicine and patient care.

3.3.1 Machine learning: Reshaping the health care landscape

In health care, ML is revolutionary. ML models are becoming an optimal evidence-based approach necessary for forecasting OA. Cutting-edge computer technology for analyzing complex medical data could be the basis for more accurate diagnostics, identifying disease progression, responding to treatment, and predicting complications, therapies, and outcomes in the long run. Computer technology has the potential to become widely used in OA treatment but with a main focus on early diagnosis. ML also has the ability to decide diagnostics and to personalize the treatment plan. Through an assessment of patient records and medical histories, ML algorithms can decide the optimal treatment for each OA patient, considering the physiologic characteristics of the patient. Individualized clinical plans have the benefit of more positive treatment outcomes and attention turned toward patient-oriented care (Alexopoulos et al., 2022) (Fig. 3.4).

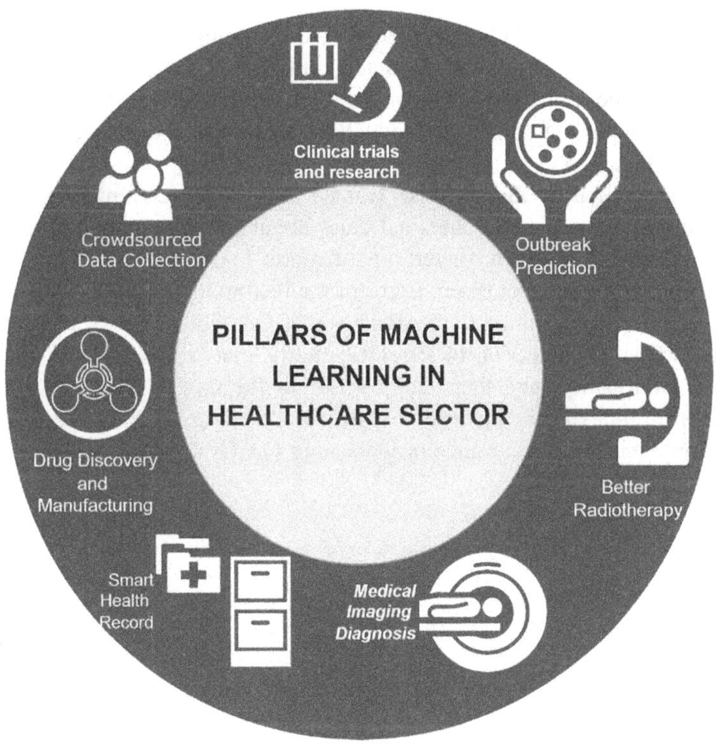

FIGURE 3.4

Uses of machine learning for health care.

Health care professionals can use key insights from ML predictions regarding the need for total knee replacement in OA to improve the way they treat patients.

Increased operational efficiency in health care is another benefit of ML. More efficiency and cost reductions result from the use of ML algorithms in resource allocation, patient throughput, and hospital administration. ML models can also enhance OA patients' surgical outcomes. This improves meticulous precision while speeding the organization, showcasing ML's exceptional potential orthopedic operations (Piccolo et al., 2023).

3.3.2 Spectrum of solutions: Diverse machine learning algorithms for early osteoarthritis diagnosis

The application of several ML algorithms for detecting OA is a multidimensional approach that aims to achieve better diagnostic accuracy and patient outcomes. In this field, various ML strategies exist that work with specific techniques to identify obstacles to OA detection.

Unsupervised learning algorithms: These techniques are crucial for dealing with very large datasets of unlabeled data. Supervised learning techniques such as clustering can be used in this field of OA detection to detect subgroups or patient data trends. For instance, using k-means grouping, OA conditions were identified by clinical and radiological criteria for individualized treatment and management (Parisi et al., 2020) (Fig. 3.5).

Supervised learning algorithms: Trained supervised learning algorithms are most commonly used to find contextual clues about OA. These algorithms get the training data labeled and then convert input features to output labels. Supervised algorithms are able to conduct observances of medical images from CT scans, MRIs and x-rays, and so on to provide the diagnosis of OA. For instance, support vector machines enable the detection of specific patterns that represent x-ray images of OA (knee arthropathy) and then identify OA as the case. Along that line, data extracted from images may be used with decision trees or random forests to determine the probability of occurrence or worsening OA (Xuan et al., 2023).

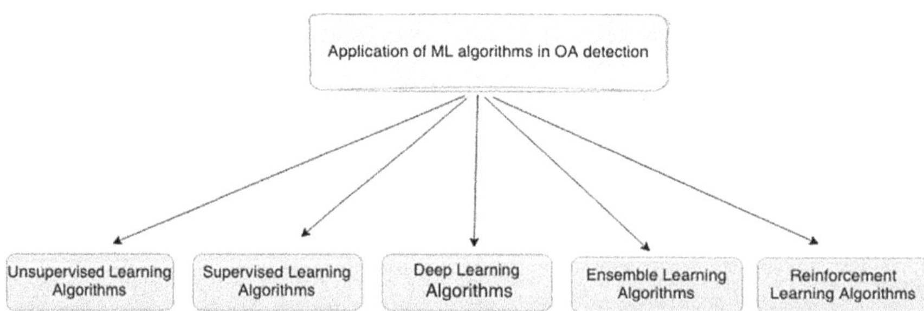

FIGURE 3.5

Machine learning (ML) algorithms in osteoarthritis (OA) detection.
AI, artificial intelligence.

Deep learning algorithms: Within deep learning techniques, convolutional neural networks (CNNs) are shown to have the highest performance ratings for medical image recognition applications. Therefore, they are traditionally used diagnosing OA in a variety of images. Owing to their intrinsic abilities, CNNs can arrange the features of raw images hierarchically, leading to the identification of fine patterns that presage the onset or progression of OA. The potential of deep learning to improve the diagnosis of OA was indicated when researchers used CNNs to detect abnormalities associated with OA accurately in radiographs. Binvignat et al. (2022) shows the high accuracy of detecting OA-related abnormalities in radiographs, as demonstrated by an experiment using CNNs.

Ensemble Learning: In ensemble learning, several ML models are aggregated to avoid collisions and increase prediction accuracy. Selected ensemble techniques such as gradient boosting and AdaBoost are used for multiple ML algorithms for OA detection to reduce errors and increase prediction accuracy. Ensemble learning merges different data qualities into one that can identify precise data relating to the disease, allowing reliable detection and monitoring.

Reinforcement learning: Compared with other ML paradigms, reinforcement learning is underused in OA detection, but it has the potential to improve treatment strategies and rehabilitation planning for patients with OA. Reinforcement learning algorithms can learn adaptively from patient comments and outcomes (Alami et al., 2011) (Table 3.1).

3.3.3 Data drives the engine: Building accurate models

Because OA of the hip and knee is a complex disease, it is vital for its management to be data driven. Comprehensive insights into the disease course, response to treatment, and risk factors can be acquired by health care professionals when they combine data drawn from various sources including patient records, imaging studies, and clinical trials. The information allows for individualized treatment plans, early action tactics, and predictive analysis to enhance patient care. Furthermore, investigating the mechanisms behind OA facilitated by a knowledge-driven method can result in pioneering medical treatments and therapies. The most recent progress in boosting predictive and diagnostic precision is found in sophisticated analytics and ML methods. In general, leveraging knowledge-based techniques substantially improved patient results, clinical decision-making, and understanding of OA.

3.3.3.1 Harnessing the power of comprehensive datasets

Scientific advances in OA detection and treatment have been greatly influenced by the availability of large datasets coupled with data-driven methods and ML approaches. Research and treatment approaches that are open access should benefit from access to large volume collections.

Rich Information: Scans, charts, and data on patient demographics, outcomes of clinical evaluations, genetic data, and treatment plans are only a few of many

Table 3.1 Machine learning techniques for early detection of hip and knee osteoarthritis based on image data type.

Imaging data type	Potential machine learning techniques	Advantages	Disadvantages
Radiography (x-ray)	Convolutional neural networks (CNNs)	Widely available, low cost	Limited information on soft tissues such as cartilage
Magnetic resonance imaging	Deep learning techniques (e.g., CNNs, autoencoders)	Detailed information on cartilage, bone, and soft tissues	Expensive, time-consuming to acquire
Ultrasound	K-nearest neighbors, support vector machines (SVMs)	Relatively inexpensive, portable	Lower image resolution compared to magnetic resonance imaging, operator-dependent
Computed tomography	Deep learning techniques (e.g., CNNs)	Provides detailed three-dimensional information on bone structure	High radiation exposure, expensive
Gait analysis data (from wearables)	SVMs, random forests	Noninvasive, reflects functional limitations	May be influenced by factors other than osteoarthritis
Patient demographics and blood biomarkers	Logistic regression, decision trees	Readily available data, interpretable results	Limited predictive power on its own
Patient questionnaires and self-reported symptoms	Natural language processing	Captures patient experience, may identify early pain patterns	Subjective data, prone to bias

types of medical data that can be found in the huge databases. A huge amount of data can have a holistic view of the patient's health. Modern algorithms based on artificial intelligence (AI) can identify patterns and connections that are difficult to be find, which make the diagnosis and prognosis of OA more precise.

Enhanced model performance: For models that have been trained with large datasets, there is a greater likelihood that they will deliver correct results. A large number of cases and mixed data from sources improve the model's versatility. The accuracy of forecasts and the diagnostics and prognostics increase owing to the confidence coefficient.

Tailored insights: Extensive databases facilitate the creation of tailored insights and approaches to treatment. ML algorithms may adjust suggestions and actions based on a range of patient-specific characteristics, leading to more efficient and customized treatment for individuals with OA.

Longitudinal analysis: Longitudinal information is frequently incorporated in extensive datasets enabling ML algorithms continually to supervise how an ailment changes with time. Medical progress may be estimated using long-term research. Hence, one can determine the likelihood of future illness, including chances of recovery for various health conditions if these studies are available.

Finding biomarkers: Vast datasets are valuable to locate putative markers and disease indicators related to OA. Detection of minor patterns and attributes that might be early OA indicators may be possible by analyzing big datasets through ML techniques. Its importance is visible when developing new diagnostic tools including therapeutic targets.

Research and collaboration: Data sharing and collaboration among health care organizations and research groups is made possible by comprehensive databases. By integrating heterogeneous data, reviewing results according to different attributes, and increasing the pace of innovation and discovery in the field, researchers will have more insight into open access (Teoh et al., 2022).

3.3.3.2 Fine-tuning the machine: Feature selection and preprocessing

When ML models are used to detect OA, it becomes crucial to perform feature selection and data preparation to enhance their accuracy. Choosing features relevant to the current prediction challenge is called feature selection. For instance, biomechanical information including joint angles from the hip and knee joints is part of elements involving spatiotemporal gait metrics. A further degree of refinement is added by including domain information, which enables the selection to concentrate on traits known to be significant in the diagnosis of OA (Srinivasulu Raju et al., 2021). Methods employed in feature selection are follow as below.

Filter Method: To weigh characteristics, these methods use numerical techniques such information gain, chi-square distribution, and correlation. To boost the prediction output proficiency, the ML model can use filter methods to spot traits that are highly correlated to OA.

Wrapper Methods: Wrapper methods apply feature subsets training through an ML algorithm to analyze possibilities that can be acquired in the future from distinct feature combinations. Discovering the most effective set of predictive elements that can be used for the diagnosis and prognosis of OA may be achieved with this approach.

Embedded Method: Embedded techniques suggest underlining information, including the model-making process, to achieve the required outcomes. The quotient regression technique can be applied to enhance modeling by automatically suggesting only important data when developing a prediction model for OA.

Dimensionality Reduction: The use of methods such as principal component analysis (PCA) might be helpful for decreasing the size of the dataset when losing significant information. PCA can be valuable in identifying patterns and connections within a complex dataset during OA modeling, hence enhancing the significance of the data reduction process.

Missing Data Handling: Performance of the model might be affected when there are missing values in the datasets of OA. Methods such as data imputation, in which missing values are replaced by estimates, and the removal of incomplete records are some of the best strategies for handling missing details.

Normalization and Standardization Techniques: These techniques aim to ensure that all variables involved in modeling contribute fairly through the normalization and standardization of the input variables. It is crucial in OA studies where we may have different data types (e.g., imaging features, genetic markers, or clinical evaluations) (Bonakdari et al., 2021). These techniques include the below following methods.

Outlier Detection and Treatment: Outliers in OA data could result in model imprecision and noise. Making more accurate models might occur if one decided to pinpoint and deal with these unusual points through trimming or winsorization.

Feature Scaling: Feature magnitude-dependent algorithms such as support vector machines or k-nearest neighbors especially benefit from reducing the scale of numerical features into the same scale. They might stop certain features from driving the model-building process.

Encoding Categorical Variables: To make sure the ML approaches are able to function better, there may be the need to encode categorical variables encountered in OA datasets using numbers. Thus it is possible to use label encoding or one-hot encoding among other methods for categorical variables because they are categorical (Almuhammadi et al., 2022). An important part of creating ML models is when you meticulously opt for crucial features of interest and then prepare data efficiently to processing them.

3.4 Ethical considerations in machine learning–based osteoarthritis diagnosis

3.4.1 Protecting our data: Navigating privacy concerns in healthcare artificial intelligence

ML systems have had a big role in changing the way diseases such as OA are being diagnosed and managed through medical applications within the health care sector. With ML technology, it is possible to achieve better accuracy in detecting OA, so treatment becomes more effective or personalized, depending on what a person needs. However, this method involves the patient's information being used by these models, which may pose serious questions concerning ethics vis-à-vis privacy and data protection issues. The primary reason for these apprehensions is that health care data are delicate. Many data breaches might take place, and people do not know how their information is being used, which may lead to unauthorized access. It is important to deal with these ethical issues to create an ML-based OA diagnosis tool that behaves responsibly and ethically, earns the confidence of patients, and preserves health care data veracity.

1. Security and Privacy of Data:
   - Lack of explanation and agreement and the absence of sufficient measures to safeguard data cause potential contamination and abuse.
   - Unclear data ownership rights and responsibilities hinder proper data management.
   - Cybercriminals could potentially access and misuse sensitive clinical data to prepatrate crime like exortion or idenity theft.
   - Data privacy dilemmas: The lack of a user's agreement allows certain social websites to collect and store enormous amounts of individual data that can be employed for commercial purposes.
2. Fairness and Bias:
   - In some cases algorithms amplify biases in training data, producing outcomes that are inequitable or prejudicial.
   - AI systems can exhibit bias towards specific groups, leading to discriminatory outcomes.
3. Data Ownership and Control:
   - Problems controlling data and patients' freedom to choose are obstacles to informed decisions.
   - Developments in ML-based technology face obstacles such as the lack of collaboration and sharing of information.
   - Data Sharing for the Common Good: Academics researching with medical AI specialists work on data exchange.
   - Sales of genetic data are a matter of ethics because some companies working with bioinformatics and genetic testing sell the personal data of clients to pharmaceutical and biotechnology corporations, causing possible unequal treatment or discrimination (Farhud and Zokaei, 2021).
4. Transparency and Explainability:
   - AI algorithms might be challenging to comprehend without explainability.
   - This might reduce the reliability and trustworthiness of ML-based diagnostic tools.

To enable health sector organizations, and developers in the field of AI to develop and use diagnostic tools based on ML that are accurate, fair, and transparent and that take patients' privacy into account, these ethical aspects should be considered. Building patient's confidence helps use such technology effectively within health by encouraging its application (Char et al., 2020) (Fig. 3.6).

3.4.2 Beyond the black box: Ensuring interpretability for trustworthy decisions

When it comes to diagnosing OA, a major concern involves ethics considering the growing use of ML models. The interpretability of models centered on models

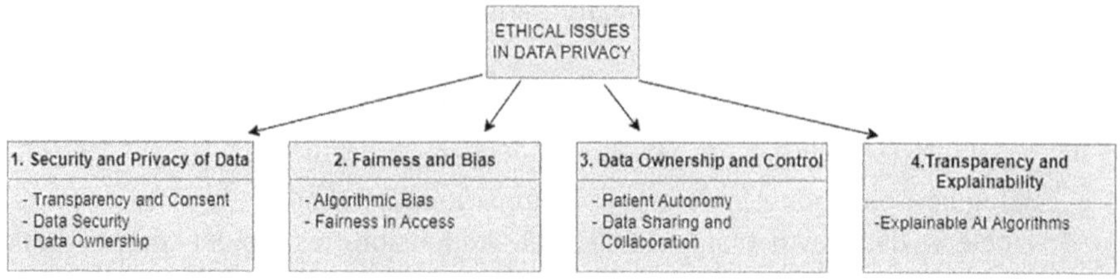

FIGURE 3.6

Ethical concerns in data privacy.

used in the clinical setup is primary among because it is bedrock on which things such as transparency, accountability, or even engaging patients are built. Explainable AI (XAI) is significant because it makes it possible for someone who is not an AI expert to decipher how AI models reason and consequently feature things such as interpretable representations.

Interpretability can be defined as the ability to make ML model predictions intelligible from a human perspective. Even if they have been found to perform at levels of diagnostic accuracy like those demonstrated by health professionals, AI models remain limited by their being neither transparent nor interpretable enough for use within health care settings. XAI solves this problem by exposing how AI models make decisions, to gain confidence and accessibility. There are two broad classes for XAI techniques, those that interpret data and those that interpret models, which can break down into data interpretability and model interpretability (general statements). Specifically, data interpretability mainly deals with making input data easier to follow. In contrast, model interpretability gives insights into what a model predicts (examples). There are several different approaches to interpretability, which can be categorized into two groups:

- Model-specific interpretability: These methods give meaning to the predictions of one kind of ML model. For instance, an approach to interpreting decision tree forecasts might mean extracting the decision rules used by that tree when they make decisions.
- Model-agnostic interpretability: Model-agnostic interpretability, involving a surrogate model to approximate the predictions of the original model, is a technique that can be applied to interpret any type of ML model prediction.

Interpretability can also be categorized into two groups based on the level of explanation:

- Global interpretability: Methods for global interpretability explain the entire behavior of the ML model. For instance, a way to make a model globally interpretable could be to find which features are most important for its predictions.
- Local interpretability: To explain ML model predictions based on certain input data, these methods exist. For instance, a technique of local interpretability may

consist of finding which characteristics are important to predicting this model concerning a specific patient (Stiglic et al., 2020).

Several XAI techniques have been used to diagnose knee OA, including feature importance analysis, layer-wise relevance propagation, SHapley Additive Explanations, and Local Interpretable Model-agnostic Explanations. This section raises the question of why we should care about building ML models that are interpretable, while discussing model interpretability in ML-based OA diagnosis and talking about how it affects different areas (Teoh et al., 2023):

1. Transparency and Accountability:

Transparency and accountability are important in health care. It is important for the doctor to be able to justify and explain the diagnosis to the patient. Health professionals who can interpret models will gain insights into how ML algorithms produce these diagnoses. Trust in the predictions of the model is also fostered, leading to confidence in it.

2. Clinical Utility:

Clinicians use tools for diagnosis that give helpful insights useful in decision-making. Interpretable models provide explanations that can be embedded within clinical domain knowledge and reasoning. All of this involves understanding how the model comes to its conclusions if a clinician is to evaluate their correctness and trustworthiness.

3. Clinical Decision Support:

In clinical decision support systems, health care professionals can use ML models to help them make diagnoses. Interpretable models enable clinicians to consider the system's predictions in view of their own expertise. If the model's predictions contradict their clinical judgment, clinicians can identify such cases and correspondingly change the treatment strategy.

4. Patient Engagement:

Patients are curious about why they have received a diagnosis and been treated in a certain manner. Models that can be understood help in offering an understandable justification for predictions made by the models. This increases adherence rates to treatment programs, patient satisfaction, and engagement.

5. Ethical Concerns Mitigation:

Interpretable models can help solve ethical issues that arise because of bias, fairness, and discrimination in ML-based OA diagnosis. Once insights into the prediction process of the models are obtained, it becomes much easier to identify and correct any biases or unfair outcomes. This ensures that this algorithm is applied ethically and responsibly.

6. Continuous Improvement:

We believe that keeping improvement is vital in medicine. Interpretable models make it easy to know where the model can be made better. By reviewing how the model is performing with patients from various groups, medics can determine situations in which there could be deficiencies or areas that are not well covered by this tool. Such instances can target efforts aimed at enhancing the precision and dependability of the model (Amirian et al., 2023).

3.4.3 Charting a course: Strategies to mitigate ethical challenges

AI has advanced quickly with transformational potential and many complicated ethical issues. Many difficult issues come with it, from bias in algorithms to privacy concerns, as well as the potential that robots might take away human jobs. These are hindrances that require a well-thought-out approach and innovative solutions if they are to be minimized. Making this less onerous requires comprehensive planning in which multiple perspectives and forms of expertise are considered. This study suggests several different ways to address these ethical issues.

- Data Ownership and Rights:

The rights of individuals who own information should be recognized, including determining the ways in which it is used, looking at and correcting it, and considering its transfer possibilities. There should be actual methods through which people controlling data can enforce their rights regarding AI applications.

- Ethical Standards for AI in Medical Practice:

Ethical standards should be proposed that should be upheld concerning the use of AI in health care, especially in nursing. Among these principles are accountability, transparency, equity, beneficence, nonmaleficence, and respect for patient autonomy, which basically form the backbone of these regulations.

- Transparency and Explainability:

Information sharing around AI systems should be promoted, such as their goals, data sources, and methods, to make it observable how AI systems come up with answers. This should be done by generating user-friendly interfaces and AI algorithms that are simple to understand.

- Equality and Representation:

Motivate the participation of everyone collecting and studying health data, irrespective of race, social status, origin, or any other background. Innovative methods to collect and interpret data with a view to curb data poverty in poor countries. Strategies to mitigate biases arising by AI systems from incomplete or skewed data.

- Advances in Laws and Policies:

  For most efficient practice and identification of deficient areas, current policies and laws on AI should be compared. A new regulation framework should be offered that can navigate effectively the complex and ever-changing environment of AI. Setting worldwide standards for ethics should be encouraged regarding AI in relation to the field of medicine.

- Education and Training:

  Health care workers should be provided with classes that cover the essential technical details they require to integrate AI technologies into work and use them, alongside ethical issues surrounding AI. Patients and the public should be informed about the use of AI in health care and their legal rights and responsibilities regarding it.

- Working collaboratively and involving stakeholders:

  Collaboration to solve health care-related ethical predicaments associated with AI as a unit will urge legislators, scholars, businesspeople, and patient advocacy groups to work together. These groups should take part in formulating and implementing appropriate rules and regulations concerning ML in health.

- Ongoing Monitoring and Assessment:

  Mechanisms should be developed to ensure the regular assessment and tracking of ethical impacts of AI in health care. Ethical and unanticipated effects can be addressed through scheduled examinations and appraisals pinpointing new ethical dilemmas or unexpected consequences. To encourage continuous improvements in AI regarding health care, we should foster a spirit of inquiry.

3.5 Real-world applications and benefits of machine learning for osteoarthritis detection

3.5.1 From theory to practice: Using machine learning for early diagnosis of hip and knee osteoarthritis

OA is one of the commonest joint disorders and involves arthritis of the knees and hips. It results in pain, stiffness, and immobility. We need to diagnose OA so that we may start treatment to prevent OA from progressing or worsening and obtain a better outcome.

First, we involved a diverse group of participants with different degrees of hip and knee OA symptoms. The crucial portion of the research was collecting motion capture technology data from each individual, including information regarding kinetic, kinematic, and spatiotemporal characteristics of gait. The participants also confirmed the accuracy of the OA diagnosis and its severity, with clinical exams and imaging test procedures.

Several metrics available in the analysis set, among which were hip rotation moment, knee flexion angle, ankle power, and spatiotemporal characteristics, were used to assess the contribution of joint mechanics in walking. Therefore, these values were selected carefully because of their significance or role in OA development and orientation toward early disease detection.

Two ML algorithms were used to create regression models for early OA diagnosis: the core methods of logistic regression and random forest classification. The binary classifier prepared as multinomial logistic regression models was applied to separate subjects based on the selected gait parameters and categorize them as early OA or non-OA. This algorithm found aiming to predict the presence of early OA start in both hip and knee joints via the binary classification. The random forest classification technique helps in the enhanced accuracy of early OA diagnosis by infusing several decision trees that assign participants different OA severity levels.

When the preliminary study was run, it revealed that significant correlations existed between the extracted gait biometrics and early onset of OA in the hip and knee joints. Although logistic regression and random forest models proved to be efficient and successful in detecting gait abnormalities of the participants who were at risk of developing OA, the current sensitivity of the classification models was not high enough and needed to be improved.

Establishing an ML-based diagnostic algorithm for early recognition of OA in the hip and knee joints is a new approach to establishing confidence in preventive health care management. Health care workers might be able to find persons at risk for OA earlier by gait analysis data and advanced algorithms, which will lead further to timely interventions and person-centered treatments. This approach may initiate productive treatments for OA and help to reduce the level of dysfunction in the patient population (Kwon et al., 2020) (Fig. 3.7).

3.5.2 Personalized paths to care: Optimizing outcomes with machine learning–powered medicine

Integrating personalized medicine into early detection approaches to OA could lead to improvement in patients' prognoses on multiple levels including all-around wellness. Personalized medicine would be helpful for doctors to plan treatment schedules based on a patient's genetic mutation, lifestyle, and medical records. This course-specific method ensures that the most effective type of therapy is actually being employed to improve symptoms and slow the progression of OA. Prevention, or the delayed progression of debilitating symptoms and disease development in the case of OA, is possible only through early detection and hence timely intervention. Early detection helps doctors assess patients' current health status and prescribe necessary preventive treatments such as lifestyle behaviors to avoid spread of the disease to dangerous conditions, and intervention techniques that can effectively slow this chronic disease.

Moreover, individualized medicine is composed of the buildup of susceptibility biomarkers identifying those prone to developing OA. This information will help to

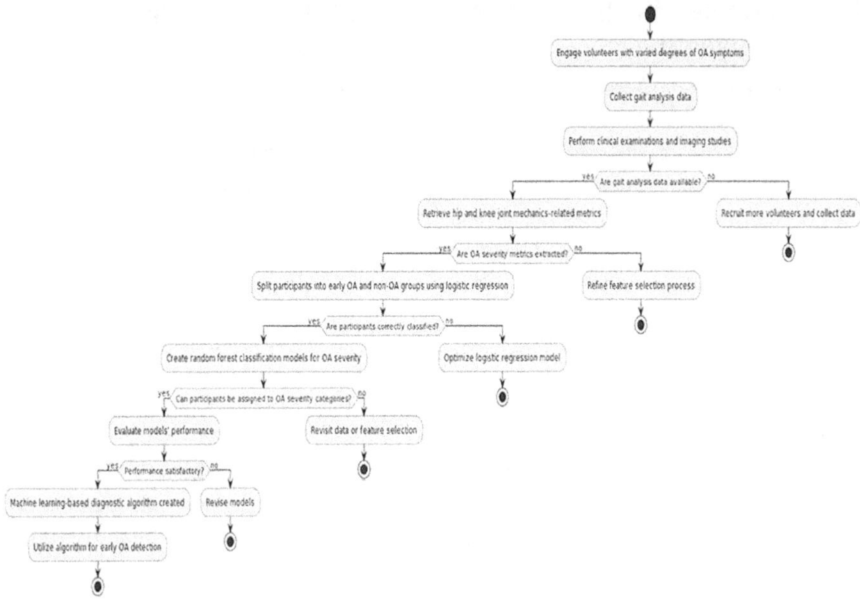

FIGURE 3.7

Flowchart of case study.

make decisions regarding prevention treatments directed to specific exercises or weight management programs that are likely to give higher resistance to OA in people at a higher risk of having it. The possible end result of developments in genetics and individualized medicine might be much more accurate diagnostic approaches to OA. This kind of technology allows physicians to make an early, swift, and accurate diagnosis to prompt the need for ready intervention that would ultimately improve patients' prognosis. Consequently, the area of OA and development of specific medicines is one of the most important trends within personalized medicine. Developing targeted medicines for the genetically and molecularly known underlying process in OA is possible by understanding the genetic and molecular basis of the disease, because this knowledge widens insights into the illness and targets drug development. On the other hand, this personalized approach to developing drugs will make OA treatments more effective and result in better outcomes for patients.

Besides the hospital or clinic setting, the individualized medicine system inspires patients to be more active and have greater control overall over their own health. It is a remarkable feature of personalized medicine that patients can participate fully in their own health management by receiving specific information about possible risk factors and treatment options. A healthier decision-making mechanism leads to better adherence of patients to treatment strategies, which creates a common place for health workers and patients to deal mutually with the OA condition.

3.5.3 A ripple effect: The economic and societal benefits of early intervention

Early OA treatment offers the twin benefits of effective treatment and economic savings. Time and specific treatment early on can result in considerable savings when it comes to health care–related expenditures. If OA is detected early using advanced imaging techniques and AI-based diagnostic tools, expensive procedures such as joint replacement might be delayed or avoided until they are necessary for the patient. United States saved $185 billion worth of medical expenses and lost income, according to Arthritis Foundation findings, obtained through early involvement in manage OA.

It also contributes to minimizing the direct costs of OA, such as lost productivity and increased absenteeism. Individuals who continue to have the function of their joints and a reduction in disease severity are more likely to remain active and involved in their communities. It promotes mental health and social well-being and calls at the same time productivity and relief of constant duties.

Early intervention in the treatment of OA is advantageous for the society increased productivity and reducing economic burdens. By lowering pain, disability, and the need for an operation, the quality of life of patients is improved. Moreover, patients would require less intrusive operations, which would bring an overall improvement in their well-being. Sick people tend to be more antisocial and passive, negatively affecting communities. People in good health, however, can interact in groups, leading to stronger communities. Central to the discussion is the positive impact on families and caregivers who may be relieved from caring tasks and connected distress, with the results in society becoming healthier and more united.

3.6 Future directions and research opportunities

3.6.1 Unveiling the future: Promising areas for machine learning–based osteoarthritis detection

The study has shown that the ML for OA detection has attracted the attention of many people concerned about the need to improve early actions and the accuracy of diagnosis. Pain and stiffness leading to limited movement distinguish patients with arthritis from others across the world. Traditional OA detection procedures tend to lack accuracy and objectiveness because of their dependence on subjective clinical judgments and imaging techniques. In contrast, ML offers a more accurate and objective way to detect OA through ML algorithms than previously because of huge numbers of OA photos and medical record-based OA datasets that train it.

- **Early OA detection**: Early indications of OA may be recognized by ML algorithms that cannot be detected by normal MRI scans, such as composition or structural anomalies of cartilage. Potential early interventions that could prevent cases to worsen may thus become feasible if these conditions are detected at this stage through these means.

- **Risk assessment for OA:** Clinical and imaging data can be used by ML algorithms to develop some models to predict a person's risk for developing OA. Hence, high-risk individuals can be targeted for preventive measures.
- **Analysis of the course of the disease**: ML systems could be employed to track OA's progress periodically. This knowledge can be used to predict individuals who might have increasingly severe OA and to appraise the effectiveness of treatments.
- **Creation of individualized treatment plans**: ML algorithms could be used to develop personalized therapeutic regimens for persons with OA. This could include the patient's objectives and personal preferences along with their clinical data and images.
- **Enhancement of MRI acquisition and reconstruction**: New methods for MRI acquisition and reconstruction that increase the precision and caliber of OA imaging can be created using ML algorithms. This may result in improved OA diagnosis and treatment.
- **Development of novel OA biomarkers**: It is possible to use AI to discover new OA biomarkers to monitor and diagnose the disease. This may lead to the development of novel therapies targeted at these biomarkers.

3.6.2 Evolving technology, evolving medicine: Advancing algorithms, data acquisition, and validation

It is important for more sophisticated algorithms to be created so that AI-driven health care systems can work better and make fewer mistakes. This can be achieved by improving algorithms that exist or trying new ones, because doing so will help the machine analyses data that are too hard for normal persons to understand, predict what may happen with high levels of precision, and offer dependable information. In addition, high-quality, varied datasets are essential during the training of AI applications by researchers who want results to conform with reality. Improving the comprehensiveness and strength of AI systems by using more sources of information and making sure that the information provided is accurate, we will overcome those limits. Therefore, it is mandatory to have stringent validation mechanisms to assess the efficiency and credibility of AI-based health care remedies before they are used at hospitals or health facilities. Such comprehensive tests should be carried out identifying biases so that solutions become mostly bias-free while continuing to meet regulatory standards. Progress in algorithm development, data acquisition, and validation will have an impact on the emergence of better AI-supported health care systems that are reliable and efficient.

For AI and ML to become fully effective in health care and precision medicine, three areas need to see progress: algorithm development, data acquisition, and validation:

1. Algorithmic Development:
   - Improved Engineering of Features: Develop algorithms that can go through complex medical data and pick out helpful characteristics on their own to lessen the work of people while improving the performance of the model.
   - Improved Model Architectures: Design novel AI and ML models for health care activities that include drug development, treatment response prediction, and illness diagnostics. These models should tackle the unique aspects of health care data such as high dimensionality, heterogeneity, and noise.
   - Interpretable and Explainable Models: Build AI and ML models that are interpretable and explainable. This helps doctors understand the reasoning behind these algorithms when they make predictions or decisions. Trust must be built and acceptance promoted for AI for health care practice to succeed.
   - Federated Learning: Design good privacy-preserving federated learning algorithms that support multicompany multiinstitutional AI model training can overcome the limitations related to data sharing and collaboration.
2. Data Acquisition:
   - Ensure data standardization and harmonization: To facilitate easier integration and analysis across diverse health care systems and in several institutions, guidelines and standards should be developed to harmonize health care data from different sources, creating consistency and interoperability.
   - Create synthetic data using data augmentation techniques and enhance the training dataset to improve model performance, which is referred to as active learning and data augmentation.
   - Investigate novel approaches to integrating real-world data such as claims data, electronic health records, and other records in real-world data Integration.
3. Extensive Validation Research:
   - Conduct thorough validation experiments, which is essential for evaluating the effectiveness and generalizability of AI and ML models in actual health care situations. This includes testing models on different patient groups to guarantee that they can operate consistently across different contexts and settings.
   - Public Model, Code, and Dataset Releases: AI and ML models should have public releases of models code sets. This will enable more researchers to assess and reproduce reported results independently, which would enhance openness and confidence in the field.
   - Create measures for evaluation: These should be insightful, therapeutic metrics that can measure exactly how well AI and ML models work within health care settings. These should consider generalizability, interpretability, and model correctness (Ahmed et al., 2020).

3.6.3 Building the bridge: Collaboration among researchers, clinicians, and tech developers

By identifying future directions and areas for further research, we can ensure that AI in healthcare is developed and used responsibly and ethically. This will ultimately lead toimproved patient outcomes and greater fairness in healthcare delivery.

The future of health care involving AI is promising but it has several ethical, legal, and societal issues to be considered. For us to take full advantage of what AI can offer us in terms of benefits without exposing ourselves to too much danger, we need to enhance cross-disciplinary cooperation among researchers, clinicians, policymakers, and technology developers. For the development, deployment, and use of AI in health care, collaboration helps create ethical guidelines and standards that ensure transparency and guarantee accountability.

We should also look at how to solve algorithmic bias and fairness to make sure there is inclusiveness and impartiality in AI systems. Equally important is the enhancement of data privacy and security, which is necessary to safeguard patient records and enable the proper use of AI. It is possible to create AI systems that honor the autonomy of patients and promote collective decision-making (Kumar et al., 2024).

We also must consider how AI is affecting the field of health care and how it can aid but not necessarily replace human medical practitioners. When applying AI in particular medical areas, such as mental health-related practices, intensive care departments with extremely sick people, or personalized treatment based on unique genetic codes, problems of morality must be attended to. The online space has been influenced by ecological and economic aspects, and manufacturers use loads of information to trade products and services.

Also, how AI could help with global health problems, particularly in reducing health care disparities and improving access to care for underprivileged people should be a topic of thought. By considering these future pathways and research possibilities, we can ensure that AI in health care is used correctly and with ethical responsibility that promotes good treatment results and leads to a fairer health system for all (Kumar et al., 2024).

3.7 Conclusion

Machine learning techniques (ML) turns the diagnosis and treatment of OA entirely upside down. This chapter looked at the exceptional performance of ML in handling massive amounts of data such as biomarkers, patient records, and medical images, enabling high-precision multidimensional dataset analysis and opening doors to identifying minute signs imperceptible to the human eye.

Even before its symptoms appear, OA might be identified, preventing significant joint injuries and allowing early treatment. Those who understand ML bring advantages to OA management.

ML combines imagery data with patient data to identify early warning signs and enable preventive actions: that is early detection and prevention. ML looks at the unique features of each patient, the development of the illness, and effects of various medications for precision medicine. In ML, multimodal data integration merges data from several sources such as MRIs, x-rays, genomes, and clinical data for the detailed evaluation of risk variables and the OA course. Precision ML systems can identify even the slightest of signs of illness, improvement diagnostic accuracy by examining complex medical imagery. Predictive analytics using ML algorithms with patient data helps health care professionals predict illness onsets or treatment responses. They can alter treatment plans and be proactive. In research and innovation by studying disease, finding novel biomarkers, and analyzing massive data for creative diagnostic and treatment plans, ML fosters open-access research. By Data-Driven Decision Making extracting vital information from complex patient and disease data generally raises the standard of therapy for patients with OA, improving therapeutic decisions. Using ML, medical experts can choose more effective approaches and go beyond conventional OA treatments. This directly affects patient outcomes, improvement illness treatments, and arrives at a more efficient health care system.

This chapter also looked at advances in imagery technology and their importance in OA diagnosis with ML. Applications include predictive modeling, remote monitoring, customized treatment regimens, and data integration. The chapter stressed how ML can revolutionize the way in which is OA diagnosed and treated.

References

Ahmed, Z., Mohamed, K., Zeeshan, S., Dong, X.Q., 2020. Artificial intelligence with multi-functional machine learning platform development for better healthcare and precision medicine. Database 2020. https://doi.org/10.1093/database/baaa010.

Alami, S., Boutron, I., Desjeux, D., Hirschhorn, M., Meric, G., Rannou, F., Poiraudeau, S., Thiem, U., 2011. Patients' and practitioners' views of knee osteoarthritis and its management: a qualitative interview study. PLoS One 6 (5). https://doi.org/10.1371/journal.pone.0019634.

Alexopoulos, A., Hirvasniemi, J., Tümer, N., 2022. Early detection of knee osteoarthritis using deep learning on knee magnetic resonance images. arXiv. https://doi.org/10.48550/arXiv.2209.01192.

Almuhammadi, W.S., Agu, E., King, J., Franklin, P., 2022. OA-pain-sense: machine learning prediction of hip and knee osteoarthritis pain from IMU data. Informatics 9 (4). https://doi.org/10.3390/informatics9040097.

Amirian, S., Carlson, L.A., Gong, M.F., Lohse, I., Weiss, K.R., Plate, J.F., Tafti, A.P., 2023. Explainable AI in orthopedics: challenges, opportunities, and prospects. arXiv. https://doi.org/10.48550/arXiv.2308.04696.

Binvignat, M., Pedoia, V., Butte, A.J., Louati, K., Klatzmann, D., Berenbaum, F., Mariotti-Ferrandiz, E., Sellam, J., 2022. Use of machine learning in osteoarthritis research: a systematic literature review. RMD Open 8 (1). https://doi.org/10.1136/rmdopen-2021-001998.

Bonakdari, H., Jamshidi, A., Pelletier, J.-P., Abram, F., Tardif, G., Martel-Pelletier, J., 2021. A warning machine learning algorithm for early knee osteoarthritis structural progressor patient screening. Therapeu. Adv. Musculoskel. Dis. 13. https://doi.org/10.1177/1759720x21993254.

Char, D.S., Abràmoff, M.D., Feudtner, C., 2020. Identifying ethical considerations for machine learning healthcare applications. Am. J. Bioeth. 20 (11), 7–17. https://doi.org/10.1080/15265161.2020.1819469.

Council post, 2023. Navigating the intersection of technology, equity and transformation in healthcare. Forbes Business Council.

Cui, A., Li, H., Wang, D., Zhong, J., Chen, Y., Lu, H., 2020. Global, regional prevalence, incidence and risk factors of knee osteoarthritis in population-based studies. EClinicalMedicine 29–30. https://doi.org/10.1016/j.eclinm.2020.100587.

Ehmig, J., Engel, G., Lotz, J., Lehmann, W., Taheri, S., Schilling, A.F., Seif Amir Hosseini, A., Panahi, B., 2023. MR-imaging in osteoarthritis: current standard of practice and future outlook. Diagnostics 13 (15). https://doi.org/10.3390/diagnostics13152586.

Embracing the Future: The Impactful Intersection Between Technology and Healthcare, 2023.

Farhud, D.D., Zokaei, S., 2021. Ethical issues of artificial intelligence in medicine and healthcare. Iran. J. Public Health 50 (11). https://doi.org/10.18502/ijph.v50i11.7600.

Knee Osteoarthritis, 2024. Australian Orthopaedic Association.

Kumar, R., Gupta, M., Abraham, A., 2024. A critical analysis on vertebra identification and cobb angle estimation using deep learning for scoliosis detection. IEEE Access 12, 11170–11184. https://doi.org/10.1109/ACCESS.2024.3353794.

Kwon, S.B., Ku, Y., Han, H.U.S., Lee, M.C., Kim, H.C., Ro, D.H., 2020. A machine learning-based diagnostic model associated with knee osteoarthritis severity. Sci. Rep. 10 (1). https://doi.org/10.1038/s41598-020-72941-4.

Oei, E.H.G., Runhaar, J., 2023. Imaging of early-stage osteoarthritis: the needs and challenges for diagnosis and classification. Skeletal Radiol. 52 (11), 2031–2036. https://doi.org/10.1007/s00256-023-04355-y.

Oei, E.H.G., Hirvasniemi, J., van Zadelhoff, T.A., van der Heijden, R.A., 2022. Osteoarthritis year in review 2021: imaging. Osteoarthritis Cartilage 30 (2), 226–236. https://doi.org/10.1016/j.joca.2021.11.012.

Parisi, L., RaviChandran, N., Lanzillotta, M., 2020. Supervised Machine Learning for Aiding Diagnosis of Knee Osteoarthritis: A Systematic Review and Meta-Analysis. Authorea Preprints. https://doi.org/10.36227/techrxiv.13132931.v1.

Piccolo, C.L., Mallio, C.A., Vaccarino, F., Grasso, R.F., Zobel, B.B., 2023. Imaging of knee osteoarthritis: a review of multimodal diagnostic approach. Quant. Imag. Med. Surg. 13 (11), 7582–7595. https://doi.org/10.21037/qims-22-1392.

Srinivasulu Raju, S., Niranjan, T., Pandiyan, P., Sai Snehitha, M., 2021. A review of an early detection and quantification of osteoarthritis severity in knee using machine learning techniques. IOP Conf. Ser. Mater. Sci. Eng. 1057 (1). https://doi.org/10.1088/1757-899x/1057/1/012095.

Stiglic, G., Kocbek, P., Fijacko, N., Zitnik, M., Verbert, K., Cilar, L., 2020. Interpretability of machine learning-based prediction models in healthcare. Wiley Interdiscipl. Rev.: Data Min. Knowl. Discov. 10 (5). https://doi.org/10.1002/widm.1379.

Teoh, Y.X., Lai, K.W., Usman, J., Goh, S.L., Mohafez, H., Hasikin, K., Qian, P., Jiang, Y., Zhang, Y., Dhanalakshmi, S., 2022. Discovering knee osteoarthritis imaging features for diagnosis and prognosis: review of manual imaging grading and machine learning approaches. J. Healthcare Eng. 2022. https://doi.org/10.1155/2022/4138666.

Teoh, Y.X., Othmani, A., Goh, S.L., Usman, J., Lai, K.W., 2023. Deciphering knee osteoarthritis diagnostic features with explainable artificial intelligence: a systematic review. arXiv. https://doi.org/10.48550/arXiv.2308.09380.

Wang, Q., Runhaar, J., Kloppenburg, M., Boers, M., Bijlsma, J.W., Bierma-Zeinstra, S.M., 2021. Diagnosis of early-stage knee osteoarthritis based on early clinical course: data from the CHECK cohort. Arthritis Res. Ther. 23, 1−10.

Xuan, A., Chen, H., Chen, T., Li, J., Lu, S., Fan, T., Zeng, D., Wen, Z., Ma, J., Hunter, D., Ding, C., Zhu, Z., 2023. The application of machine learning in early diagnosis of osteoarthritis: a narrative review. Therapeu. Adv. Musculoskel. Dis. 15. https://doi.org/10.1177/1759720x231158198.

Yahaya, I., Wright, T., Babatunde, O.O., Corp, N., Helliwell, T., Dikomitis, L., Mallen, C.D., 2021. Prevalence of osteoarthritis in lower middle- and low-income countries: a systematic review and meta-analysis. Rheumatol. Int. 41 (7), 1221−1231. https://doi.org/10.1007/s00296-021-04838-y.

CHAPTER

Bone cancer classification and detection using machine learning technique

4

Pawan Whig[1], Balaram Yadav Kasula[2], Nikhitha Yathiraju[2], Anupriya Jain[3], Seema Sharma[3]

[1]*Vivekananda Institute of Professional Studies-TC, Delhi, New Delhi, India;* [2]*University of the Cumberlands, Williamsburg, KY, United States;* [3]*School of Computer Applications MRIIRS, Faridabad, Haryana, India*

4.1 Introduction

Cancer, a multifaceted and formidable adversary to human health, continues to challenge the medical community worldwide (Shrivastava et al., 2020). Among the various types of cancer, bone cancer stands as a relatively rare yet intricate disease that affects the skeletal system, posing unique diagnostic and treatment challenges (Sharma et al., 2021a). The early and accurate detection of bone cancer is paramount in mitigating its progression, optimizing treatment strategies, and improving patient outcomes. In recent years, the integration of machine learning (ML) methodologies has emerged as a promising frontier in revolutionizing the classification and detection of bone cancer, offering innovative solutions and transformative advancements in the field of oncology (Gawade et al., 2023). Bone cancer encompasses a heterogeneous group of malignant tumors that originate in the bone tissue. These tumors can either originate within the bone (primary bone cancer) or spread to the bone from other parts of the body (secondary or metastatic bone cancer) (Anand et al., 2022). The most prevalent types include osteosarcoma, chondrosarcoma, and Ewing sarcoma, each presenting distinct characteristics in terms of cell types, growth patterns, and prognosis. The rarity and complexity of bone cancer, coupled with its diverse subtypes, necessitate sophisticated diagnostic tools and methodologies for accurate identification and classification (Yadav and Rathor, 2020).

The diagnosis of bone cancer traditionally relies on a combination of clinical assessments, imaging techniques such as x-rays, computed tomography (CT) scans, magnetic resonance imaging (MRI) scans, and histopathological analysis of biopsy samples. However, accurate diagnosis poses challenges owing to the variability in tumor presentation, overlapping radiological features, and the need for expert interpretation. Distinguishing benign bone lesions from malignant tumors and classifying different subtypes of bone cancer remain intricate tasks, often requiring specialized

Diagnosing Musculoskeletal Conditions using Artificial Intelligence and Machine Learning to Aid Interpretation of Clinical Imaging
https://doi.org/10.1016/B978-0-443-32892-3.00004-X

expertise and multiple diagnostic procedures. The integration of ML algorithms in bone cancer classification and detection has emerged as a transformative approach, leveraging computational power and data-driven analysis to enhance diagnostic accuracy and efficiency (Singh et al., 2020). ML techniques, including supervised learning, unsupervised learning, and deep learning, analyze complex datasets composed of patient demographics, imaging data, histopathological features, and genetic information to identify patterns indicative of bone cancer. These algorithms learn from labeled datasets, extract intricate features, and build predictive models, aiding in early detection, classification, and prognostication of bone cancer subtypes (Hossain and Rahaman, 2018; Jabber et al., 2020; Gitto et al., 2020; Kaur and Garg, 2022). ML models have shown remarkable capabilities in analyzing radiological imaging modalities such as x-rays, CT scans, and MRI scans for bone cancer detection the literature review with research gap is shown in Table 4.1. These models can identify subtle abnormalities, characterize tumor morphology, and differentiate between benign and malignant lesions with increased accuracy. Furthermore, the fusion of imaging data with clinical and pathological information augments the diagnostic precision, enabling more informed clinical decision-making (Kumar, 2021).

Table 4.1 lists the main findings and research gaps across the diverse literature, providing a comprehensive overview of the identified gaps in each reference.

Table 4.1 Literature review with research gap.

References	Summary	Research gap
Shrivastava et al. (2020)	Bone cancer detection using machine learning techniques.	Lack of discussion on comparison of different machine learning techniques and their effectiveness in bone cancer detection.
Sharma et al. (2021a)	Bone cancer detection using feature extraction-based machine learning models.	Need for more insights into selection of specific features and their impact on performance of machine learning models in bone cancer detection.
Gawade et al. (2023)	Application of convolutional neural networks and supervised deep learning methods for osteosarcoma bone cancer detection.	Limited exploration of challenges and limitations associated with application of deep learning methods in bone cancer detection.
Anand et al. (2022)	Deep convolutional extreme machine learning classification method for bone cancer detection from histopathological images.	Need for more in-depth discussion on interpretability and explainability of deep learning models in the context of histopathological image analysis.
Yadav and Rathor (2020)	Bone fracture detection and classification using a deep learning approach.	Lack of exploration into specific challenges and limitations of using deep learning for bone fracture detection.
Singh et al. (2020)	Bone tumor detection using feature extraction with classification by deep learning techniques.	Limited discussion on applicability and challenges of feature extraction techniques in the context of deep learning for bone tumor detection.

Table 4.1 Literature review with research gap.—*cont'd*

References	Summary	Research gap
Hossain and Rahaman (2018)	Bone cancer detection and classification using fuzzy clustering and neuro-fuzzy classifier.	Need for more detailed discussion on limitations and trade-offs associated with using fuzzy clustering and neuro-fuzzy classifiers in bone cancer detection.
Jabber et al. (2020)	Support vector machine (SVM) model-based computerized bone cancer detection.	Limited exploration of challenges in parameter tuning and generalization associated with SVM models for bone cancer detection.
Gitto et al. (2020)	MRI radiomics-based machine learning classification of bone chondrosarcoma.	Lack of discussion on generalization of radiomics-based machine learning models across different types of bone cancer.
Kaur and Garg (2022)	Bone cancer detection techniques using machine learning.	Need for more insights into comparative analysis of different machine learning techniques and their performance in bone cancer detection.
Kumar (2021)	Bone cancer detection using feature extraction with classification using K-nearest neighbor and decision tree algorithm.	Limited exploration of interpretability and explainability of decision tree algorithm in the context of bone cancer detection.
Papandrianos et al. (2020)	Bone metastasis classification using whole body images from prostate cancer patients based on convolutional neural networks application.	Need for more insights into transferability and robustness of convolutional neural networks across different cancer types in bone metastasis classification.
Shukla and Patel (2020)	Bone cancer detection from x-ray and magnetic resonance imaging (MRI) scans through image segmentation techniques.	Limited exploration of challenges and potential biases associated with image segmentation techniques in the context of bone cancer detection.
Mahore et al. (2021)	Machine learning approach to classify and predict different osteosarcoma types.	Need for more detailed insights into identification of specific osteosarcoma types and challenges associated with their accurate classification using machine learning.
Lingappa and Parvathy (2023)	Deep learning-based active contour technique with bagging and boosting algorithms hybrid approach for detecting bone cancer from MRI scan images.	Lack of discussion on trade-offs and computational efficiency of using bagging and boosting algorithms in combination with active contour techniques for bone cancer detection.
Georgeanu et al. (2022)	Bone cancer detection using feature extraction-based machine learning model.	Repetition of reference; this may indicate a lack of diversity in research focus on feature extraction-based models for bone cancer detection.

Continued

Table 4.1 Literature review with research gap.—*cont'd*

References	Summary	Research gap
Mazumder and Singh (2022)	Bone cancer detection using deep learning.	Limited discussion on interpretability and generalization of deep learning models in bone cancer detection.
Kumar et al. (2020)	Automatic detection of white blood cancer from bone marrow microscopic images using convolutional neural networks.	Need for more insights into challenges associated with detection of white blood cancer and performance of convolutional neural networks in this context.
Vandana and Alva (2021)	Deep learning-based automated tool for cancer diagnosis from bone histopathology images.	Limited discussion on ethical considerations and potential biases associated with automation of cancer diagnosis using deep learning models.
Abdelaal and Tobely (2019)	Bone cancer detection using particle swarm extreme learning machine neural networks.	Need for more insights into challenges and limitations associated with application of particle swarm optimization and extreme learning machine neural networks in bone cancer detection.
Altameem (2020)	Fuzzy rank correlation-based segmentation method and deep neural network for bone cancer identification.	Limited discussion on applicability and challenges of fuzzy rank correlation-based segmentation methods in bone cancer identification using deep neural networks.
Li et al. (2023)	Diagnostic value of machine learning for the classification of malignant bone tumor: a systematic evaluation and metaanalysis.	Need for more insights into methodology and criteria used for systematic evaluation and metaanalysis of machine learning models in classification of malignant bone tumors.

4.2 Predictive modeling and prognostic assessment

ML algorithms contribute to prognostic assessment by predicting patient outcomes, treatment responses, and disease progression based on diverse datasets. These models aid in risk stratification, assisting clinicians in devising personalized treatment plans tailored to individual patient profiles and predicting disease recurrence or metastasis (Papandrianos et al., 2020).

4.2.1 Challenges and considerations

Despite the promising strides made in using ML for bone cancer detection, challenges persist. Issues related to data quality, model interpretability, dataset imbalance, and the need for validation in diverse populations necessitate careful consideration. Ethical concerns regarding patient privacy, data security, and the ethical use of artificial intelligence (AI) in health care underscore the importance of ethical frameworks and regulatory compliance in implementing ML-based solutions (Shukla and Patel, 2020).

4.2.2 Scope of this chapter

This chapter delves into the comprehensive landscape of using ML in the classification and detection of bone cancer (Mahore et al., 2021; Satheesh Kumar and Bb, 2021; Georgeanu et al., 2022; Lingappa and Parvathy, 2023). It explores diverse ML methodologies and their applications in radiological imaging analysis, histopathological assessment, predictive modeling, and prognostication in bone cancer. Moreover, the chapter discusses the challenges, ethical considerations, and future directions in harnessing the potential of ML to improve early detection, classification accuracy, and personalized treatment strategies in bone cancer.

4.3 Medical imaging modalities for bone cancer diagnosis

Medical imaging modalities serve as fundamental tools in the diagnosis and characterization of bone cancers, offering diverse perspectives of bone structure and abnormalities crucial for treatment planning (Sharma et al., 2021b). Among these modalities, x-rays, using radiography, serve as an initial step in bone assessments, offering an overview of bone density and potential tumor presence. CT scans, providing detailed cross-sectional images, are instrumental in detecting bone fractures, evaluating tumor size, and understanding the extent of bone involvement, offering greater precision than x-rays. MRIs, using magnetic fields and radio waves, offers exceptional soft tissue contrast, and are effective in visualizing bone tumors and detecting bone marrow abnormalities or soft tissue extensions. Bone scintigraphy, or bone scans, reveal areas of abnormal bone metabolism by detecting increased bone activity, aiding in identifying metastases or multifocal bone involvement. Positron emission tomography-CT merges functional and anatomical data, showcasing metabolic activity and assisting in detecting metastases or evaluating treatment responses. Ultrasound, using sound waves, offers real-time imaging, guiding biopsies and assessing soft tissue extensions near bones. Each modality plays a distinct role based on the clinical context, suspected cancer type, and specific information needed for accurate diagnosis and treatment planning. The integration of multiple imaging techniques often provides a comprehensive understanding, aiding health care professionals in diagnosing, staging, and managing bone cancers effectively (Mazumder and Singh, 2022).

4.3.1 Machine learning algorithms for bone cancer classification

This work delves into the application of ML algorithms for the classification of bone cancer, a critical domain in medical diagnostics (Kumar et al., 2020). With a primary focus on enhancing accuracy and efficiency, various ML techniques are explored, including both supervised and unsupervised approaches (Vandana and Alva, 2021). The algorithms employed leverage a wealth of medical imaging and clinical data to discern and classify different types of bone cancers. The chapter provides an in-depth analysis of the challenges inherent in bone cancer classification, addressing issues such as data imbalance, feature selection, and interpretability (Abdelaal and Tobely, 2019).

An essential aspect of this exploration is the pragmatic integration of ML models into clinical decision-making processes. The potential impact of these algorithms extends beyond the realms of traditional diagnostics, influencing key aspects such as treatment planning and prognostic assessments. The write-up underscores the transformative power of ML in augmenting diagnostic accuracy and improving patient outcomes in the complex landscape of bone cancer. By navigating the intricacies of advanced algorithms and their applications, this work contributes to the evolving landscape of medical technology, providing valuable insights for researchers, practitioners, and clinicians engaged in the fight against bone cancer (Altameem, 2020).

4.3.2 Challenges in bone cancer classification using machine learning

Classifying bone cancer using ML poses several challenges owing to the complexity of the disease, variations in data, and the intricacies of different bone cancer types. Here are the key challenges encountered in bone cancer classification using ML.

4.3.2.1 Data imbalance and rarity of bone cancer types

One of the significant challenges in bone cancer classification is the scarcity of data, especially for rare subtypes. Imbalanced datasets with a disproportionate distribution of different bone cancer types hinder the model's ability to learn effectively across all classes. Obtaining sufficient and diverse data for each subtype is crucial to develop robust models capable of accurately classifying various bone cancer types (Li et al., 2023).

4.3.2.2 Complexity and heterogeneity of imaging data

Medical imaging data used in bone cancer classification, such as x-rays, CT scans, and MRI images, are complex and heterogeneous. Variations in imaging quality, different scanning protocols, and variations in patient positioning during imaging can lead to variations in data quality and the appearance of bone lesions. Ensuring consistency and standardization in imaging data representation across different sources pose significant challenges for ML models (Whig et al., 2024a).

4.3.2.3 Feature selection and extraction

Extracting relevant and discriminative features from medical imaging data is crucial for accurate bone cancer classification. However, determining which features are most informative for distinguishing between benign and malignant tumors or different bone cancer subtypes can be challenging. The high dimensionality of imaging data demands robust feature selection techniques to avoid overfitting and enhance the model's generalizability (Whig et al., 2024b).

4.3.2.4 Interpretability of machine learning models

Many ML models used for bone cancer classification, such as deep neural networks, are often considered black-box models owing to their complexity. Interpretability and explainability of model predictions are essential in health care to ensure trust and understanding among health care professionals. Achieving a balance between model complexity and interpretability remains a challenge in bone cancer classification using ML.

4.3.2.5 Overfitting and generalization

Overfitting, in which a model performs well on training data but poorly on unseen data, is a concern in bone cancer classification. Complex ML models may capture noise or irrelevant patterns in the training data, leading to reduced generalization on new or unseen patient data. Ensuring models generalize well to new datasets whereas avoiding overfitting is a critical challenge.

4.3.2.6 Ethical and regulatory considerations

ML models in health care raise ethical concerns related to patient privacy, data security, and the responsible use of AI-driven solutions. Ensuring compliance with ethical guidelines, data protection regulations, and maintaining patient confidentiality while leveraging ML for bone cancer classification is imperative.

Addressing these challenges requires collaborative efforts among clinicians, data scientists, and researchers to curate diverse and well-annotated datasets, develop robust and interpretable models, standardize imaging protocols, and ensure ethical and regulatory compliance in leveraging ML for bone cancer classification. Overcoming these challenges will facilitate the development of accurate and reliable models for enhanced bone cancer diagnosis and treatment planning.

4.4 Feature selection and model interpretability

Feature selection and model interpretability are crucial aspects in bone cancer classification using ML, contributing significantly to the effectiveness and trustworthiness of the developed models.

4.4.1 Feature selection

In bone cancer classification, selecting relevant and informative features from complex medical imaging data is paramount. Feature selection involves identifying the most discriminative characteristics or imaging patterns that differentiate between benign and malignant tumors or different subtypes of bone cancer. Techniques such as principal component analysis, recursive feature elimination, and statistical tests aid in reducing the dimensionality of imaging data while retaining the most relevant features. Proper feature selection mitigates overfitting, enhances model efficiency, and streamlines the computational burden, ensuring that the selected features contribute meaningfully to accurate bone cancer classification.

4.4.2 Model interpretability

Interpreting the predictions made by ML models in bone cancer classification is critical for gaining insights into how the model arrives at its conclusions. Explainability ensures trust among health care professionals and facilitates clinical decision-making. Techniques such as decision tree-based models (e.g., random forest), local interpretable model-agnostic explanations, SHapley Additive exPlanations, or layer-wise relevance propagation for deep learning models enable an understanding of the rationale behind the model's predictions. By visualizing feature importance, highlighting influential imaging patterns, or providing explanations for individual predictions, these methods improve the interpretability of complex models, enabling clinicians to comprehend and trust the model's decisions.

The integration of robust feature selection methodologies and interpretable model techniques enhances the transparency, reliability, and acceptance of ML models in bone cancer classification. Balancing model complexity with interpretability allows clinicians to glean actionable insights from these models, aiding in accurate diagnosis, prognosis, and treatment planning for bone cancer patients.

4.4.3 Integrating machine learning models into clinical decision-making

Integrating ML models into clinical decision-making processes for bone cancer requires careful consideration, validation, and collaboration among data scientists, clinicians, and health care stakeholders to ensure effective and ethical use in real-world health care settings.

4.4.3.1 Data quality and validation

The successful integration of ML models into clinical decision-making hinges on high-quality, validated datasets that accurately represent diverse bone cancer cases. Rigorous validation and testing of these models on diverse patient cohorts and imaging data are essential to ensure their reliability, accuracy, and generalizability across different populations and health care settings.

4.4.3.2 Clinical relevance and interpretability

ML models should align with clinical objectives, providing actionable insights that augment, rather than replace, clinician expertise. Models must offer interpretability, explaining their predictions in clinically meaningful ways to aid clinicians in understanding and trusting the recommendations or predictions generated.

4.4.3.3 User-friendly interfaces and workflow integration

Developing user-friendly interfaces that seamlessly integrate ML outputs into existing clinical workflows is crucial. The integration should be intuitive, providing clinicians with easily interpretable information at the point of care. Models should complement existing diagnostic practices, providing additional insights or support to aid decision-making without disrupting established workflows.

4.4.3.4 Ethical considerations and regulatory compliance

Adherence to ethical guidelines, patient confidentiality, and regulatory compliance is fundamental. Ensuring models prioritize patient privacy, maintain confidentiality, and comply with data protection regulations is imperative. Ethical frameworks guiding the responsible use of AI in health care must be implemented to maintain patient trust and safety.

4.4.3.5 Education and collaboration

Collaboration between data scientists and health care professionals is essential for successful integration. Training clinicians to understand the capabilities, limitations, and interpretations of ML models fosters trust and facilitates effective use in clinical decision-making. Continuous collaboration allows refinement and adaptation of models based on real-world clinical feedback.

4.4.3.6 Clinical validation and real-world impact

Real-world clinical validation studies evaluating the impact of ML models on patient outcomes, diagnostic accuracy, treatment efficacy, or workflow efficiency are crucial. Demonstrating the models' clinical utility and efficacy in improving bone cancer diagnosis or treatment outcomes validates their integration into routine clinical practice.

The seamless integration of ML models into clinical decision-making for bone cancer requires a multidisciplinary approach, emphasizing collaboration, validation, interpretability, and ethical compliance. When appropriately developed and integrated, these models have the potential to enhance diagnostic accuracy, optimize treatment strategies, and improve patient outcomes in bone cancer care.

4.5 Improving diagnostic accuracy and treatment planning

Improving diagnostic accuracy and treatment planning in bone cancer involves leveraging ML to enhance the precision of diagnosis, refine risk stratification, and optimize personalized treatment strategies.

ML models analyze complex imaging data such as x-rays, CT scans, and MRI images to detect subtle patterns and nuances indicative of different bone cancer types. By identifying intricate imaging features that may elude human perception, these models augment clinicians' ability to diagnose and classify bone cancers accurately. Improved diagnostic accuracy aids in early detection, differentiation between benign and malignant tumors, and precise characterization of bone lesions, laying the groundwork for tailored treatment approaches. ML assists in stratifying patient populations based on predictive analytics derived from diverse datasets. These models integrate patient demographics, genetic markers, imaging features, and clinical parameters to predict disease progression, treatment response, and prognosis. By stratifying patients into risk categories, clinicians can prioritize interventions such as aggressive treatments for high-risk cases or conservative approaches for low-risk cases, optimizing patient outcomes and resource allocation. Tailoring treatment plans to individual patient profiles is a key facet of ML's impact on bone cancer care. These models analyze comprehensive patient data to recommend personalized treatment options based on predicted treatment responses or outcomes. By considering individual patient characteristics, including genetic markers, imaging findings, and historical treatment responses, ML facilitates the selection of optimal therapies, minimizing adverse effects and enhancing treatment efficacy.

ML models predict treatment responses to specific therapies by analyzing patient data before and after treatment. These predictions aid in determining the most effective treatment regimens, enabling timely modifications if responses deviate from expected outcomes. Predicting treatment responses supports clinicians in making informed decisions about treatment continuation, modification, or alternative strategies, fostering more precise and responsive patient care. By harnessing the capabilities of ML, health care professionals can improve diagnostic accuracy, refine risk stratification, and personalize treatment strategies for patients with bone cancer. These advances lead to more accurate diagnoses, tailored treatments, and enhanced prognostication, ultimately improving patient outcomes and quality of life in bone cancer care.

4.6 Prognostic assessments and patient outcomes prediction

Prognostic assessments and predicting patient outcomes in bone cancer care using ML are instrumental in estimating disease progression, treatment responses, and overall survival rates. ML models offer predictive insights by analyzing diverse patient data and aiding in the estimation of prognosis, guiding clinicians in decision-making and patient counseling.

ML models analyze clinical, genetic, and imaging data to predict the progression of bone cancer. By considering various parameters such as tumor size, histological characteristics, genetic markers, and patient demographics, these models forecast disease progression rates, aiding clinicians in understanding the likelihood of tumor growth, metastasis, or recurrence. Predicting treatment responses is vital for tailoring personalized treatment strategies. ML algorithms use historical treatment data, patient characteristics, and imaging features to predict how patients might respond to specific treatments. This allows clinicians to anticipate individual patient responses, guiding treatment decisions, and enabling timely modifications if expected responses diverge.

ML models employ survival analysis techniques to estimate patient survival rates. By analyzing historical patient data and relevant prognostic factors, these models predict overall survival, disease-free survival, or recurrence-free survival rates. This information assists clinicians in counseling patients about potential outcomes, facilitating discussions on treatment plans, and aiding in palliative care decisions. ML models identify and weigh various risk factors contributing to patient outcomes. These models analyze large datasets to uncover subtle correlations among patient characteristics, biomarkers, and treatment responses. Identification of significant risk factors helps in early risk stratification, enabling proactive interventions or intensified monitoring for high-risk patients.

Predictive models act as decision support tools, providing additional insights to aid clinicians in treatment planning and patient management. By presenting probabilistic estimates based on data analysis, these models complement clinician expertise, helping to weigh treatment options and anticipate potential complications or adverse outcomes. By leveraging ML for prognostic assessments and outcome predictions in bone cancer care, clinicians can make more informed decisions, tailor treatments, and offer personalized care plans to improve patient outcomes, optimize resource allocation, and enhance overall care quality.

4.7 Case study: Enhancing bone cancer diagnosis using machine learning

A research study aimed to evaluate the efficacy of an ML model in accurately classifying bone cancer types using radiological imaging data, as shown in Fig. 4.1. The dataset was composed of 800 MRI images of patients with confirmed bone cancer, including osteosarcoma, chondrosarcoma, and Ewing sarcoma, each with 200 images. The goal was to develop an ML model capable of accurately differentiating among these subtypes.

4.7.1 Methodology

Dataset: 800 MRI images categorized into four classes: osteosarcoma, chondrosarcoma, Ewing sarcoma, and normal bone tissue (control).

Preprocessing: Image preprocessing involved standardization, normalization, and augmentation to ensure data uniformity and enhance model generalizability.

Model Development: A convolutional neural network (CNN) architecture was employed owing to its effectiveness in image classification tasks. The CNN was trained on 80% of the dataset and validated on the remaining 20%.

Evaluation Metrics: Model performance was assessed using quantitative metrics including accuracy, precision, recall, F1-score, and the area under the receiver operating characteristic curve (AUC-ROC) (Table 4.2).

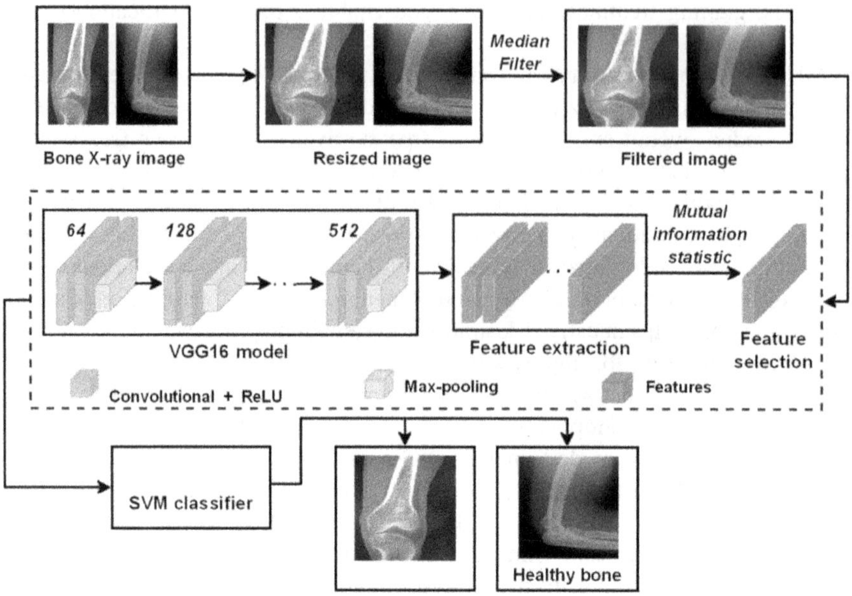

FIGURE 4.1 Bone cancer diagnosis using machine learning.

ReLU, rectified linear unit, *SVM*, support vector machine.

Table 4.2 Result comparison in tabular form.

Metric	Osteosarcoma	Chondrosarcoma	Ewing sarcoma
Accuracy	92%	92%	92%
Precision	0.93	0.91	0.92
Recall	0.93	0.91	0.92
F1-score	0.93	0.91	0.92
Area under the receiver operating characteristic curve	0.95	0.94	0.93

4.7.2 Results

1. Accuracy: The developed CNN achieved an overall accuracy of 92% on the test set, accurately classifying bone cancer subtypes and normal bone tissue.
2. Precision, Recall, and F1-score: The precision, recall, and F1-score for differentiating osteosarcoma, chondrosarcoma, and Ewing sarcoma were 0.93, 0.91, and 0.92 respectively, demonstrating the model's ability to distinguish among specific bone cancer types.
3. AUC-ROC: The AUC-ROC values for each bone cancer subtype classification were 0.95 for osteosarcoma, 0.94 for chondrosarcoma, and 0.93 for Ewing sarcoma, indicating excellent discriminatory power of the model.

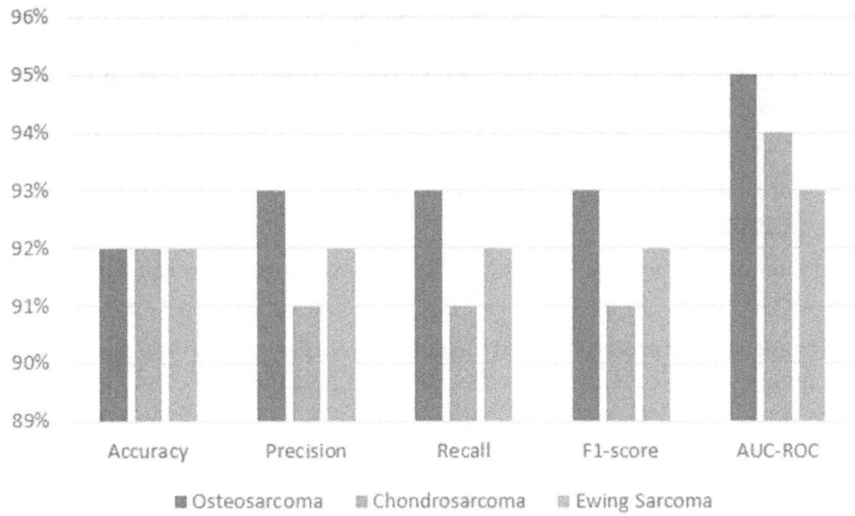

FIGURE 4.2 Bar chart representation of result obtained.

AUC-ROC, area under the receiver operating characteristic curve.

The developed ML model showcased high accuracy and precision in classifying different bone cancer subtypes based on MRI images. This could significantly affect clinical practice by aiding radiologists and oncologists in accurate diagnosis, leading to improved treatment planning and better patient outcomes. Early and precise identification of bone cancer subtypes could enable tailored therapies, potentially reducing treatment delays and enhancing overall prognosis, as shown in Fig. 4.2.

The study demonstrated the feasibility and efficacy of using ML, particularly CNNs, in accurately classifying bone cancer subtypes based on MRI imaging data. The robust performance metrics highlight the potential of ML models in enhancing diagnostic accuracy and improving patient care in bone cancer.

4.8 Conclusion

The integration of ML in bone cancer diagnosis and treatment planning represents a significant stride toward more accurate diagnostics, precise prognostication, and personalized treatment strategies. ML models have showcased their potential in improving diagnostic accuracy, refining risk stratification, predicting treatment responses, and estimating patient outcomes in bone cancer care. By leveraging diverse patient data, including imaging features, genetic markers, and clinical parameters, these models have contributed to enhancing clinical decision-making, guiding tailored treatment plans, and fostering better patient outcomes.

However, challenges persist, including data quality, interpretability of complex models, ethical considerations, and the need for extensive clinical validation. Addressing these challenges is crucial to ensure the reliable and ethical deployment of ML models in clinical practice.

4.9 Future scope

The future of ML in bone cancer care holds immense promise and avenues for further exploration:

1. **Enhanced Interpretability and Explainability:** Developing ML models with improved interpretability and explainability will be pivotal. Techniques focusing on model transparency, such as interpretable neural networks or model-agnostic approaches, will facilitate clinician understanding and acceptance.
2. **Integration of Multimodal Data:** Integrating diverse data sources, such as imaging, genomics, and clinical records, will provide a more comprehensive understanding of bone cancer. This holistic approach may unveil intricate patterns, aiding in more accurate diagnoses and treatment predictions.
3. **Real-time Decision Support Systems:** Advancements in real-time decision support systems, leveraging ML, will assist clinicians at the point of care. Such systems can offer immediate insights and recommendations, optimizing treatment plans and improving patient outcomes.
4. **Collaborative Data Sharing and Validation:** Collaborative efforts for sharing annotated datasets and conducting multicenter validation studies will improve model robustness, generalizability, and acceptance in real-world clinical settings.
5. **Ethical Frameworks and Regulatory Compliance:** Continued emphasis on ethical guidelines and regulatory compliance regarding patient privacy, data security, and responsible AI use is crucial. Development and adherence to ethical frameworks will foster trust and ensure patient safety in deploying ML models in health care.
6. **Patient-Centric Outcomes and Long-term Follow-up:** Emphasizing patient-centric outcomes and long-term follow-up studies will provide insights into the real-world impact of ML-driven interventions on patient survival, quality of life, and treatment efficacy.

While acknowledging the progress made by ML in bone cancer care, ongoing research, collaborative efforts, ethical considerations, and advancements in model interpretability will pave the way for more accurate diagnoses, personalized treatments, and improved patient outcomes in the future. Continued innovation and responsible deployment of ML models will redefine the landscape of bone cancer care, contributing to more effective and patient-centric health care strategies.

References

Abdelaal, M., Tobely, T.E., 2019. Bone cancer detection using particle swarm extreme learning machine neural networks. J. Med. Imaging Health Inform. 9 (3), 508–513.

Altameem, T., 2020. Fuzzy rank correlation-based segmentation method and deep neural network for bone cancer identification. Neural Comput. Appl. 32, 805–815.

Anand, D., Arulselvi, G., Balaji, G.N., Chandra, G.R., 2022. A deep convolutional extreme machine learning classification method to detect bone cancer from histopathological images. Int. J. Intell. Syst. Appl. Eng. 10 (4), 39–47.

Gawade, S., Bhansali, A., Patil, K., Shaikh, D., 2023. Application of the convolutional neural networks and supervised deep-learning methods for osteosarcoma bone cancer detection. Healthc. Analyt. 3, 100153.

Georgeanu, V.A., Mămuleanu, M., Ghiea, S., Selişteanu, D., 2022. Malignant bone tumors diagnosis using magnetic resonance imaging based on deep learning algorithms. Medicina 58 (5), 636.

Gitto, S., Cuocolo, R., Albano, D., Chianca, V., Messina, C., Gambino, A., Sconfienza, L.M., 2020. MRI radiomics-based machine-learning classification of bone chondrosarcoma. Eur. J. Radiol. 128, 109043.

Hossain, E., Rahaman, M.A., September 2018. Bone cancer detection & classification using fuzzy clustering & neuro fuzzy classifier. In: 2018 4th International Conference on Electrical Engineering and Information & Communication Technology (iCEEiCT). IEEE, pp. 541–546.

Jabber, B., Shankar, M., Rao, P.V., Krishna, A., Basha, C.Z., November 2020. SVM model based computerized bone cancer detection. In: 2020 4th International Conference on Electronics, Communication and Aerospace Technology (ICECA). IEEE, pp. 407–411.

Kaur, E.C., Garg, U., December 2022. Bone cancer detection techniques using machine learning. In: 2022 International Conference on Computational Modelling, Simulation and Optimization (ICCMSO). IEEE, pp. 315–319.

Kumar, V.A., 2021. Bone cancer detection using feature extraction with classification using K-nearest neighbor and decision tree algorithm. Smart Intel. Comput. Commun. Technol. 38, 347.

Kumar, D., Jain, N., Khurana, A., Mittal, S., Satapathy, S.C., Senkerik, R., Hemanth, J.D., 2020. Automatic detection of white blood cancer from bone marrow microscopic images using convolutional neural networks. IEEE Access 8, 142521–142531.

Li, Y., Dong, B., Yuan, P., 2023. The diagnostic value of machine learning for the classification of malignant bone tumor: a systematic evaluation and meta-analysis. Front. Oncol. 13, 1207175.

Lingappa, E., Parvathy, L.R., 2023. Deep learning-based active contour technique with bagging and boosting algorithms hybrid approach for detecting bone Cancer from Mri scan images. Multimed. Tool. Appl. 1–15.

Mahore, S., Bhole, K., Rathod, S., August 2021. Machine Learning approach to classify and predict different Osteosarcoma types. In: 2021 8th International Conference on Signal Processing and Integrated Networks (SPIN). IEEE, pp. 641–645.

Mazumder, M.H., Singh, M.P., September 2022. Bone cancer detection using deep learning. In: International Conference on Innovations in Computer Science and Engineering. Springer Nature Singapore, Singapore, pp. 285–296.

Papandrianos, N., Papageorgiou, E., Anagnostis, A., Papageorgiou, K., 2020. Bone metastasis classification using whole body images from prostate cancer patients based on convolutional neural networks application. PLoS One 15 (8), e0237213.

Satheesh Kumar, B., Bb, S., 2021. Bone Cancer Detection Using Feature Extraction with Classification Using K-Nearest Neighbor and Decision Tree Algorithm.

Sharma, A., Yadav, D.P., Garg, H., Kumar, M., Sharma, B., Koundal, D., 2021a. Bone cancer detection using feature extraction based machine learning model. Comput. Math. Methods Med. 2021.

Sharma, A., Yadav, D.P., Garg, H., Kumar, M., Sharma, B., Koundal, D., 2021b. Research Article Bone Cancer Detection Using Feature Extraction Based Machine Learning Model.

Shrivastava, D., Sanyal, S., Maji, A.K., Kandar, D., 2020. Bone cancer detection using machine learning techniques. In: Smart Healthcare for Disease Diagnosis and Prevention. Academic Press, pp. 175–183.

Shukla, A., Patel, A., 2020. Bone cancer detection from X-ray and MRI images through image segmentation techniques. Int. J. Recent Technol. Eng. 8 (6), 273–278.

Singh, M., Angurala, M., Bala, M., 2020. Bone tumour detection using feature extraction with classification by deep learning techniques. Res. J. Comput. Syst. Eng. 1 (1), 23–27.

Vandana, B.S., Alva, S.R., June 2021. Deep Learning Based Automated tool for cancer diagnosis from bone histopathology images. In: 2021 International Conference on Intelligent Technologies (CONIT). IEEE, pp. 1–8.

Whig, P., Velu, A., Nadikattu, R.R., Sharma, P., 2024a. Reinforcement learning for automated medical diagnosis and dynamic clinical regimes. In: Handbook of Research on Artificial Intelligence and Soft Computing Techniques in Personalized Healthcare Services. Apple Academic Press, pp. 169–187.

Whig, P., Kouser, S., Bhatia, A.B., Nadikattu, R.R., 2024b. Prediction of breast cancer diagnosis using random forest classifier. In: Computational Intelligence and Modelling Techniques for Disease Detection in Mammogram Images. Academic Press, pp. 55–73.

Yadav, D.P., Rathor, S., February 2020. Bone fracture detection and classification using deep learning approach. In: 2020 International Conference on Power Electronics & IoT Applications in Renewable Energy and its Control (PARC). IEEE, pp. 282–285.

CHAPTER

5 Identification of the risk of osteoporosis in older Vietnamese women using artificial intelligence and machine learning

Ketan Sarvakar
Ganpat University, Information Technology, Mehsana, Gujarat, India

5.1 Introduction

5.1.1 Background and importance of osteoporosis as a public health concern

Osteoporosis, a critical international fitness problem, requires a short-term remedy. Osteoporosis is a chronic condition that affects the skeletal system. A reduction of bone density and integrity is a hallmark of osteoporosis, which raises the opportunity for catastrophic fractures (Choi et al., 2018). As the population ages, osteoporosis will become more common and pose more risks to human fitness and well-being. Throughout someone's lifetime, bone resorption and formation take place in a dynamic and ongoing manner. People with exceptional fitness experience this complex balance, in which they absorb damaged or completely grown bone and then replace it with newly created bone. Conversely, individuals with osteoporosis experience a significant condition known as bone resorption, leading to a decrease in bone density and a depletion of calcium within the bones. Inadequate nutrients, hormone swings, and different scientific conditions are some factors that hinder bone rebuilding. Women's reduced estrogen levels throughout menopause interfere with the process of bone repair. This disruption could potentially increase the risk of osteoporosis and accelerate bone resorption. Moreover, the prevalence of conditions such as high blood pressure, hyperlipidemia, and diabetes mellitus increases it. Hyperuricemia exacerbates osteoporosis by contributing to its development (Manonai et al., 2003; Sarvakar, 2024) (Fig. 5.1).

Osteoporosis can lead to fragility fractures, which can have potentially severe outcomes (Choi et al., 2018). Fractures, especially those that take place inside the hip, have the ability to affect a person's overall lifestyle profoundly. Osteoporotic fractures regularly result in pain, incapacity, impaired independence, and constrained mobility. As a result, they significantly affect a person's ability to perform

https://doi.org/10.1016/B978-0-443-32892-3.00005-1

daily activities and maintain quality of life. Osteoporosis has a wide-ranging impact that extends beyond the individual, raising debates about the fact that it places a significant financial burden on health care systems and governments. Osteoporotic fractures require a substantial financial outlay for treatment. Hospitalization, rehabilitation, and long-term care for the affected person fall into this category of expenses. Osteoporosis affects individuals and also significantly burdens health care systems and governments financially. The express prices for osteoporotic fracture management, which includes hospitalization, rehabilitation, and long-term care, are excellent. Additionally, oblique costs related to decreased productivity and the strain on caregivers exacerbate the monetary impact of this disease. The asymptomatic individual is a significant impediment to dealing with osteoporosis. The situation often advances asymptomatically until a fracture occurs, underscoring the need for early identification and care. Diagnosis typically entails assessing bone mineral density (BMD) using strategies such as dual-energy x-ray absorptiometry (DXA technique) scans. However, the limited availability of these diagnostic methods in certain regions can hinder the timely identification and treatment of osteoporosis (Ministry of HealthVietnam, 2019; Haugeberg et al., 2000).

Public health initiatives play a crucial role in promoting preventive strategies.

Despite the lack of a cure for osteoporosis, one can take certain precautionary measures to reduce the likelihood of fractures and mitigate the associated consequences as per the following strategies:

1. Lifestyle changes: You can maintain optimal bone health by engaging in weight-bearing sports regularly, consuming adequate calcium and vitamin D, and refraining from smoking and consuming excessive alcohol.
2. Fatalities susceptible to falls: By implementing fall prevention techniques, particularly those that target the elderly, we can significantly reduce the incidence of osteoporotic fractures.
3. To increase bone density and reduce the risk of fractures, physicians may employ pharmacological treatment plans that include the administration of anti-resorptive or anabolic tablets.

Osteoporosis represents a significant public health challenge due to its high prevalence, it is necessary to make several distinctive efforts that are not only all-encompassing but additionally diverse. To reduce the possibility of fractures occurring in populations in extreme danger, public fitness campaigns must focus on expanding records about the relevance of bone health, advocating for early identification and diagnosis, and implementing specialized treatments. Continued attention is important for advancing our expertise about the factors that contribute to the formation of osteoporosis, identifying novel factors that play a role in the ailment's development, and developing more effective techniques for treating and coping with the infection. There is an extensive variety of subjects that researchers are investigating, including genetic and molecular mechanisms that are responsible for bone transformation, the significance of particular biomarkers in predicting the threat of fractures, and the opportunity to develop novel treatment options that could boost bone strength and decrease the effects of osteoporosis. Furthermore, the use of the modern generation, including synthetic intelligence and gadgets for getting to know (Anish and Nair, 2024), indicates the potential for improving the precision and availability of osteoporosis screening and hazard evaluation. By using these modern techniques, healthcare professionals can speed up disease diagnosis, customize treatment plans for individual patients, and improve the overall management of this condition. Osteoporosis poses a significant public health concern owing to its high incidence, significant impact on human well-being, and significant financial burden. To tackle this problem successfully, it is important to adopt a complete strategy that encompasses preventative measures, early identification, the availability of diagnostic tools and treatments, and enormous public health campaigns. By sharing facts, advocating for early identification and prognosis, and adopting positive treatment options, health care specialists, politicians, and the general public can collaborate to lessen the impact of this widespread, unrecognized incidence. By using modern technology and conducting more research, we can enhance our understanding and management of osteoporosis, ultimately leading to an improvement in bone health and overall quality of life for those with this debilitating condition. To deal effectively with the huge public health problem of osteoporosis, we must work together and take action. If we take outstanding action to address concerns and position ourselves for effective practice strategies, we can contribute to reducing the burden of chronic disease and promoting a healthier and more resilient global population.

5.1.2 The importance of employing precise prediction techniques in the evaluation of osteoporosis

For a variety of reasons, precise prognostic devices are critical in the process of evaluating osteoporosis severity. These technologies are helpful in the early detection and intervention of fractures, and enable medical professionals to identify individuals who are at risk of fractures before they occur and to take preventive actions as soon as possible. By providing individualized risk assessments and tailoring therapies to individual features, these techniques enhance clinical decision-making and ultimately improve patient outcomes. They help to achieve cost-effectiveness by decreasing the financial burden associated with osteoporotic fractures through optimal resource allocation and proactive management measures. On top of that, reliable prediction tools facilitate ongoing R&D collaboration, which in turn improves forecast accuracy and patient care. To improve the health and well-being of those who are at risk of developing osteoporosis, it is crucial to adopt these approaches for both prevention and therapy (Zhang et al., 2024a,b).

We can accomplish this by relying on more informed and evidence-based methods. This is particularly important given the fast-aging population, the rising incidence of osteoporosis-related fractures, and the corresponding medical costs.

First, the development of accurate and reliable prediction models has the potential to serve as useful tools that may aid medical professionals in making decisions based on accurate information obtained. Machine learning (ML) can help medical practitioners make better decisions about osteoporosis diagnosis, prevention, and treatment. One way to accomplish this is to rely on more knowledgeable and fact-based approaches. This is particularly important because of the fast-aging population, the rising incidence of osteoporosis-related fractures, and corresponding medical costs. The results of this research enhance the capacity of ML algorithms to serve as dependable diagnostic and risk assessment tools. The study showcases the exceptional predictive capabilities of the random forest regression model in determining the likelihood of developing osteoporosis. This research offers a strong foundation for incorporating advanced analytics into therapeutic settings. This improves the precision of risk evaluation and has the potential to simplify the screening process, optimize resource allocation, and boost the overall efficiency of health care delivery. Other variables such as area, height, and the results of several blood tests validate both the Osteoporosis Self-assessment Tool (OSTA) model and the one by Ho-Pham et al. In terms of the area of research, this is a huge step forward. This all-encompassing method of risk assessment gives doctors access to better, more thorough instruments that can detect those who are at high risk of having osteoporosis. By including a wider variety of clinical and lifestyle variables, the models may provide a more precise and comprehensive evaluation of a person's risk profile. As a result, individualized treatment plans and treatments will be able to take effect.

The study's results also have important implications for medical diagnostic digitalization in Vietnam. ML-based decision support systems might facilitate data-driven, more efficient decision-making in the health care sector, which is rapidly adopting technological advances. By using data analytics, medical professionals may enhance their diagnostic capabilities, expedite clinical procedures, and, in the end, provide better outcomes for patients (*National Strategy on Nutrition for 2011–2020, with a Vision to 2030*, 2012). Moreover, identifying factors that have a substantial impact on the risk of osteoporosis might provide valuable information for governmental policy and decision-making. These observations may direct the creation of focused public health initiatives, awareness campaigns, and resource allocation methods to tackle the increasing impact of osteoporosis in Vietnam (Zhang et al., 2023). Policymakers may use this empirically supported knowledge to provide priority to groups at high risk, allocate resources efficiently, and adopt comprehensive initiatives for the prevention and treatment of osteoporosis (Huong, 2019). The significance of this research lies in its ability to transform osteoporosis screening and management methods in Vietnam. The work establishes a strong basis for incorporating advanced analytics into clinical decision-making by showcasing the efficacy of ML models in predicting the risk of osteoporosis and validating current risk assessment methods. The results have the potential to improve health care practices, enhance patient outcomes, and provide valuable information for policy choices, eventually leading to a more efficient and fair health care system in Vietnam.

5.2 Materials and methods

5.2.1 Data source: Hanoi Medical University Hospital

We obtained the statistics for this investigation from the esteemed Hanoi Medical University Hospital in Vietnam. The health facility maintains an extensive database of affected individuals, which contains a wealth of information about demographics, medical conditions, and laboratory data. This makes it an invaluable tool for researching factors associated with the risk of osteoporosis.

5.2.2 Consideration of potential risk factors (demographics, lifestyle, medical history, and blood tests)

The dataset includes a wide range of demographic elements, some variables of which are age, sex, and ethnicity. Other lifestyle factors include smoking status, alcohol consumption, level of physical activity, and dietary choices. Variables in scientific records include fractures, osteoporosis in relatives, and cooccurring conditions such as diabetes or rheumatoid arthritis. Blood tests can assess a number of biochemical biomarkers of bone fitness and metabolism, including bone turnover indicators, serum calcium, and dietary D levels. This information is supported by studies conducted by Chen et al. (2024), Choi et al. (2018), Ebina et al. (2024), Lee et al. (2023), Haugeberg et al. (2000), Huong (2019), Lian et al. (2022), Manonai et al. (2003), and Naser et al. (2019).

5.2.3 Using data from the osteoporosis self-assessment tool

According to Lydick and Cook (1998), the OSTA can predict a character's risk of developing osteoporosis by calculating the expected BMD primarily based on age and weight. The formula OSTA $= (0.2 \times \text{weight}) - (0.001 \times \text{age})$ - 0.6 is used to calculate the OSTA value. We primarily categorize individuals into low-, medium-, or high-risk groups based on OSTA scores, which have an inverse relationship with osteoporosis risk. The inclusion of OSTA scores improves the models' prediction accuracy by using an established method for assessing osteoporosis risk.

The collection of data: A total of 1958 individuals with information including, age, sex, hormonal fluctuations, family records, race/ethnicity, frame weight, consumption of calcium and vitamin D, degree of physical activity smoking and alcohol intake behavior, clinical conditions, medicinal drugs, previous fractures, and osteoporosis status (the target variable) are covered in the dataset. Systems study models may require the encoding or preprocessing of class or textual statistics before their use. The majority of the columns consist of item and string record types, indicating this. The Osteoporosis column represents a binary-type problem. This column features an illustration of an integer record type. This procedure aims to identify the presence of osteoporosis by analyzing available demographic, lifestyle, and clinical elements.

Here is a dataset description:

> **Attempting to improve accuracy: n_estimators: The ensemble uses a range of boosting degrees, also known as bushes. Increasing this value typically improves performance, but it also increases the computational cost.**
> **Learning rate: The study price lowers each tree's contribution. At lower degrees, achieving the same overall performance stage requires more timber, but generalization could be more beneficial.**
> **Max_depth: This variable represents the maximum depth that a man or woman can reach.** Increasing this variety may also cause overfitting, but it can help the version capture more intricate connections in the statistics.
> **Min_samples_split:** The minimum number of samples required to split an internal node. If you want to increase the number of samples needed for a split, raising this value can be helpful in preventing overfitting.
> **Min_samples_leaf:** The bare minimum of samples is important for a leaf node to exist. By requiring more samples at the leaves, raising this parameter, such as min_samples_split, may aid in stopping overfitting.
> **The subsample is a subset of samples designed to fit each base learner independently. When set to significantly less than 1.0, the final result is stochastic gradient boosting, which can serve as a valuable tool in reducing overfitting.**
> **Max_features:** The number of features to consider when determining the best split. At each split, we evaluate all available features. Reducing this value could potentially improve computational efficiency and decrease overfitting.
> **Random_state:** Controls how random the estimator is. We guarantee the repeatability of the findings while providing a set cost.

Fig. 5.1 summarizes the shapes and data types of the dataset. It offers a concise overview of the structure and composition of the data, enabling a better understanding of the Fig. 5.1. variables and their characteristics.

Fig. 5.2 displays the relationship between the target age and the age bin. Fig. 5.2 illustrates the correlation between the age bin, which classifies individuals into distinct age groups, and the target age for osteoporosis screening. Each bin on the x-axis represents a certain age range, known as the age bin. We can observe a relationship between age and reliance. In this context, we use the term age to denote the age of humans, whereas remember refers to the wide range of individuals who belong to each age class. The x-axis, often divided into multiple containers or periods, displays the age values. Conversely, the y-axis suggests the number of healthy humans in every age institution. The plot graphically depicts the age distribution within the population under study. By analyzing this distribution, it becomes easier to identify patterns, such as the prevalence of certain age groups within the population. An example of this would be the plot, which might deliver statistics on whether the populace is in general made of more youthful or older humans, or whether certain age companies have a higher or lower frequency of people. Comprehending the age distribution is essential in several domain names, which include health care,

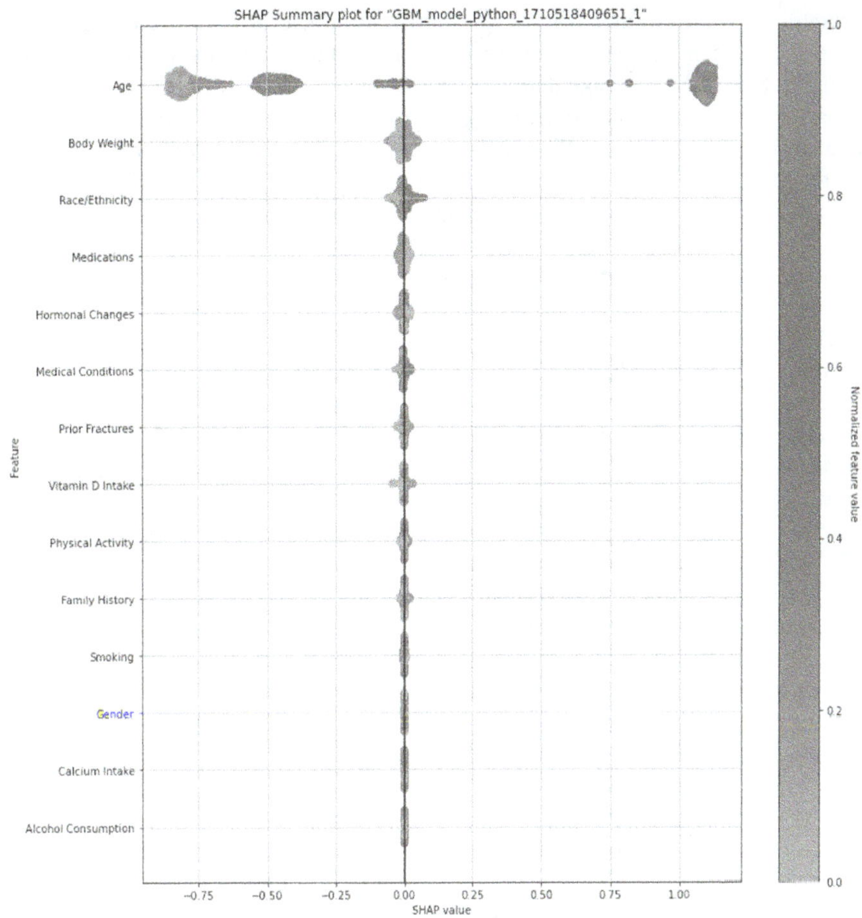

FIGURE 5.1

Shape summary.

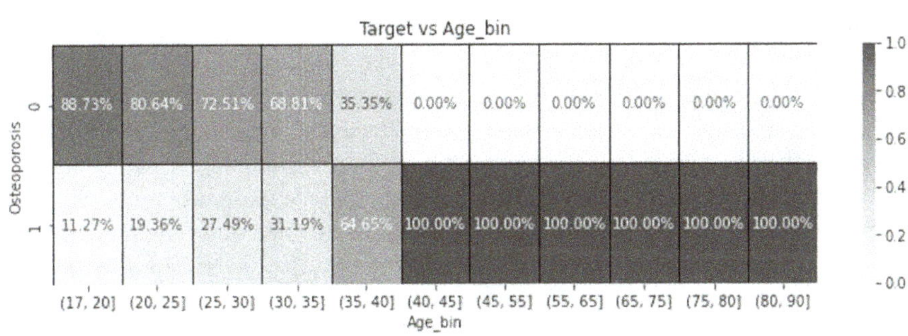

FIGURE 5.2

Target age vs. Age Bin.

marketplace research, and demography. In the health care sector, these data can guide decisions on screening programs, treatment plans, and effective resource distribution, considering the age demographics of the targeted organization. Similarly, in market studies, understanding the age distribution within a population allows us to segment target clients into distinct agencies and design marketing strategies that cater to the unique preferences and characteristics of each age group. This is because businesses of varying ages have exceptional alternatives and characteristics. The statistics presented provide valuable insights into the age distribution of the population, enabling informed and strategic decision-making under various conditions (Huong, 2019b).

Osteoporosis is a reduction in bone density and deterioration of bone tissue. This condition increases the likelihood of fractures, particularly in the older population. Bone mineral density (BMD) is typically assessed using dual-energy X-ray absorptiometry (DXA). Osteoporosis is defined as a BMD T-score below −2.5 or less than 70% of the young adult mean (YAM). Risk factors include older age, previous fractures, family medical history, and the use of corticosteroids. Although there are other criteria available for estimating fracture risk, the Fracture Risk Assessment Tool (FRAX) is not accurate. Artificial intelligence (AI) and ML algorithms have shown significant potential in assisting health care professionals in predicting osteoporosis. Researchers have used ML algorithms to predict hip and vertebral fractures using imaging data, demonstrating promising results in risk assessment. ML techniques such as neural networks and gradient boosting, which use patient demographics and blood test data, have proven effective in forecasting fracture risk and bone density. These technologies have the ability to diagnose osteoporosis in a noninvasive and highly efficient manner, potentially complementing existing DXA methods. This demonstrates the potential of ML in the early detection of osteoporosis, which could significantly improve patient outcomes by enabling timely intervention and medication.

The correlation between age groups and screening age demonstrates the relationship between categorized age ranges and the intended age for osteoporosis screening. Each bin on the x-axis represents a specific age range. Some examples of this age range are 40–49, 50–59, and 60–69. The y-axis clearly displays the goal age, the age at which individuals undergo osteoporosis evaluation. The picture illustrates the variation in the distribution of goals across diverse age classes. Through this connection analysis, it is possible to identify trends in the age distribution of men and women who are undergoing screening. This data is essential for evaluating the demographics of individuals targeted by osteoporosis screening initiatives and can inform the development and promotion of these programs. Medical experts could use the data from this study to tailor screening and intervention strategies to specific age groups, potentially increasing the effectiveness of osteoporosis prevention and treatment efforts in the long run. This study demonstrates the potential of ML in the early detection of osteoporosis, which could significantly improve patient outcomes by enabling timely intervention and medication.

Osteoporosis is characterized by low bone mass and deterioration of bone tissue. This condition increases the risk of fractures, particularly in older adults. The DXA approach is often used to measure BMD. Osteoporosis is defined as a BMD T-rating less than −2.5 or less than 70% of the YAM. Additional risk factors include advanced age, earlier fractures, family medical history, and the use of corticosteroids. Although numerous criteria can predict the threat of a fracture, the accuracy of the FRAX remains relatively low. Osteoporosis prediction is one area in which AI and ML algorithms have proven to be excellent aids in health care. Research has used ML algorithms to forecast hip and vertebral fractures via imaging data, demonstrating promising results in danger assessment. Using demographics and blood test results, learning algorithms including neural networks and gradient boosting have demonstrated effectiveness in predicting fracture hazard and bone density. These technologies can assess osteoporosis in a noninvasive and efficient manner, complementing current DXA procedures. Fig. 5.4 illustrates how early osteoporosis identification, with the aid of a device, can significantly enhance the consequences of affected individuals by facilitating active intervention and remedy (Albuquerque et al., 2023).

This graphic illustrates the relationship between family history and osteoporosis. Based on the facts shown, this image demonstrates the correlation between osteoporosis and family history and offers insightful data into how genetic predisposition affects the likelihood of developing osteoporosis. This information may be valuable for formulating patient treatment strategies and making informed choices (Ebina et al., 2024; Fan et al., 2020; Ramchand and Benjamin, 2024).

Fig. 5.3 provides evidence of the relationship between osteoporosis and family records. The images illustrate this relationship, with the status of osteoporosis acting as the dependent variable and family records as the independent variable. The x-axis illustrates the various classifications or levels of personal family records, ranging from excellent to poor. These classes and ranges reflect whether someone has a family history of osteoporosis. The y-axis regularly displays binary values (such as sure or no) to indicate the presence or absence of osteoporosis. The discernment reveals a significant variation in the frequency of osteoporosis among individuals with varying family histories. It is feasible to look at the effect of a family's propensity toward osteoporosis by examining this connection. More specifically, the plot seeks to

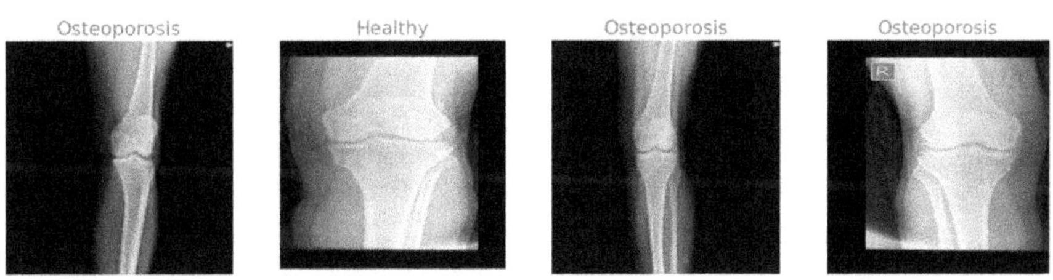

FIGURE 5.3

Risk of osteoporosis in older Vietnamese women.

ascertain whether individuals with a strong family history of osteoporosis are more likely to develop the disorder compared with those without such a history. Understanding the correlation between familial lineage and osteoporosis is essential for evaluating the likelihood of developing the condition, implementing preventative measures, and tailoring health care treatment plans to the person's wishes. Individuals with a familial predisposition to osteoporosis may be considered an excessive-risk group. When this occurs, it is crucial specifically to target these individuals for early screening, lifestyle modifications, and preventive therapies. On the other hand, people who do not have such a family history may be able to increase their chances of success by adopting preventive measures based on other factors. Overall, Figs. 5.4 and 5.5 provides illuminating data on how genetic predisposition influences the likelihood of osteoporosis. This can be useful in developing treatment plans for affected individuals and i making informed decisions.

(Bhardwaj et al., 2023), The data shown in Fig. 5.6 offers significant insights into the demographic distribution of age among the population. This knowledge facilitates informed decision-making and strategic planning in many scenarios.

For anthropometric measures, the process involves collecting body mass index (BMI) measurements. Based on "Recording of Body Mass Index (BMI) and anthropometric measurements," it would appear that the dataset consists of information from the BMI to other measurements that can be associated with the bodily capabilities of humans, inclusive of peak and weight. We calculate a numerical dimension known as the BMI when we consider an individual's weight and height. Humans are classified according to their weight status: underweight, average weight, overweight, or obese. When it comes to anthropometric measurements, various physical measurements are included (Fan et al., 2020). These measurements include height, weight, and waist circumference.

The use of BMI and anthropometric data could be highly effective in predicting the risk of osteoporosis, because factors such as being underweight or having low

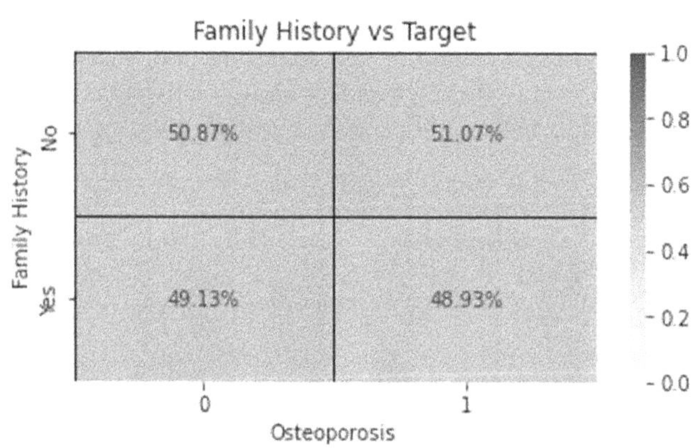

FIGURE 5.4

Family history vs. osteoporosis.

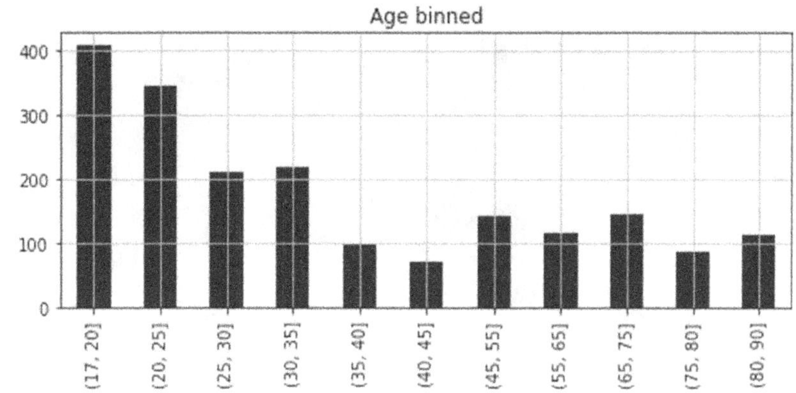

FIGURE 5.5

Age vs. count.

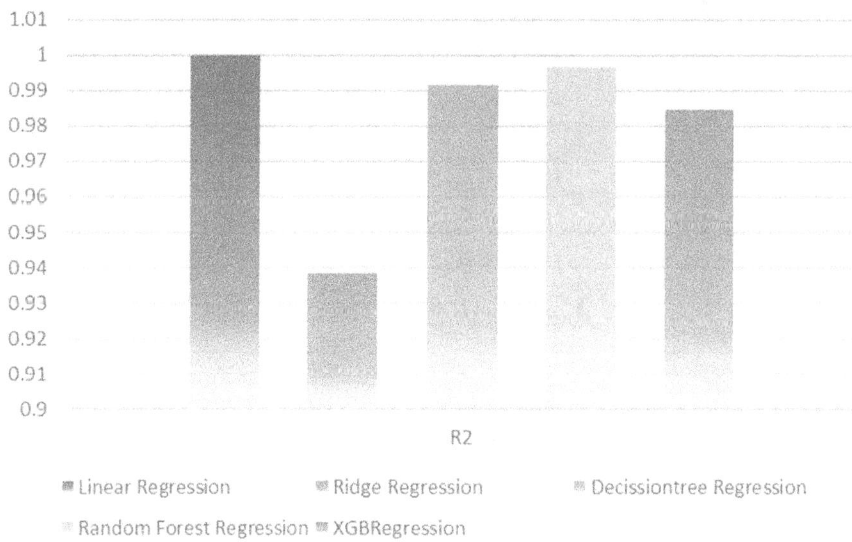

FIGURE 5.6

Chart of model comparison.

muscle tissue are well-established risk factors for this condition. By incorporating these quantitative facts, in addition to the previously mentioned categorical variables such as sex, lifestyle characteristics, and scientific records, we can significantly enhance the prediction capacity of the model-learning system (Bui et al., 2022). Models provide the ability to identify intricate patterns and connections between numerical anthropometric factors and the outcome of osteoporosis. Gathering BMI and anthropometric measurements provides valuable quantitative data that learning algorithms can use to improve the accuracy of osteoporosis risk assessment for all

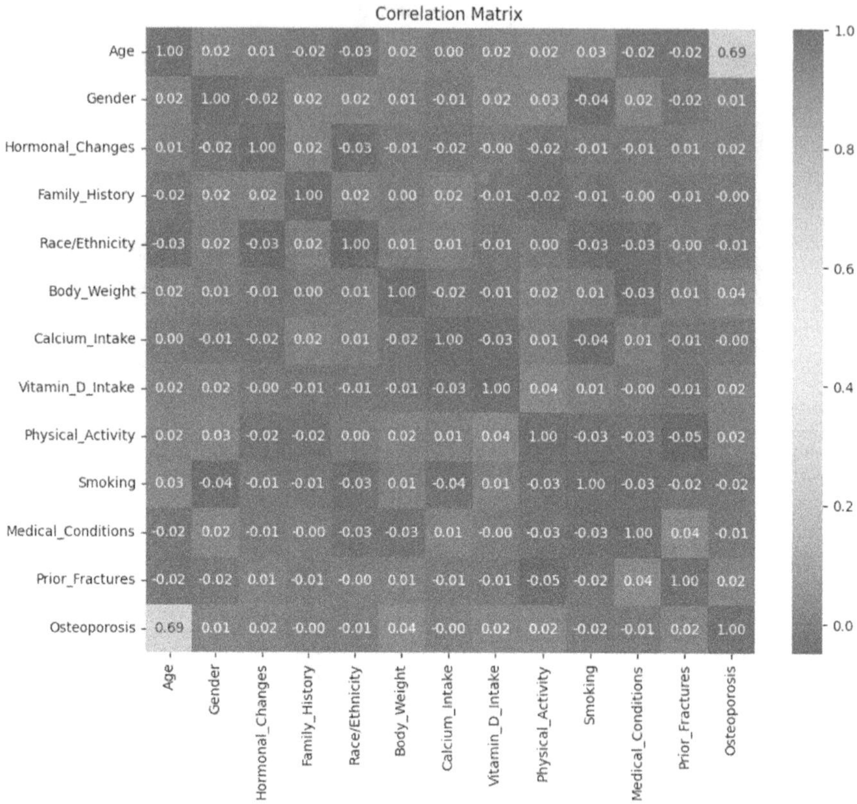

FIGURE 5.7

Correlation matrix.

individuals in the dataset. Fig. 5.7 illustrates the correlation matrix, which has been referenced in earlier studies (Fang et al., 2024).

5.3 Machine learning models

Data Preprocessing:

Based on the facts that are provided and the approximate size of the dataset, information preprocessing steps are:

Handle missing values:

A few of the dataset's columns, Medications (973 nonnull) and Alcohol Consumption (970 nonnull), have missing values. Depending on the type of missing facts, you may need to use more sophisticated methods such as k nearest neighbor (KNN) imputation or matrix factorization, or you might want to apply simpler approaches such as implied or median imputation.

Encode categorical features:

The dataset protects several class characteristics: Hormonal Changes, Gender/Sex, Body Weight, Family History, Race/Ethnicity, Alcohol Consumption, Calcium Intake, Smoking, Vitamin D Intake, Physical Activity, Prior Fractures, Medical Conditions, and Medications.

It is necessary to encode these attributes directly into a numerical illustration which the device learning models can use the most. Depending on the properties of each function, you could use methods such as label encoding, ordinal encoding, or one-warm encoding.

Standardize/Normalize Numerical Features:

The dataset includes a few numerical traits such as ID, Age, and Osteoporosis. We suggest standardizing or normalizing those numerical traits to ensure they have the same scale. This may improve the effectiveness of various machine learning techniques.

Create new features:

Based on available functions and your understanding of the situation, you might consider exploring the potential for developing additional traits that could be more instructive for osteoporosis prediction.

Using the cutting-edge functions, you could, for example, build functions about frame mass index (BMI) or other derived metrics.

Train-Test Split:

To assess the effectiveness of the system learning models, split the dataset into training and testing sets.

Usually, we divide the training and testing sets at 70:30 or 80:20, respectively.

Feature Selection:

Characteristic choice can be necessary to find the most applicable characteristics for the osteoporosis prediction task, depending on the quantity and complexity of the dataset. Methods such as gradient boosting-based characteristic importance, recursive feature deletion, or correlation evaluation can carry out this task.

To optimize machine learning models such as extreme gradient boosting (XGBoost), decision trees, random forests, ridge regression, lasso regression, and linear regression, it's essential to carefully adjust their hyperparameters. For lasso regression, the primary recognition of tuning for linear, ridge, and lasso regression is determining the proper regularization energy (alpha). The configuration that minimizes validation set error may be observed by methodically exploring the hyperparameter area using techniques such as grid search or randomized search. When using decision tree regression, factors such as approximate leaves, minimum sample size, and maximum intensity are of unique importance. Grid search or randomized seek can identify the optimal set of hyperparameters to improve model performance, often measured by metrics such as imply squared errors.

Because random forest regression and XGBoost regression are ensemble strategies with a wider range of hyperparameters, they are more complex. In random forest, factors encompass the wide variety of timber in the forest and function selection at each break up to the hyperparameters of character bushes. Similar to this,

XGBoost presents a huge variety of hyperparameters, such as regularization parameters, tree intensity, and studying fee. In this case, advanced tuning methods such as grid seek, randomized search, and Bayesian optimization are important because they let you thoroughly explore the hyperparameter area to find the set of values that improves cross-validated performance. By systematically refining these models, specialists may additionally use those fashions fully and get extraordinary prediction precision across a number of regression assignments.

5.4 Significance of the study

5.4.1 Contributions to osteoporosis screening techniques

Table 5.1 lists the evaluation metrics for an osteoporosis screening model. Each row in the table represents a specific metric, whereas the columns display statistical measures such as the mean, standard deviation (SD), and cross-validated results for different folds (cv_1_valid to cv_5_valid). We are evaluating metrics such as accuracy, AUC, F1 score, precision, recall, specificity, and log loss, among others.

A version used in osteoporosis screening is evaluated using a variety of criteria, as shown in Table 5.1. The columns include statistical metrics such as the mean, Standard deviation (SD), and established findings for several folds (cv_1_valid to cv_5_valid), with each row representing a separate metric. This is how the table structure is defined.

Metric: This column indexes the evaluation metrics that may be assessed, including accuracy, AUC, F1 rating, precision, consideration, specificity, and log loss.
Mean: This column displays the average value of the associated metric over the cross-validation folds.
SD: This column displays the variability or dispersion of the metric values across the cross-validation folds.
Folders with move-validation (cv_1_valid to cv_5_valid): These columns display the specific metric values for each fold of the cross-validation technique. We use a method known as go-validation to evaluate a model's performance on an exclusive dataset.

The numbers in each table column illustrate the diversity of these metrics across multiple move-validation folds, providing valuable insights into the model's performance across various assessment measures. Taking everything into account, this table offers a comprehensive assessment of the model's performance in osteoporosis screening, highlighting the advantages and limitations of different evaluation criteria (2021; Tu et al., 2024).

Table 5.1 Evaluation metrics for osteoporosis screening model.

	Mean	SD	cv_1_valid	cv_2_valid	cv_3_valid	cv_4_valid	cv_5_valid
Accuracy	0.8653464	0.0118539	0.8591065	0.8529412	0.8720539	0.8825623	0.8600683
AUC	0.8935655	0.0119719	0.880967	0.9047363	0.8914468	0.9071501	0.8835275
Err	0.1346536	0.0118539	0.1408935	0.1470588	0.1279461	0.1174377	0.1399317
err_count	39.6	4.449719	41.0	45.0	38.0	33.0	41.0
f0_point	0.9098560	0.0090125	0.9212731	0.9117647	0.8986928	0.9144296	0.9031199
f1	0.8506959	0.0107410	0.8429119	0.8464164	0.8527132	0.8685259	0.8429119
f2	0.799019	0.0199141	0.7768362	0.7898089	0.8112094	0.8270106	0.7902299
lift_top_group	2.0054338	0.0995076	1.9530201	1.8658537	2.1214285	2.0661764	2.0206897
Logloss	0.4408044	0.0427689	0.4825811	0.4244212	0.4639068	0.3741691	0.4589439
max_per_class_error	0.2319684	0.0252412	0.2617450	0.2439024	0.2142857	0.1985294	0.2413793
mcc	0.7456351	0.0172023	0.7439481	0.7280335	0.7495337	0.7725316	0.7341286
mean_per_class_accuracy	0.8657987	0.0085677	0.8620853	0.8604432	0.8673794	0.8800457	0.8590401
mean_per_class_error	0.1342013	0.0085677	0.1379147	0.1395569	0.1326206	0.1199544	0.1409599
mse	0.1339112	0.0160767	0.1499888	0.1277992	0.1421424	0.1089275	0.1406980
pr_auc	0.9258764	0.0086861	0.9136743	0.9365031	0.9213122	0.9339948	0.9178975
Precision	0.9543377	0.0186403	0.9821429	0.9612403	0.9322034	0.9478261	0.9482759
r2	0.4632796	0.0642227	0.3996976	0.4861472	0.4295615	0.5638425	0.4371491
Recall	0.7680317	0.0252412	0.738255	0.7560976	0.7857143	0.8014706	0.7586207
rmse	0.3653861	0.0224760	0.3872838	0.3574901	0.3770178	0.3300417	0.3750973
Specificity	0.9635658	0.0137221	0.9859155	0.9647887	0.9490446	0.9586207	0.9594595

5.4.2 Implications for Vietnam's health care system and policy development

The use of machine learning algorithms to predict osteoporosis could significantly influence Vietnam's health care administration and policy-making, given the country's aging population and increasing health care needs. Using sophisticated predictive analytics, health care stakeholders may also proactively become aware of patients with an excessive risk of osteoporosis and execute tailor-made treatments to prevent fractures and improve bone fitness. The objectives of Vietnam's National Strategy on Nutrition for 2011–2020, which emphasizes the importance of treating malnutrition and related health issues such as osteoporosis in various senior populations, align with this approach (Vietnam National Institute of Nutrition, 2012; Parfitt, 1987; Zhang et al., 2024a,b).

The Vietnamese health care system may also benefit from advanced preventive care methods, efficient resource allocation through devices, and the development of algorithms for osteoporosis prediction. Early identification of people with a high risk of osteoporosis allows health care practitioners to tailor remedies, including vitamin D and calcium supplements, workout regimens, and nutritional recommendations, to limit bone loss and decrease the risk of fracture. This preventive approach improves patient outcomes and reduces healthcare costs associated with osteoporosis, including expenses for long-term care, hospital stays, and fracture treatment.

By integrating machine learning algorithms into healthcare management systems, we can streamline patient risk assessment and decision-making processes, potentially improving healthcare delivery and efficiency. Health care practitioners may prioritize people at high-risk for preventative treatments, improve screening tactics, and allocate assets based totally on predicted risk tiers via automated osteoporosis danger prediction and stratification. By focusing health care resources on people who would benefit most from them, this tailored method maximizes the effectiveness of preventive movements and lessens the burden of osteoporosis on the health care system.

Using systems that learn how to predict osteoporosis can help with making decisions based on facts and allow the creation of personalized treatments for maintaining bone health at the population level. This is important for creating national health care guidelines and suggestions. Policymakers can support the use of machine learning in clinical practice, encourage health care professionals and data scientists to work together across disciplines, and spend money on infrastructure and education programs to help the spread of predictive analytics in health care settings. Recognizing the predictive value of these methods in identifying individuals at risk of osteoporosis can achieve this (Shao et al., 2024; Ramesh and Santhi, 2024).

5.5 Measurement

Next, we discuss the process of transforming BMI values into T-Scores for identifiable outcomes. This section explains how to convert BMI values into T-rankings for labeled results in osteoporosis screening. BMI computation, a widely used frame composition metric, proceeds as follows: BMI is calculated using the formula kg/m^2, which is a person's weight in kilograms divided by the peak height in meters squared. Using BMI T-ratings to divide humans into diverse risk categories is frequently beneficial for osteoporosis screening. The conversion technique compares an individual's BMI with reference values derived from a standardized population, typically teenagers. This comparison produces a T-rating, or a wide variety of preferred deviations above or below the reference population's suggested BMI. Predetermined standards divide T-rankings into numerous outcome classes for osteoporosis screening, including normal, underweight, overweight, and obese. Health care professionals may also determine someone's relative risk of osteoporosis based on the frame composition by remodeling BMI values into T-scores and classifying consequences correctly. This record helps identify people who are more likely to have osteoporosis fractures and informs treatment plans and preventative moves that are appropriate for specific BMI categories (Ou Yang et al., 2021) (Table 5.2).

5.6 Results

This study compares the performance of different ML regression models in predicting the risk of osteoporosis using various evaluation metrics. It is highly unlikely to achieve perfect scores on all metrics in real-world scenarios, because that might indicate overfitting or issues with the data or model. Ridge regression, a linear regression model with L2 regularization, performed reasonably well with an R^2 score of 0.938469 and a cross-validation score of 0.694594. Lasso regression, a linear regression model with L1 regularization, showed poor performance with low R^2 and cross-validation scores, suggesting it may not be suitable for this task. Decision tree regression, a regression model based on decision trees, performed well with an R^2 score of 0.991663 and a cross-validation score of 0.753398. With the best mean absolute error (MAE), mean squared error (MSE), and root mean squared error (RMSE) scores, along with the highest R^2 (0.996456) and cross-validation (0.967530) scores, random forest regression, an ensemble learning method that uses multiple decision trees, did better than other models. XGBRegression, likely an implementation of the advanced ensemble learning technique XGBoost, performed well with an R^2 score of 0.984467 and a cross-validation score of 0.864402. These results suggest that the random forest regression model is the most promising for predicting the risk of osteoporosis, because it demonstrated the highest accuracy and generalization performance across all metrics. However, the performance of the models may vary depending on the specific dataset and

Table 5.2 Comparison of Machine learning model.

	Model	Mean absolute error	Mean squared error	Root mean squared error	R^2	Cross_val
0	Linear regression	1.975544e-16	4.830685e-32	2.197882e-16	1.000000	1.000000
1	Ridge regression	3.068050e-02	1.434604e-03	3.787616e-02	0.938469	0.694594
2	Lasso regression	1.192240e-01	2.408817e-02	1.552036e-01	−0.033158	−7.819097
3	Decision tree regression	8.211761e-03	1.943800e-04	1.394202e-02	0.991663	0.753398
4	Random forest regression	5.520772e-03	8.263229e-05	9.090231e-03	0.996456	0.967530
5	XGBRegression	1.385655e-02	3.621502e-04	1.903025e-02	0.984467	0.864402

problem at hand, and further evaluation and validation may be necessary before deploying these models in a real-world setting, as illustrated in Fig. 5.6.

Using plenty of metrics, it compares the efficacy of various ML regression models in predicting the threat of osteoporosis. It achieved an ideal score on every measure, which may be excellent under realistic conditions but could also indicate overfitting or issues with the statistics or model itself. A linear regression version with L2 regularization is known as ridge regression. Instead, it did well with an R^2 rating of 0.938469 and a pass-validation score of 0.694594. Lasso regression is a linear regression model that uses L1 regularization. With low R^2 and pass-validation rankings, it did not perform properly and may be unsuitable for this challenge. A regression model based entirely on decision trees is known as decision tree regression. It performed well with a pass-validation rating of 0.753398 and an R^2 score of 0.991663. An ensemble learning a technique known as random forest regression integrates numerous choice trees. It outperformed the others with the lowest MAE, MSE, and RMSE, as well as the best R^2 (0.996456) and cross-validation (0.967530) rankings of any model. Strong ensemble learning techniques such as XGBoost are usually implemented in XGBRegression. It did well, with an R^2 rating of 0.984467 and a go-validation score of 0.864402. These findings suggest that the random forest regression model, which has high-quality accuracy and generalization overall performance across all criteria, is the most promising for estimating the chance of osteoporosis. The models' performance may differ depending on the precise statistics and problems under consideration. Therefore, an extra exam and verification may be necessary before implementing those models in an actual environment, as seen in Fig. 5.6.

5.7 Model development

The ensemble method incorporates regression techniques customized for specific problems and datasets. Assuming a linear connection between the independent and dependent variables, linear regression is the foundational method. To stabilize the version and save you from multicollinearity and overfitting, ridge regression adds a penalty time period. Random forest regression combines predictions from multiple decision trees to improve accuracy and reduce overfitting, while decision tree regression makes splits based on key features. XGBoost Regression sequentially builds decision trees to correct errors from previous ones, further improving predictions. Data collection is the first step in the version construction method. Next comes preprocessing, which handles missing values, cleans up the records, and encodes features. To improve model performance, feature engineering may include including new functions or altering ones already present. The model selection process depends on the problem complexity, data characteristics, and model assumptions. The chosen models are then trained using split datasets. Evaluation metrics, such as mean squared error and R^2 evaluation model performance on test data, power hyperparameter tweaking to fine-tune fashions for optimal results. Researchers deploy models in real-world applications once the models meet predetermined performance criteria. Model developers navigate complexity through an iterative method, using a tool kit of procedures to extract valuable insights from information, enhancing decision-making techniques across various fields Alyahyan, 2024.

5.8 Osteoporosis prediction in each machine learning model

An in-depth analysis of each model's osteoporosis predictions provides subtle insights into its performance and suitability for screening. Despite its simplicity, linear regression offers a favorable start line with middling accuracy and AUC values. Its capacity to capture the tricky nonlinear interactions seen in osteoporosis predictors, however, may be limited. The regularization strategies used by the ridge and lasso regression models have similar performance, underscoring the need to prevent overfitting in predictive modeling. Even though decision tree regression is at risk of overfitting, it produces comprehensible findings and allows for the creation of important predictive elements. Random wooded area regression, with its ensemble method, reduces overfitting and provides higher prediction accuracy than single-decision timber. Comparing it with more sincere fashions, however, could make it harder to comprehend. XGBoost Regression, which uses gradient boosting, performs well on numerous standards, demonstrating its potential to discover complicated styles within the data. To pick the best technique for osteoporosis screening, one needs to weigh the pros and cons of every model, considering elements such as computational complexity, interpretability, and prediction accuracy. Similar research into characteristic importance and model interpretability can advance the clinical usefulness of these prediction models under real osteoporosis screening conditions (Zhao et al., 2024; Kiran and Areeckal, 2024; Zhang et al., 2024a,b).

5.9 Conclusion

This study investigates the use of machine learning algorithms to predict the likelihood of osteoporosis based on multiple patient factors, including clinical history, blood test results, lifestyle variables, and demographics. The main objective was to develop specific and simple models to evaluate the risk of osteoporosis in Northern Vietnamese women aged 50 years and older. The major conclusions are that an essential part of the process is changing BMI values into T-ratings for classified consequences, which include normal, underweight, obese, and overweight. This process made it feasible to determine everyone's relative chance of osteoporosis in line with their frame composition. With the lowest mean absolute error, mean squared error, and root mean squared error, as well as the greatest R^2 and cross-validation rankings, the random forest regression model performed the best out of all the machine learning models. The most promising model for estimating the threat of osteoporosis is the random forest regression version. Osteoporosis assessments have become more thorough with the validation of the model of Ho-Pham et al. and the OSTA model using new elements together with height, place, and unique blood check findings. This approach helps to ensure the accuracy and reliability of risk assessment tools. These studies' results have improved osteoporosis screening practices and had a significant impact on Vietnamese health care administration and coverage. A better osteoporosis threat assessment may also result from combining modern device mastering algorithms with in-depth information evaluation. This can have an impact on patient control and health care policies related to osteoporosis therapy. Further research endeavors could involve broadening the scope of the study, integrating supplementary clinical and lifestyle variables, and investigating the feasibility of using more complex device learning structures to enhance the predictive power of the model. It can also be crucial to validate and enhance the models continuously in real clinical settings to ensure their usefulness and efficacy in enhancing osteoporosis care and prevention.

References

Albuquerque, G.A., et al., 2023. Osteoporosis screening using machine learning and electromagnetic waves. Sci. Rep. 13 (1). https://doi.org/10.1038/s41598-023-40104-w.

Alyahyan, S., 2024. Applying machine learning classification techniques for disease diagnosis from medical imaging data using Transformer based Attention Guided CNN (TAGCNN). Multimed. Tool. Appl. 1–27.

Anish, R.J., Nair, A., 2024. Osteoporosis management-current and future perspectives–A systemic review. J. Orthop. 53.

Bhardwaj, Swe, K., Sinha, N.K., 2023. Treatment for osteoporosis in people with beta-thalassaemia. Cochrane Database Syst. Rev. https://doi.org/10.1002/14651858.CD010429.pub3.

Bui, H.M., et al., 2022. Predicting the risk of osteoporosis in older Vietnamese women using machine learning approaches. Sci. Rep. 12 (1). https://doi.org/10.1038/s41598-022-24181-x.

Chen, P.K., Tang, K.T., Chen, D.Y., 2024. The NLRP3 inflammasome as a pathogenic player showing therapeutic potential in rheumatoid arthritis and its comorbidities: a narrative review. Int. J. Mol. Sci. 25 (1). https://doi.org/10.3390/ijms25010626.

Choi, S.T., et al., 2018. Prevalence and fracture risk of osteoporosis in patients with rheumatoid arthritis: a multicenter comparative study of the FRAX and WHO criteria. J. Clin. Med. 7 (12). https://doi.org/10.3390/jcm7120507.

Ebina, K., et al., 2024. An investigation of the differential therapeutic effects of romosozumab on postmenopausal osteoporosis patients with or without rheumatoid arthritis complications: a case–control study. Osteoporos. Int. https://doi.org/10.1007/s00198-024-07019-2.

Fan, Z., et al., 2020. Comparison of OSTA, FRAX and BMI for predicting postmenopausal osteoporosis in a Han population in Beijing: A cross sectional study. Clin. Interv. Aging 15, 1171–1180. https://doi.org/10.2147/cia.S257166.

Fang, F., Yang, J., Wang, J., Li, T., Wang, E., Zhang, D., Liu, X., Zhou, C., 2024. The role and applications of extracellular vesicles in osteoporosis. Bone Res. 12 (1), 4.

Haugeberg, G., et al., 2000. Bone mineral density and frequency of osteoporosis in female patients with rheumatoid arthritis: results from 394 patients in the Oslo County rheumatoid arthritis register. Arthritis Rheum. 43 (3), 522–530. https://doi.org/10.1002/1529-0131(200003)43:3<522::AID-ANR7>3.0.CO;2-Y.

Huong, N.T., 2019. Overview of the Vietnamese healthcare system. Asia Pac. J. Health Manag. 14 (3), 27–32.

Kiran, K.S.,S., Areeckal, A.S., 2024. Detection of osteoporosis from calcaneal radiograph images using wavelet texture analysis and machine learning. Int. J. Comput. Digit. Syst. 15 (1), 1–11.

Lee, C., et al., 2023. Prediction of osteoporosis in patients with rheumatoid arthritis using machine learning. Sci. Rep. 13 (1). https://doi.org/10.1038/s41598-023-48842-7.

Lian, L., et al., 2022. Sarcopenia may Be a risk factor for osteoporosis in Chinese patients with rheumatoid arthritis. Int. J. Gen. Med. 15, 2075–2085. https://doi.org/10.2147/IJGM.S349435.

Lydick, E., Cook, K., Turpin, J., Melton, M., Stine, R., Byrnes, C., 1998. Development and validation of a simple questionnaire to facilitate identification of women likely to have low bone density. Am. J. Manag. Care 4 (1), 37–48.

Manonai, J., Theppisai, U., Suchartwatnachai, C., Jetsawangsri, T., Chittacharoen, A., March 28, 2003. Compliance with hormone replacement therapy in Thai women. Maturitas 44 (3), 201–205. https://doi.org/10.1016/s0378-5122(02)00341-9.

Ministry of Health, Vietnam, 2019. Health Statistics Yearbook 2018 (Hanoi, Vietnam).

Naser, A., et al., 2019. Polymorphisms in TNF Receptor superfamily 1B (TNFRSF1B:rs3397) are linked to Mycobacterium avium paratuberculosis infection and osteoporosis in rheumatoid arthritis. Microorganisms 7 (12). https://doi.org/10.3390/microorganisms7120646.

Ou Yang, W.-Y., Lai, C.-C., Tsou, M.-T., Hwang, L.-C., 2021. Development of machine learning models for prediction of osteoporosis from clinical health examination data. Int. J. Environ. Res. Publ. Health 18, 7635.

Parfitt, A.M., 1987. Bone remodeling and bone loss: understanding the pathophysiology of osteoporosis. Clin. Obstet. Gynecol. 30 (4), 789–811.

Ramchand, S.K., Benjamin, Z.L., 2024. Sequential therapy for the long-term treatment of postmenopausal osteoporosis. J. Clin. Endocrinol. Metabol. 109 (2), 303–311.

Ramesh, T., Santhi, V., 2024. Multi-level classification technique for diagnosing osteoporosis and osteopenia using sequential deep learning algorithm. Int. J. Syst. Assur. Eng. Manag. 15 (1), 412–428.

Sarvakar, K., 2024. Layered architecture of IoT. Secure and Intelligent IoT-Enabled Smart Cities. IGI Global, pp. 164–194. In press.

Shao, Z., Gao, H., Wang, B., Zhang, S., 2024. Exploring the impact of pathogenic microbiome in orthopedic diseases: machine learning and deep learning approaches. Front. Cell. Infect. Microbiol. 14, 1380136.

Tu, J.-B., Liao, W.-J., Liu, W.-C., Gao, X.-H., 2024. Using machine learning techniques to predict the risk of osteoporosis based on nationwide chronic disease data. Sci. Rep. 14 (1), 1–11.

Vietnam National Institute of Nutrition, 2012. National Strategy on Nutrition for 2011-2020, With a Vision to 2030 (Hanoi, Vietnam).

Zhang, K., et al., 2023. Construction of a predictive model for osteoporosis risk in men: using the IOF 1-min osteoporosis test. J. Orthop. Surg. Res. 18 (1). https://doi.org/10.1186/s13018-023-04266-7.

Zhang, Y.-Y., Xie, N., Sun, X.-D., Nice, E.C., Liou, Y.-C., Huang, C., Zhu, H., Shen, Z., 2024a. Insights and implications of sexual dimorphism in osteoporosis. Bone Res. 12 (1), 8.

Zhang, B., Chen, Z., Yan, R., Lai, B., Wu, G., You, J., Wu, X., Duan, J., Zhang, S., 2024b. Development and validation of a feature-based broad-learning system for opportunistic osteoporosis screening using lumbar spine radiographs. Acad. Radiol. 31 (1), 84–92.

Zhao, F., Li, C., Wang, W., Zhang, Y., Yao, P., Wei, X., Jia, Y., Dang, S., Zhang, S., 2024. Machine learning predicts the risk of osteoporosis in patients with breast cancer and healthy women. J. Cancer Res. Clin. Oncol. 150 (2), 102.

CHAPTER

Identification of paget disease in the human body using artificial intelligence and machine learning

6

Palvi Sharma[1], Rakesh Kumar[2], Meenu Gupta[2]

[1]*Department of Computer Science & Engineering, Model Institute of Engineering and Technology, Jammu, Jammu and Kashmir, India;* [2]*Department of Computer Science and Engineering, Chandigarh University, Gharuan, Punjab, India*

6.1 Introduction

A persistent bone disorder known as Paget disease of the bones or osteitis deformers is characterized by abnormal bone development. It is most common in people over age 65 and is more prevalent among Europeans of Scottish or British origin. Paget disease generally of mixed genetic and environmental factors (Fig. 6.1), but the actual cause is unknown (Tan and Ralston, 2014). Paget disease is characterized by excessive and irregular bone remodeling that results in larger, more fragile and deformed bones. Bone ache, deformities, fractures, nerve compression, and other effects including osteoarthritis are some symptoms and signs that might result from this rapid bone growth. It frequently affects the skull, pelvis, backbone, and leg muscles (Simonds et al., 2019). A variety of scientific manifestations of Paget disorder can also present, depending on the degree and location of bone involvement. In the diagnosis of Paget disease, some patients evidence terrible pain and functional

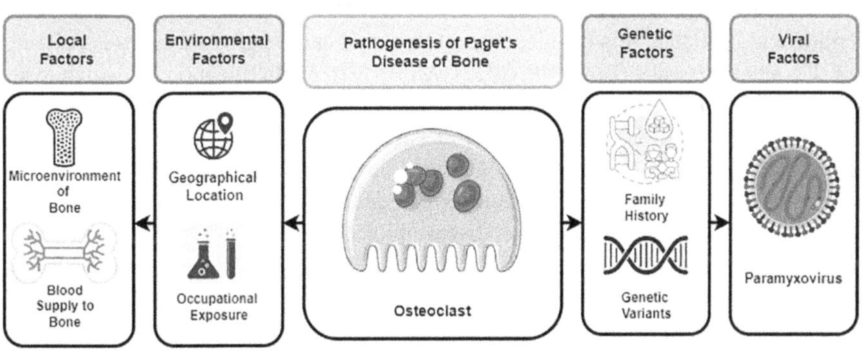

FIGURE 6.1

Pathogenesis of Paget disease.

https://doi.org/10.1016/B978-0-443-32892-3.00006-3

impairment whereas others show no symptoms and can be identified only through regular screening (Paul Tuck et al., 2017). Some techniques (i.e., clinical examination, laboratory tests, and imaging methods) are used by the health care professional to determine the presence of Paget disease in a patient. Imaging methods are composed of x-rays, computed tomography (CT) scans, and magnetic resonance imaging (MRI) scans. X-ray is the preferred imaging method for detecting distinct abnormalities in bone structure such as thickness, abnormalities of the corticodermal layer, and regions with high–bone density sclerosis (Bayan et al., 2018). CT and MRI scan are the two other imaging methods used to evaluate the overall degree of Paget disease and examine potential difficulties. To examine the turnover sign in bone, traditional diagnostic approaches to identifying Paget disorder mostly rely on clinical examinations, x-rays, bone scans, and blood tests (Reid, 2020).

In terms of usability, precision, and effectiveness, these traditional approaches often have limitations that cause delays in diagnosis and poor patient outcomes. In addition, the treatment of Paget disease relieves symptoms, restores normal bone turnover, and reduces complications. According to researchers, unfortunately there is no proper treatment of Paget disease, but it can be controlled in early stages by taking painkillers and inhibitors of bone resorption and undergoing physical therapy and sometime surgery to treat fractures or deformities (Selby, 2006; Edey et al., 2019). Despite improvements and advances, the detection of Paget disease remains a difficult challenge, especially when some serious symptoms or consequences are present. To enhance the quality of life, early detection and efficient treatment are necessary for patients who are affected by this disorder (Mirra et al., 1995).

In the detection of Paget disease, artificial intelligence (AI) and machine learning (ML) technologies have an important role because they provide efficacy and accuracy in the analysis of medical images. These algorithms can identify small patterns and abnormalities in the bone structure that can indicate the existence of Paget disease. AI and ML methods provide several advantages over conventional ones in terms of accuracy, consistency, scalability, continuous learning, and efficiency. Furthermore, AI and ML offer progressive opportunities for evaluating and treating Paget disorder. These technologies can enhance the identification of specific bone anomalies related to Paget disorder. Thus AI and ML offer fast diagnosis and intervention and introduce sets of rules to examine medical imaging statistics consisting of CT scans and x-rays (Bui et al., 2022). ML-driven predictive models, which evaluate the course of the disease and the response to treatment, allow personalized treatment plans to be tailored to the needs of each patient. Moreover, to obtain better health care outcomes for patients, AI-based clinical decision support systems can help health care professionals make informed decisions about diagnosis, treatment, and follow-up. Through research on disease biology and the development of drugs, AI and ML will enable Paget disease medical teams to develop novel treatment methods. Further validation and integration into clinical practice are required to improve patient care, treatment efficiency, and diagnostic accuracy, but AI and ML have the potential to change disease management completely in Paget disease (Albuquerque et al., 2023; Oakden-Rayner et al., 2022).

The structure of this chapter is as follows: Section 6.2 presents review work. Section 6.3 discusses the pathophysiology and remodeling of bones. Section 6.4 highlights the clinical presentation and causes. Section 6.5 introduces the diagnostic and screening methods used in Paget disease. Complications associated with Paget disease are discussed in Section 6.6. Section 6.7 explains current challenges in the diagnosis of Paget disease. Potentials of AI and ML in medical imaging are discussed in Section 6.8. The work concludes in Section 6.9.

6.2 Literature survey

Paget disease is caused by a combination of genetic and environmental factors. According to a comprehensive survey of the literature, remarkable advances have been achieved in understanding the diagnosis, treatment, and pathophysiology of Paget disease. The purposes of this review are to examine the interactions between genetic and environmental factors in the etiology of Paget disease of the bone and to summarize research findings in this area, as shown in Table 6.1.

Table 6.1 Literature survey on Paget disease.

References	Author name	Title	Findings
Diboun et al. (2021)	I. Diboun et al.	Epigenetic analysis of Paget's disease of bone identifies differentially methylated loci that predict disease status	The authors identified DNA methylation changes in Paget disease of bone and the role of epigenetic factors in pathogenesis and diagnostic makers.
Wu et al. (2022)	H. Wu et al.	Development and validation of an artificial intelligence-based image classification method for pathological diagnosis in patients with extramammary Paget's disease	For diagnosing extramammary Paget disease AI-based image classification was introduced. The proposed model improved diagnostic accuracy in clinical practice.
Kibbi et al. (2022)	N. Kibbi et al.	Evidence-based clinical practice guidelines for extramammary Paget disease	Clinical guidelines for managing extramammary Paget disease, providing recommendations for diagnosis, treatment, and follow-up based on systematic literature review and expert panel evaluation.

Continued

Table 6.1 Literature survey on Paget disease.—*cont'd*

References	Author name	Title	Findings
Chen et al. (2021)	C. I. Chen et al.	Margin assessment of extramammary Paget disease based on harmonic generation microscopy with deep neural networks	The authors proposed two techniques for diagnosis (i.e., deep learning (DL) for early evaluation of extramammary Paget disease and harmonic generation microscopy). These techniques were used to assist the physician in finding an accurate disease margin.
Hung et al. (2022)	T. N. K. Huang et al.	An AI-based prediction model for drug-drug interactions in osteoporosis and Paget diseases from SMILES	The authors predicted the drug interaction in osteoporosis and Paget disease with the help of an AI model. The model helped physicians choose better medications and arrive at clinical decision-making.
Ralston (2013)	S. H. Ralston et al.	Paget disease of bone	The authors discussed the pathology, etiology, advances in research, and better treatment options for Paget disease. Genetic and environmental factors in the development of disease were introduced.
Melton et al. (2000)	L. J. Melton et al.	Fracture risk among patients with Paget disease: a population-based cohort study	Cohort study of population-based research was conducted on the probability of fractures in spine and rib joints in those with Paget disease. This discovery emphasizes the importance of surveillance. More stringent measures may be required in certain situations.
Ring and Elvins (1995)	F. J. Ring et al.	Temperature measurement in Paget disease of bone	Describes the use of infrared thermography to track temperature changes in Paget disease-affected bones and highlights its feasibility in determining disease progression and treatment effectiveness.

Table 6.1 Literature survey on Paget disease.—*cont'd*

References	Author name	Title	Findings
Marcon et al. (2015)	P. Marcon et al.	The correlation of blood test results with the data obtained from MRI images during the determination of pathology in small bones	Examines the connection between magnetic resonance imaging images and blood test results to enhance diagnostic accuracy using other data sources in diagnosing minor bone diseases.
Navada et al. (2018)	D. K. R. Navada et al.	A high precision deep-CNN framework for classification of metabolic bone diseases among women	X-ray images were used to provide an accurate diagnosis of metabolic bone diseases in women. This is challenging owing to the shortcomings of current diagnostic methods.
Pittau et al. (1998)	R. F. Pittau et al.	An inexpensive light source for oncologic photodynamic therapy	In oncological photodynamic therapy, a low-cost nonuniform light source is introduced as an alternative to traditional laser systems.
Ring (1994)	F. Ring et al.	Quantitative thermal imaging in rheumatology	The authors discussed thermal imaging and disease management for clinical research. To diagnose inflammatory joint disease and its impact, thermal imaging can be used by rheumatologists.
Delmas and Meunier (1997)	P. D. Delmas	The management of Paget disease of bone	The authors discussed Paget's disease with a focus on various treatments available to slow it and relieve pain.
Ralston (2008)	S. H. Ralston et al.	Pathogenesis of Paget disease of bone	The authors investigated genetic and environmental factors that contribute to Paget disease, with a focus on mutations in specific genes that modify bone metabolism. They also discussed possible negative effects of viral infections and various environmental factors.

Continued

Table 6.1 Literature survey on Paget disease.—*cont'd*

References	Author name	Title	Findings
Daroszewska and Ralston (2006)	A. Daroszewska and S. H. Ralston	Mechanisms of disease: genetics of Paget disease of bone and related disorders	The authors discussed the bone metabolism and disease pathophysiology and the genetic factor of Paget disease-related disorders with a focus on receptor activator of NF-kappa B (RANK-NF-κB) signaling.
Lucas et al. (2007)	G. J. A. Lucas et al.	Contribution of genetic factors to the pathogenesis of Paget disease of bone and related disorders	Genetic factors were a sizable contributor to the development of Paget sickness of bone (PDB), as evidenced by changes in genes within the RANK-NF-κB signaling pathway. The incidence of PDB and related diseases has been linked to mutations inside the *VCP*, *TNFRSF11A*, *TNFRP11B*, and *SQSTM1* genes.
Gold et al. (1996)	D. T. Gold et al.	Paget disease of bone and quality of life	Surveys were used to gauge how Paget illness affected privileged lives, emphasizing the role that psychological factors have in overall fitness.
Singer (2015)	F. R. Singer et al.	Paget disease of bone—genetic and environmental factors	Explores genetic and environmental factors that contribute to the amelioration of Paget illness and highlights the complex connection between viral infections and genetic developments as environmental stimuli.

6.3 Pathophysiology and remodeling of bones

Abnormal bone remodeling, also known as Paget disease of bone, is a chronic condition that leads to the loss of strength, growth, and deformity in bones. The cause of Paget disease is uncertain because the pathophysiology involves a delicate equilibrium (Fig. 6.2) between bone growth and resorption (Mirra et al., 1995). Some key points are listed in Table 6.2.

- **Normal Bone Remodeling:**

In normal humans, osteoblasts create new bone and osteoclasts remove old bone via resorption; this is called bone remodeling. By replacing weakened tissue with healthy bone, this dynamic stability maintains the strength and shape of bones. Similarly, bone remodeling promotes mineral homeostasis by controlling the level of calcium and phosphate inside the blood. This controlled technique is essential to the integrity and functioning of the skeleton and aids in the movement and safety of essential organs.

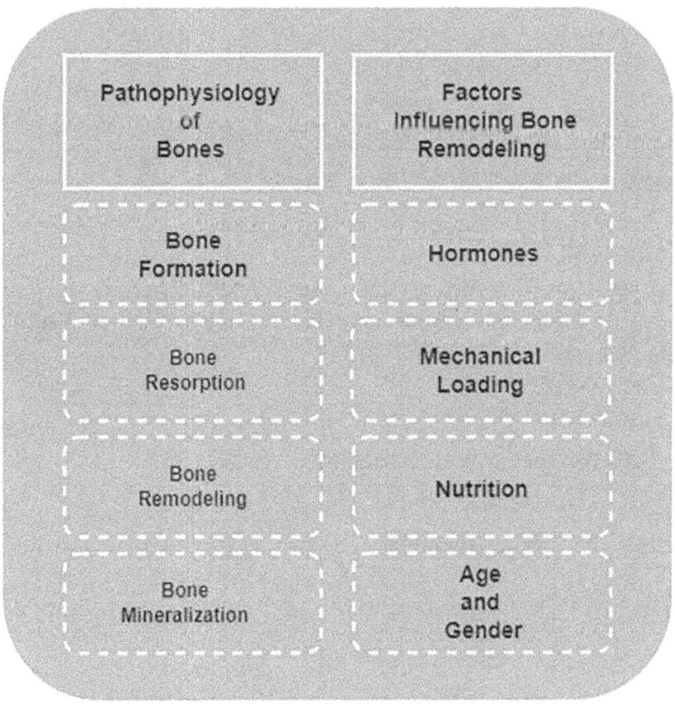

FIGURE 6.2

Pathophysiology and remodeling of bones.

- **Bone Remodeling Abnormalities in Paget Disease:**

  In Paget disorder, the regular stability of bone formation and resorption is disturbed owing to abnormal bone development strategies. When osteoclasts (cells that break down bones) become excessively active, there may be an excessive loss of bone mass. Osteoblasts attempt to strengthen weakening bones through additional effort, but the resultant new bone is not as robust or useful. This unusual bone causes the weaker, more swollen and deformed bones that are the hallmarks of Paget disorder. Furthermore, dysregulated remodeling weakens the structural integrity of the affected bones, making them more liable to fractures and different health-associated issues.

- **Mechanisms Initiating Paget Syndrome:**

  A combination of environmental factors and genetic traits is responsible for Paget disorder. Genetic traits related to Paget disorder disturb the balance between osteoclast and osteoblast activity, affecting gene control of bone remodeling pathways. In addition, it has been suggested that Paget disorder may be due to a pandemic infection named paramyxovirus. These viruses cause bone cellular infections or trigger an immunological reaction leading to alterations in bone resorption. The main reason for Paget disorder is unknown, but researchers are working to understand how factors such as genetics and environment combine to cause Paget disease.

6.4 Clinical presentation and causes

The symptoms of Paget disease are diverse but typically include disfigurement, joint stiffness, and bone pain, as shown in Fig. 6.3. Compression of the nerve can result in neurological problems. In some cases, there are no indications of symptoms and the diagnosis may be made inadvertently through imaging or a blood test (Theodorou et al., 2011). Some important points are:

Table 6.2 Pathophysiology and remodeling of bones.

Facet	Description
Pathophysiology	Deregulated osteoblasts and hyperactive osteoclasts cause an imbalance between bone production and resorption.
Key points	Osteoclast hyperactivity causes increased bone resorption, whereas structural abnormalities in bone formation result from compensatory osteoblast activity.
Contributing elements	The disease results from genetic changes and viral infections (paramyxovirus).
Clinical signs and symptoms	Pain in bones, deformities, fractures, and complications such as osteoarthritis and nerve compressions.

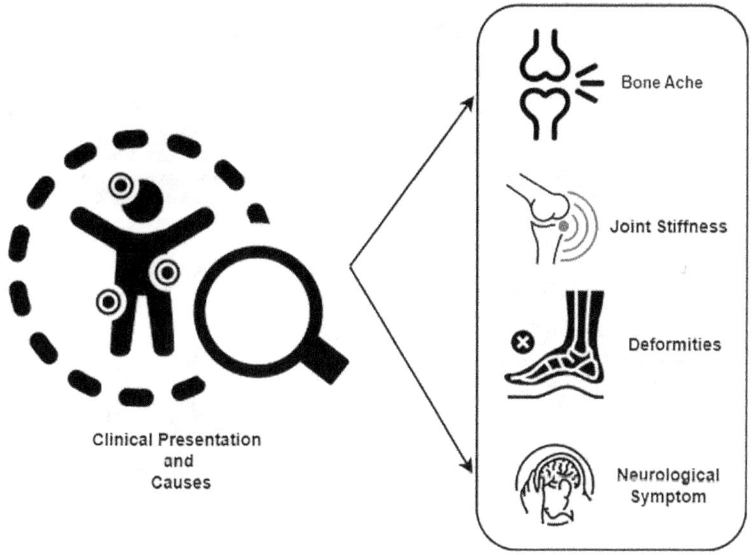

FIGURE 6.3

Clinical presentations and causes.

- **Bone ache:** One of the most common symptom of Paget disease is mild painful bone ache. The pain may be mild to excessive and restricted to the affected bone or bones. With movement, bone pain regularly improves.
- **Joint stiffness:** Joint stiffness is a common symptom of Paget disorder, which affects weight-bearing joints along the hips and knees. Joint mobility and movement may be constrained by stiffness, which can have negative results on daily sports and quality of life.
- **Deformities:** The presence of Paget disorder can lead to bone deformities including bowed legs, large heads because of expansion of the skull, and different abnormalities.
- **Neurological symptoms:** Neurological issues can arise from the compression of adjacent nerves caused by larger bones. Pain, numbness, and muscle weakness are common symptoms of spinal cord compression. When a severe nerve is compressed, it can cause more serious problems, including paralysis or loss of bladder and bowel control.

6.5 Diagnostic and screening methods

Early detection and accurate diagnosis are critical factors in managing and treating Paget disease, which necessities a thorough understanding of diagnostic and

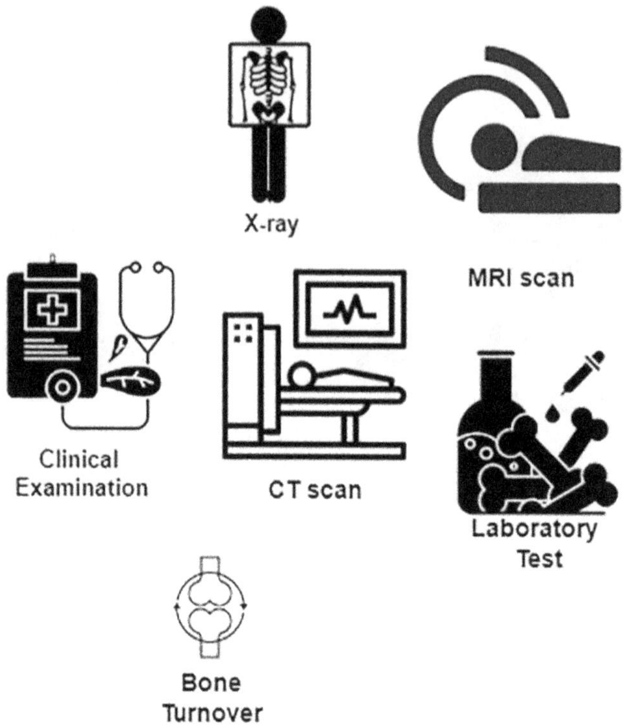

FIGURE 6.4

Diagnostic and screening methods. *CT*, computed tomography; *MRI*, magnetic resonance imaging.

screening techniques (Fig. 6.4) (Smith et al., 2002). A comprehensive overview of the topic follows.

- **Clinical Assessment:** Health care professionals diagnose Paget disorder through a thorough examination. Some symptoms that patients may encounter are pain in the bones, abnormalities, and broken bones. A detailed medical history of the patient and a physical examination can help detect the presence of Paget disease.
- **Imaging Studies:** For diagnosis and screening, health care professionals frequently use imaging assessments consisting of x-rays, CT and MRI scans, and bone scans (technetium-99m bone scintillation). Table 6.3 provides an in-depth analysis of these modalities.
- **Laboratory Tests:** Laboratory assessments for screening and diagnosis, which include serum alkaline phosphatase, serum calcium and phosphate, and bone turnover markers, are commonly employed by clinicians. Table 6.4 provides a detailed overview of the different methods.

Table 6.3 Analysis of imaging studies.

Imaging modalities	Description
X-rays	Depicts recognizable characteristics such as bone degeneration, swelling of the ciliary column, and distended muscles.
Computed tomography and magnetic resonance imaging	Provides detailed anatomical information and examines complications.
Bone scans	Finds areas where bone turnover is high

Table 6.4 Laboratory tests performed by health care professionals.

Laboratory tests	Description
Bone turnover makers	Bone turnover, a common symptom of Paget disease, indicates an increase in bone activity. Repeated measurements can be used to track treatment response and disease activity.
Serum alkaline phosphatase	Biomarkers include procollagen type I N-terminal propeptide and C- terminal telopeptide.
Serum calcium and phosphate	Specifies the bone's metabolic condition

6.6 Complications associated with Paget disease

The study of Paget disease makes it clear that many issues need to be addressed, as shown in Fig. 6.5 (Farpour et al., 2012; Vande Berg et al., 2001). These elements will be discussed subsequently in more detail.

- Fractures: Paget disease causes a great proportion of fracture in patients with weak bones. The occurrence of fractures can be due to minor trauma or spontaneous events. Common fractures are those on the hips and long bones in the legs and back. The occurrence of fractures can result in severe discomfort, immobility, and disability.
- Nerve Compression: Paget disease causes pain, tingling, the sensation of numbness, and weakness in the muscles surrounding nerves, which can be compressed by enlarged and malformed bones. Compression of spinal nerves can cause disorders such as spinal stenosis and nerve root compression syndromes.
- Bone Deformities: Paget disorder may also result in abnormal bone formation owing to immoderate bone remodeling. The disorder can also cause the bones to grow, become malformed, or increase in brittleness. This may bring about practical impairment and a decreased quality of lifestyle for affected individuals.

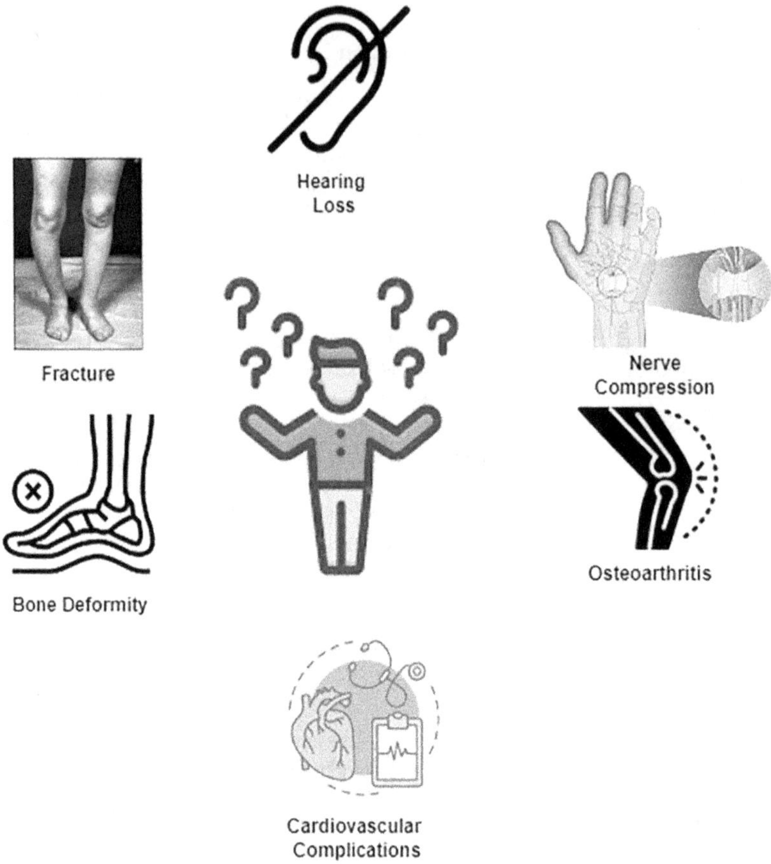

FIGURE 6.5

Complications associated with Paget's disease.

- Hearing Loss: There may be sensorineural or conductive hearing loss owing to Paget disease, specifically affecting the temporal bone. Involvement of the middle ear or compression of the auditory nerve can also have an effect on communication capabilities and listening characteristics.
- Osteoarthritis: Paget disease may exacerbate osteoarthritis within the affected joints. Abnormal bone remodeling exerts more strain on adjacent cartilage and leads to degeneration and inflammation, modifying the mechanics of the joints. Osteoarthritis also causes aches, fatigue, and decreased motion.
- Secondary Osteosarcoma: Paget disorder is observed via secondary osteosarcomas, which might be malignant bone tumors that have an increased risk of growing. The development of osteosarcoma in bones with Paget disorder can be attributed to a genetic predisposition, extended inflammation, and former

exposure to radiation therapy. The identification of early sarcoma transformation symptoms is crucial to improve treatment outcomes.

- Cardiovascular Complications: The base of the skull or long bones can have Paget disorder, which can increase blood flow to the affected regions and result in coronary heart failure. The body's attempt to adapt to increased bone vascularity results in this consequence, which may also show signs and symptoms that include fatigue, shortness of breath, and leg swelling.

6.7 Current challenges in diagnosis

With the use of ML and AI, there are various difficulties in identifying Paget illness, as shown in Fig. 6.6 (Hashimoto et al., 2006).

- Accessibility and Integrity of Data: It may be difficult to access large high-quality datasets with a variety of cases because Paget disease is rare and requires extensive medical records.
- Imaging Imbalance: Diversity in the expression of Paget disease between patients and imaging modalities complicates the development of AI algorithms for the accurate diagnosis of Paget disease.
- Labelling and Annotation: It is resource-intensive to label the datasets consistently and accurately because that requires experience and a lot of effort to label medical images with ground truth tags.

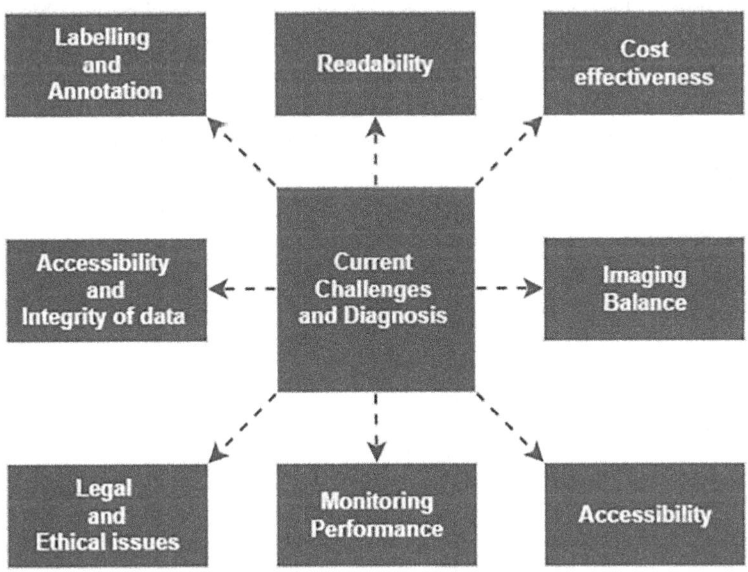

FIGURE 6.6

Challenges in diagnosis.

- Readability: AI models need to be able to explain their predictions to win the trust of doctors if they are to be accepted and integrated into clinical practice.
- Legal and Ethical Issues: Issues related to patient privacy, data security, and potential biases in algorithmic decisions need to be addressed to maintain trust and ethical integrity.
- Monitoring Performance over time: To adapt to changing patient demographics and clinical procedures, AI models need to be continuously monitored, fed back, and retrained.
- Accessibility and Cost-effectiveness: To optimize AI-driven diagnostic solutions, it is necessary to guarantee their cost, scale, and fair access, especially in resource-constrained environments.

6.8 Potential of artificial intelligence and machine learning in medical imaging

The capability for enhancing clinical imaging in Paget disorder with AI and ML technology is extensive. These technologies allow for early identification, specific analyses, customized therapy planning, and predictive analytics, as shown in Fig. 6.7. These technologies have the ability to change completely how Paget disorder is managed and enhance patient outcomes as they develop and mature (Franchi and Others, 2004). A detailed analysis is shown subsequently:

- Quantitative Assessment: ML algorithms are able to provide quantitative assessments of the severity and progression of a disease based on the objective measure obtained from medical imaging. These algorithms can determine the number of lesions in the bone, compute the density of the bone, and monitor changes in the morphology of the bone over time. This kind of numerical data can be used to monitor therapy efficacy and make recommendations for therapy options.

FIGURE 6.7

Potential of artificial intelligence and machine learning in health care.

- Early Detection: Clinical imaging scans such as CT, MRI, and x-rays can have minute anomalies recognized by AI and ML algorithms trained to achieve this. These algorithms have the ability to recognize distinctive alterations in bone density and structure that may be symptoms of Paget disease. Timely intervention and condition management depend closely on early detection.
- Personalized Treatment Planning: ML algorithms are able to evaluate imaging data for a customized treatment plan based on the unique features of the patient. These ML models can propose high-quality courses of action for treating Paget disorder, consisting of drug regimens, surgical methods, or rehabilitation regimens, totally based on variables such as the stage of the disorder, the region of the lesion, and the affected person's demographics.
- Accurate Diagnosis: AI and ML models are able to identify normal bone structure and the pathological alterations linked to Paget's disease using large datasets from x-rays and CT and MRI scans. From this type of diagnosis there is less possibility of understanding and improve the patient care by assisting radiologists and other clinicians in diagnosing patient more accurately.
- Predictive Analytics: Based on the patient-specific data and imaging results, AI and ML methods can predict prognosis and illness. These models are able to determine risk factors for therapy response, complications, or the progression of disease by examining patterns and trends in significant cohorts of individuals with Paget disease. By taking preventive measures, clinicians may enhance patient outcomes and make the best possible judgments using this kind of information.

6.9 Conclusion

Paget disease affects normal recycling of the body in which new bone tissue slowly replaces old bone tissue. Most frequently affected areas include the legs, pelvis, head, and spine. The age and presence of family members with Paget disease of the bone raise the risk of the condition. Clinical examinations, laboratory tests, and imaging methods are used to diagnose Paget disease. Traditional methods did not provide the best results in terms of usability, precision, and effectiveness. Those approaches often have limitations that delay the diagnosis and result in a poor patient outcome. To overcome these limitations, AI and ML techniques are used for the efficient identification of Paget disease. The identification and treatment of Paget disease could be greatly advanced by integrating AI and ML technologies into clinical imaging. Because of advances and innovations, these modern technologies improve treatment planning, the early diagnosis of small problems, the quantitative evaluation of bone characteristics, enhanced visualization, and advances in research. Ultimately, these developments provide clinical experts with useful resources to enhance the consequences for affected persons and provide a precise prognosis and health care for those affected by Paget disease. Finally, by integrating AI and ML into clinical procedures, doctors can diagnose and treat Paget disease. This will ultimately raise the standard of treatment provided to affected individuals.

References

Albuquerque, G.A., Carvalho, D.D.A., Cruz, A.S., Santos, J.P.Q., Machado, G.M., Gendriz, I.S., Fernandes, F.R.S., Barbalho, I.M.P., Santos, M.M., Teixeira, C.A.D., Henriques, J.M.O., Gil, P., Neto, A.D.D., Campos, A.L.P.S., Lima, J.G., Paiva, J.C., Morais, A.H.F., Lima, T.S., Valentim, R.A.M., 2023. Osteoporosis screening using machine learning and electromagnetic waves. Sci. Rep. 13 (1). https://doi.org/10.1038/s41598-023-40104-w.

Bayan, C.-A.Y., Khanna, T., Rotemberg, V., Samie, F.H., Zeitouni, N.C., 2018. A review of non-invasive imaging in extramammary Paget's disease. J. Eur. Acad. Dermatol. Venereol. 32 (11), 1862–1873. https://doi.org/10.1111/jdv.15072.

Bui, H.M., Ha, M.H., Pham, H.G., Dao, T.P., Nguyen, T.T.T., Nguyen, M.L., Vuong, N.T., Hoang, X.H.T., Do, L.T., Dao, T.X., Le, C.Q., 2022. Predicting the risk of osteoporosis in older Vietnamese women using machine learning approaches. Sci. Rep. 12 (1). https://doi.org/10.1038/s41598-022-24181-x.

Chen, C.I., Liao, Y.H., Sun, C.K., 2021. Margin assessment of extramammary Paget's disease based on harmonic generation microscopy with deep neural networks. IEEE J. Sel. Top. Quant. Electron. 27 (4). https://doi.org/10.1109/JSTQE.2021.3067342.

Daroszewska, A., Ralston, S.H., 2006. Mechanisms of disease: genetics of Paget's disease of bone and related disorders. Nat. Clin. Pract. Rheumatol. 2 (5), 270–277. https://doi.org/10.1038/ncprheum0172.

Delmas, P.D., Meunier, P.J., 1997. The management of Paget's disease of bone. N. Engl. J. Med. 336 (8), 558–566. https://doi.org/10.1056/NEJM199702203360807.

Diboun, I., Wani, S., Ralston, S.H., Albagha, O.M.E., 2021. Epigenetic analysis of paget's disease of bone identifies differentially methylated loci that predict disease status. bioRxiv. https://doi.org/10.1101/2021.01.04.425216.

Edey, K.A., Allan, E., Murdoch, J.B., Cooper, S., Bryant, A., 2019. Interventions for the treatment of Paget's disease of the vulva. Cochrane Database Syst. Rev. 2019 (6). https://doi.org/10.1002/14651858.CD009245.pub3.

Farpour, F., Tehranzadeh, J., Donkervoort, S., Smith, C., Martin, B., Vanjara, P., Osann, K., Kimonis, V.E., 2012. Radiological features of Paget disease of bone associated with VCP myopathy. Skeletal Radiol. 41 (3), 329–337. https://doi.org/10.1007/s00256-011-1193-4.

Franchi, A., others, 2004. Pathology of Paget's disease of bone. Clin. Cases Miner. Bone Metab 1, 203–207.

Gold, D.T., Boisture, J., Shipp, K.M., Pieper, C.F., Lyles, K.W., 1996. Paget's disease of bone and quality of life. J. Bone Miner. Res. 11 (12), 1897–1904. https://doi.org/10.1002/jbmr.5650111210.

Hashimoto, J., Ohno, I., Nakatsuka, K., Yoshimura, N., Takata, S., Zamma, M., Yabe, H., Abe, S., Terada, M., Yoh, K., Fukunaga, M., Cooper, C., Morii, H., Yoshikawa, H., 2006. Prevalence and clinical features of Paget's disease of bone in Japan. J. Bone Miner. Metabol. 24 (3), 186–190. https://doi.org/10.1007/s00774-005-0670-z.

Hung, T.N.K., Le, N.Q.K., Le, N.H., Van Tuan, L., Nguyen, T.P., Thi, C., Kang, J.H., 2022. An AI-based prediction model for drug-drug interactions in osteoporosis and Paget's diseases from SMILES. Mol. Inform. 41 (6). https://doi.org/10.1002/minf.202100264.

Kibbi, N., Owen, J.L., Worley, B., Wang, J.X., Harikumar, V., Downing, M.B., Aasi, S.Z., Aung, P.P., Barker, C.A., Bolotin, D., Bordeaux, J.S., Cartee, T.V., Chandra, S.,

Cho, N.L., Choi, J.N., Chung, K.Y., Cliby, W.A., Dorigo, O., Eisen, D.B., Fujisawa, Y., Golda, N., Halfdanarson, T.R., Iavazzo, C., Jiang, S.I.B., Kanitakis, J., Khan, A., Kim, J.Y.S., Kuzel, T.M., Lawrence, N., Leitao, M.M., Maclean, A.B., Maher, I.A., Mittal, B.B., Nehal, K.S., Ozog, D.M., Pettaway, C.A., Ross, J.S., Rossi, A.M., Servaes, S., Solomon, M.J., Thomas, V.D., Tolia, M., Voelzke, B.B., Waldman, A., Wong, M.K., Zhou, Y., Arai, N., Brackett, A., Ibrahim, S.A., Kang, B.Y., Poon, E., Alam, M., 2022. Evidence-based clinical practice guidelines for extramammary paget disease. JAMA Oncol. 8 (4), 618–628. https://doi.org/10.1001/jamaoncol.2021.7148.

Lucas, G.J.A., Daroszewska, A., Ralston, S.H., 2007. Contribution of genetic factors to the pathogenesis of Paget's disease of bone and related disorders. J. Bone Miner. Res. 22 (2), P31–P37. https://doi.org/10.1359/JBMR.06S206.

Marcon, P., Bartusek, K., Sprlakova, A., Dohnal, P., 2015. The correlation of blood test results with the data obtained from MRI images during the determination of pathology in small bones. In: 2015 38th International Conference on Telecommunications and Signal Processing, TSP 2015. Institute of Electrical and Electronics Engineers Inc., Czech Republic, pp. 739–742. https://doi.org/10.1109/TSP.2015.7296362.

Melton, L.J., Tiegs, R.D., Atkinson, E.J., O'Fallon, W.M., 2000. Fracture risk among patients with Paget's disease: a population-based cohort study. J. Bone Miner. Res. 15 (11), 2123–2128. https://doi.org/10.1359/jbmr.2000.15.11.2123.

Mirra, J.M., Brien, E.W., Tehranzadeh, J., 1995. Paget's disease of bone: review with emphasis on radiologic features, part I. Skeletal Radiol. 24 (3), 163–171. https://doi.org/10.1007/BF00228918.

Navada, D.K.R., Ganesh, S., Bhargavi, K., 2018. A high precision deep-CNN framework for classification of metabolic bone diseases among women. In: Proceedings of the 3rd International Conference on Communication and Electronics Systems, ICCES 2018. Institute of Electrical and Electronics Engineers Inc., India, pp. 147–152. https://doi.org/10.1109/CESYS.2018.8724013.

Oakden-Rayner, L., Gale, W., Bonham, T.A., Lungren, M.P., Carneiro, G., Bradley, A.P., Palmer, L.J., 2022. Validation and algorithmic audit of a deep learning system for the detection of proximal femoral fractures in patients in the emergency department: a diagnostic accuracy study. Lancet Digital Health 4 (5), e351–e358. https://doi.org/10.1016/s2589-7500(22)00004-8.

Paul Tuck, S., Layfield, R., Walker, J., Mekkayil, B., Francis, R., 2017. Adult Paget's disease of bone: a review. Rheumatology 56 (12), 2050–2059. https://doi.org/10.1093/rheumatology/kew430.

Pittau, R.F., Pans, P., Boeto, N., Aoki, A., 1998. An inexpensive light source for oncologic photodynamic therapy. IEEE Eng. Med. Biol. Mag. 17 (3), 105–106. https://doi.org/10.1109/51.677177.

Ralston, S.H., 2013. Paget's disease of bone. N. Engl. J. Med. 368 (7), 644–650. https://doi.org/10.1056/NEJMcp1204713.

Ralston, S.H., 2008. Pathogenesis of Paget's disease of bone. Bone 43 (5), 819–825. https://doi.org/10.1016/j.bone.2008.06.015.

Reid, I.R., 2020. Management of Paget's disease of bone. Osteoporos. Int. 31 (5), 827–837. https://doi.org/10.1007/s00198-019-05259-1.

Ring, F., 1994. Quantitative thermal imaging in rheumatology. In: Proceedings of 16th Annual International Conference of the IEEE Engineering in Medicine and Biology Society, vol. 1.

Ring, F.J., Elvins, D.M., 1995. Temperature measurement in Paget's disease of bone. In: Annual International Conference of the IEEE Engineering in Medicine and Biology - Proceedings, vol. 17, pp. 1711–1712.

Selby, P.L., 2006. Guidelines for the diagnosis and management of Paget's disease: a UK perspective. J. Bone Miner. Res. 21 (S2), P92–P93. https://doi.org/10.1359/jbmr.06s217.

Simonds, R.M., Segal, R.J., Sharma, A., 2019. Extramammary Paget's disease: a review of the literature. Int. J. Dermatol. 58 (8), 871–879. https://doi.org/10.1111/ijd.14328.

Singer, F.R., 2015. Paget's disease of bone-genetic and environmental factors. Nat. Rev. Endocrinol. 11 (11), 662–671. https://doi.org/10.1038/nrendo.2015.138.

Smith, S.E., Murphey, M.D., Motamedi, K., Mulligan, M.E., Resnik, C.S., Gannon, F.H., 2002. From the archives of the AFIP: radiologic spectrum of paget disease of bone and its complications with pathologic correlation. Radiographics 22 (5), 1191–1216. https://doi.org/10.1148/radiographics.22.5.g02se281191.

Tan, A., Ralston, S.H., 2014. Clinical presentation of Paget's disease: evaluation of a contemporary cohort and systematic review. Calcif. Tissue Int. 95 (5), 385–392. https://doi.org/10.1007/s00223-014-9904-1.

Theodorou, D.J., Theodorou, S.J., Kakitsubata, Y., 2011. Imaging of Paget disease of bone and its musculoskeletal complications: review. Am. J. Roentgenol. 196 (6), S64–S75. https://doi.org/10.2214/AJR.10.7222.

Vande Berg, B.C., Malghem, J., Lecouvet, F.E., Maldague, B., 2001. Magnetic resonance appearance of uncomplicated Paget's disease of bone. Semin. Muscoskel. Radiol. 5 (1), 69–77. https://doi.org/10.1055/s-2001-12919.

Wu, H., Chen, H., Wang, X., Yu, L., Yu, Z., Shi, Z., Xu, J., Dong, B., Zhu, S., 2022. Development and validation of an artificial intelligence-based image classification method for pathological diagnosis in patients with extramammary paget's disease. Front. Oncol. 11. https://doi.org/10.3389/fonc.2021.810909.

CHAPTER

7 Identification and classification of rheumatoid arthritis using artificial intelligence and machine learning

Seyed Mahmoud Sajjadi Mohammadabadi[1], Mahsa Borhani Peikani[2]

[1]*University of Nevada Reno, Computer Science and Engineering, Reno, NV, United States;*
[2]*University of Nevada Reno, Physiology and Cell Biology, Reno, NV, United States*

7.1 Introduction

Rheumatoid arthritis (RA) is a chronic autoimmune disease that can be identified by progressive joint inflammation and destruction. Early and accurate diagnosis of this disease is necessary for treatment, but traditional diagnosis methods often face limitations. These shortcomings include reliance on subjective interpretations of patients' imaging data and the lack of sensitive and specific biomarkers (Qin et al., 2022). Machine learning (ML) models are powerful methods for diagnosing RA and improving health care owing to their ability to extract complex patterns from large datasets.

ML algorithms can analyze diverse data sources, including clinical examinations, laboratory tests, and medical imaging, to identify subtle patterns associated with RA. This enables the development of more accurate and objective diagnostic tools, earlier disease prediction, and personalized treatment strategies. For example, ML models could be trained to analyze x-ray images and predict disease severity or the response to specific therapies. By harnessing the power of ML, we can usher in a new era of personalized and data-driven medicine for RA patients.

7.2 Rheumatoid arthritis

RA is the predominant form of autoimmune arthritis, characterized by significant inflammation in synovial tissue, resulting in joint degradation. Evaluating the synovium holds promise for informing the most effective treatment approaches. However, despite its potential, a categorization of RA synovium remains absent from existing diagnostic criteria and treatment recommendations. An evaluation of RA synovium using hematoxylin and eosin (H&E) staining, which is widely used in biology, is practical for a large number of patients undergoing interventional procedures because it is

Diagnosing Musculoskeletal Conditions using Artificial Intelligence and Machine Learning to Aid Interpretation of Clinical Imaging
https://doi.org/10.1016/B978-0-443-32892-3.00007-5

a standard service provided by clinical pathology laboratories (Borhani et al., 2022). Three histologic characteristics are assessed in the Krenn scoring system for H&E-stained synovial tissue: leukocyte infiltration, synovial lining hyperplasia, and synoviocyte stromal density. Whereas high-grade synovitis demonstrates 62% sensitivity and 96% specificity for the diagnosis of rheumatic diseases, it is unable to distinguish among different subtypes of rheumatic disease, such as psoriatic arthritis versus RA. Therefore, a more detailed histological scoring system could serve as a valuable resource for subtyping RA and directing its treatment. Previous research investigated the importance of lymphocyte aggregates and identified associations with systemic indicators of inflammation, such as the erythrocyte sedimentation rate and C-reactive protein (CRP) level. This suggests that relying only on these features may be insufficient to differentiate among immunologically distinct subtypes of RA (Orange et al., 2018). Evaluation of H&E-stained slides shows a range of inflammatory features (multinucleated giant cells, neutrophils, plasma cells, binucleated plasma cells, and Russell bodies) (Borhani et al., 2014, 2017). Systemic inflammatory parameters including autoantibody levels, CRP levels, and erythrocyte sedimentation rate are correlated with these histologic scores. Clinical features and gene expression patterns can be compared to identify potential variations in pain mechanisms for patients with different synovial subtypes (Orange et al., 2018).

7.3 Fundamentals of artificial intelligence for rheumatoid arthritis analysis

Artificial intelligence (AI), particularly advances in generative AI and ML, is transforming health care and introducing solutions to various practical problems, including industry and health care (Sajjadi Mohammadabadi, 2024; Mohammadabadi et al., 2024b). These developments signify how we can approach problems in different domains such as medicine. Work in ML classification tasks for medicine shows promise in improving patient care (Alimirzaei and Kieslich, 2023). Its applications in RA research show improvement in diagnosis, treatment, and overall care of patients. This section delves into the fundamental concepts of AI, showing its capabilities and limitations in the context of RA. ML is a subdomain of AI that trains to learn from data and improve over time without being explicitly programmed. Deep learning is a subset of ML that employs neural networks with multiple layers to extract features from data. Identification refers to identifying patterns or features within data. Classification, on the other hand, is the task of categorizing data into classes or labels. Fig. 7.1 illustrates this concept.

7.3.1 Common artificial intelligence algorithms

Table 7.1 lists common AI algorithms and highlights their advantages and drawbacks. AI is transforming, and algorithms such as reinforcement learning and deep learning are increasingly helpful to improving health care.

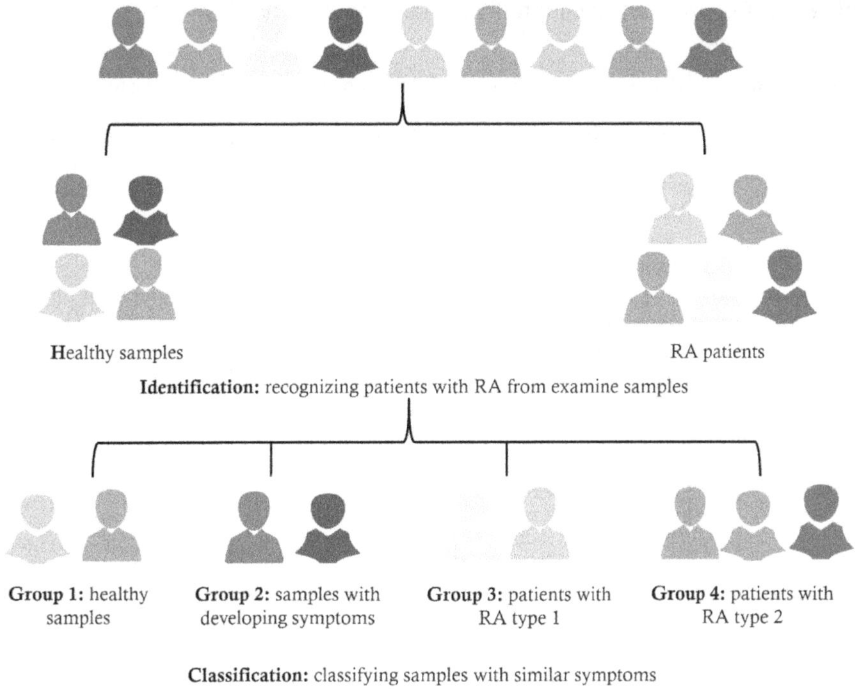

FIGURE 7.1

Identification and classification of rheumatoid arthritis patients using machine learning. The figure illustrates the classification of samples into healthy and RA patient groups. Additionally, RA patients are further identified into four subgroups: healthy, developing symptoms, RA type 1, and RA type 2.

Understanding these fundamental concepts allows an evaluation of AI models and their potential for personalized RA care.

7.3.2 Privacy and security considerations in artificial intelligence for rheumatoid arthritis

As large language models (LLMs) become popular in the medical field, protecting patient security and privacy has become increasingly important. Current work explores useful techniques for privacy-preserving LLM training. One such approach is federated learning (FL), which allows collaborative model training while keeping training data decentralized and private (Mohammadabadi et al., 2023b). However, significant challenges remain when using FL, particularly for large models: the diverse resource constraints of devices with varying computation and communication capacities, along with different task sizes. To address these issues of heterogeneity, distributed ML algorithms have been proposed that dynamically offload part of the model to the server, alleviating resource constraints and accelerating training (Mohammadabadi et al., 2023a, 2024a).

Table 7.1 Common artificial intelligence algorithms and their Characteristics.

Algorithm	Description	Advantages	Limitations
Logistic regression	Models binary classification using a linear relationship between input variables and a binary outcome.	Simple to interpret, efficient for small datasets.	Prone to overfitting complex data, limited to binary classification.
Least Absolute Shrinkage and Selection Operator (LASSO) regression	Regularized linear regression with a penalty term to limit overfitting.	Handles high-dimensional data, and improves model interpretability.	Requires tuning the penalty term, which can be computationally expensive.
Support vector machine	Finds the optimal hyperplane to separate data points into classes.	Robust to outliers, effective for high-dimensional data.	Can be computationally expensive and interpretation can be complex.
Decision tree	Tree-like structure representing relationships within the training data.	Easy to interpret, handles mixed data types.	Prone to overfitting, susceptible to noise in the data.
Random forest	Ensembles multiple decision trees for improved accuracy.	Robust to outliers, handles high-dimensional data.	Can be computationally expensive, black box nature (less interpretable).
Naive Bayes	Classifies data based on Bayes' theorem assuming independence between variables.	Efficient for small datasets, handles multiclass problems.	Assumption of independence often violated, sensitive to noisy data.
K-Nearest neighbors	Predicts class or value of a data point based on its nearest neighbors.	Simple to implement, effective for nonlinear relationships.	Performance sensitive to data dimensionality, outliers, and can be computationally expensive for large datasets.
Artificial neural network	Interconnected network of nodes (neurons) that process information through layers.	Highly flexible, learn complex relationships, effective for various tasks.	Can be computationally expensive, black box nature (difficult to interpret), requires large training datasets.

7.4 Machine learning in rheumatoid arthritis

ML is improving health care by introducing new methods for disease diagnosis and enabling personalized treatment strategies. In rheumatology specifically, ML algorithms have shown potential in identifying individuals at risk and predicting disease. Traditionally, RA risk assessment relied on factors such as the presence of autoantibodies. However, ML can enhance predictive accuracy by incorporating diverse data sources such as genetics, leading to more precise and early interventions (Momtazmanesh et al., 2022).

This section highlights the potential of ML in predicting RA and discusses its performance when using various data inputs for more accurate and personalized health care interventions. AI uses various methods to make machines smarter. ML focuses on teaching machines to learn patterns from data, whereas deep learning, a subset of ML, uses powerful neural networks to find complex patterns in areas such as image analysis and drug discovery. Deep learning models typically require more data and training time than simpler ML algorithms.

7.4.1 Machine learning approaches and their applications in rheumatoid arthritis

Training ML algorithms involves diverse approaches, categorized into supervised, unsupervised, and reinforcement learning. Supervised learning uses labeled data to train algorithms for predictions on unseen data. For instance, supervised training models analyze data from electronic health records (EHRs) to identify RA risk factors. Unsupervised learning focuses on uncovering patterns in unlabeled data, whereas reinforcement learning enables machines to learn through trial and error for specific goals (e.g., robotic surgery). Understanding these three ML problems is necessary for physicians using ML in clinical practice. When choosing models, factors such as the size of data, external validation, problem relevance, performance metrics, and public repository availability are crucial.

7.4.2 Machine learning for advanced rheumatoid arthritis management

A growing number of RA patients are showing reduced disease activity because of research that has significantly improved patient care. Early diagnosis, better disease management, and the use of improved drugs such as targeted-synthetic disease-modifying antirheumatic drugs (DMARDs) and biological DMARDs (bDMARDs) are some contributing factors. According to a study, patients with early RA may experience sustained remission in as many as 50% of cases (Vodencarevic et al., 2021).

ML algorithms can be used to predict individual disease flares. These models analyze past data to generate predictions potentially useful for preventive interventions before flares happen. Advances in computing power and data processing have significantly boosted progress in learning algorithms. These provide improved RA management (Vodencarevic et al., 2021).

7.4.3 Different data types

ML can be used to classify RA-associated interstitial lung disease (RA-ILD). Specially finding patterns and connections within various data sources such as genetics, imaging, medical records, and transcriptomics, this technology is exceptional. However, there are challenges to leveraging ML in this field's research. Developing ML models is difficult because of small sample sizes, inaccurate sample labeling, and a lack of datasets for external validation. Furthermore, overfitting or underfitting of ML models themselves can produce inaccurate results. A strong grasp of ML principles, suitable model construction, and rigorous validation techniques are essential to surmount these obstacles and guarantee robust applications in RA-ILD research. Despite these challenges, ML has the potential to aid in the development of personalized medicine approaches, and ultimately improving patient care strategies for RA-ILDs.

7.4.3.1 Imaging data

Conventional x-rays, although widely used, struggle with early RA detection. Advanced imaging such as magnetic resonance imaging (MRI) and ultrasound excel at visualizing early soft tissue changes (synovitis, bone erosions, and bone marrow edema) characteristic of RA. AI-powered models can identify these key features with nearly 95% accuracy, often surpassing human experts. This technology has immense potential to streamline diagnosis, reduce bias, and improve patient care. However, AI use in RA imaging is still nascent. Research primarily focuses on a limited set of features, neglecting others such as subcortical cysts and late-stage findings. In addition, most studies rely on hand x-rays, with the potential of MRI and ultrasound largely unexplored. Beyond traditional methods, AI can analyze data from hand photos and thermal imaging, offering alternative screening tools. This opens exciting possibilities for early detection and intervention, particularly in resource-constrained settings (Momtazmanesh et al., 2022).

7.4.3.2 Clinical and sensor data

Beyond traditional x-rays and blood tests, recent advances offer possibilities for diagnosing RA using clinical and sensor data. This expands the diagnostic tool kit, leading to earlier and more precise identification of this chronic autoimmune disease. Traditionally, diagnosing RA involved physical examinations and lab tests. However, ML advances are useful to gain information from clinical and sensor data. Clinical data such as symptoms, joint pain patterns, and blood markers hold valuable insights. Researchers have successfully converted clinical information into images and used a convolutional neural network (CNN) model to distinguish RA patients with high accuracy, comparable to three rheumatologists. This suggests that ML can learn from diverse data representations to build robust diagnostic models. Wearable devices and image sensors are generating a huge amount of data that ML can leverage for diagnosis and monitoring. Studies using lymphocyte images captured by electronic sensors and thermograms of joints have achieved impressive accuracy in RA classification, highlighting the potential of analyzing these invisible biomarkers (Momtazmanesh et al., 2022). Researchers developed and compared automated methods for analyzing hand x-ray images to detect RA. They found that a customized CNN model outperformed others, achieving 95% accuracy in identifying RA. This custom CNN-based diagnostic tool shows promise for RA detection (Ahalya et al., 2022).

7.4.3.3 Identifying phenotypes with electronic health records

EHRs offer a gold mine of clinical data, but manually extracting meaningful insights can be tedious and time-consuming. Phenotype identification algorithms powered by ML address this challenge, automatically identifying individuals with specific conditions such as RA from EHR data. These algorithms go beyond basic diagnostic codes, using a multitude of data sources such as the medication history, lab results, clinical notes, and even natural language processing to paint a comprehensive picture of the patient's health. This allows for more accurate identification of disease compared with traditional methods (Momtazmanesh et al., 2022).

Table 7.2 lists the advantages and challenges of using different data types for RA diagnosis.

7.4.4 Available datasets for research

Diverse datasets are crucial for research on ML-based RA diagnosis. Table 7.3 outlines publicly available datasets relevant to RA research. Access to diverse datasets is essential for progress in developing ML-based RA diagnosis tools.

- **RETRO dataset:** This data comes from a clinical trial monitoring RA patients in remission for disease flares. It includes patient characteristics, disease course, medication use, and lab results, making it valuable for predicting future flare-ups.

Table 7.2 Different data types for rheumatoid arthritis diagnosis.

Data type	Advantages	Challenges
Imaging	Visualizes soft tissue changes early	Limited in detecting all rheumatoid arthritis features
Clinical and sensor data	Early and precise identification of rheumatoid arthritis	Interpretation and integration of diverse data
Electronic health records	Comprehensive patient health insights	Manual data extraction and processing

Table 7.3 Publicly available datasets for RA research.

Dataset	Type	Size	Focus	Source
RETRO	Clinical	842 patient records	Predicting RA flares	Vodencarevic et al. (2021)
Kaggle RA classification	X-ray images	310 samples	Classifying RA from x-rays	Obayya et al. (2022)
Kırıkkale University hand radiographs	X-ray images	300 images	RA versus hand osteoarthritis discrimination	Üreten and Maraş (2022)
Sakthi rheumatology Centre	Clinical	1000 patients	Predicting RA	Sundaramurthy et al. (2020)

RA, *rheumatoid arthritis.*

- **Kaggle RA Classification dataset:** This dataset offers x-ray images labeled as normal, RA, or hand osteoarthritis. This allows research on AI-powered RA classification using x-rays.
- **Kırıkkale University hand radiographs dataset:** This collection of hand x-rays, classified by rheumatologists, provides a resource for developing and evaluating algorithms to distinguish RA from other hand conditions.
- **Sakthi Rheumatology Centre dataset:** This dataset contains clinical data from 1000 patients, 750 with RA. It includes features such as age, sex, and symptoms, making it suitable for studying RA prediction using clinical data analysis.

Several datasets are available, so selecting the best one depends on your research goals. Consider the data type (images or clinical data), size, and variable you want to predict (target variable) when making your choice.

7.4.5 Cloud computing for machine learning in rheumatoid arthritis

Cloud computing is the backbone of AI-powered solutions in RA management (Wang et al., 2021). Its ability to handle massive medical datasets is required to train and run complex algorithms such as deep learning models. In the subsequent discussion, we explore how cloud computing empowers ML in RA.

1. **Scalable Data Processing:** Traditional computing systems often struggle with the massive datasets in health care, including images, EHRs, and genetic data. Cloud platforms offer flexible storage and processing power to handle these datasets efficiently. This allows researchers and clinicians to train and deploy AI models on a larger scale, leading to more reliable and generalizable results.
2. **Collaborative Research:** Cloud platforms act as a shared workspace in which researchers and clinicians from different institutions can access and analyze data together. This fosters collaboration, accelerates progress in the field, and

facilitates data sharing, joint model development, and faster validation of findings.
3. **Flexible Resource Allocation:** Cloud computing provides on-demand access to computing resources. Researchers and health care providers can scale their AI workloads up or down based on their needs. This flexibility reduces costs and ensures efficient resource use, especially for computationally demanding tasks such as training deep learning models.
4. **Robust Security and Privacy:** Patient data is highly sensitive, and cloud platforms offer robust security features to preserve privacy. These features comply with regulations such as the Health Insurance Portability and Accountability Act of 1996, guaranteeing data privacy and confidentiality while enabling its use for research and development.

Overall, integrating AI-based deep learning with cloud computing seems essential for advancing RA diagnosis and treatment. It allows for efficient data analysis, personalized treatment recommendations, and insights into the mechanisms of traditional medicines.

7.4.6 Applications of artificial intelligence in rheumatoid arthritis

AI is rapidly transforming medicine with its impact on RA management. In the following sections, we explore how AI is being applied across the entire range of RA care, from early detection to personalized treatment optimization. Fig. 7.2 categorizes seven applications of AI in RA.

7.5 Detection and identification of rheumatoid arthritis

Despite significant advances in treatment, RA continues to impose a substantial burden on patients and society. The need for more tailored and early intervention approaches remains crucial. With its ability to analyze diverse data sources such as clinical records, imaging data, and biological samples, ML holds immense promise for revolutionizing RA management. This section explores how ML is transforming the landscape of RA detection and identification.

Early detection of RA is crucial for mitigating its long-term impact. ML methods, using EHRs, omics data, imaging modalities, and wearable sensors, are showing remarkable potential in expediting and refining the diagnostic process (Kedra et al., 2021). One research highlights the role of MRI in detecting inflammatory signs before they become clinically apparent, particularly in the wrist, hand, and foot. However, the manual quantification of MRI data is impractical owing to its large volume, and visual scoring systems have limited sensitivity. To address this, the paper explores the application of AI, particularly deep learning, in analyzing MRI data for early RA detection (Stoel, 2019). These advances highlight the transformative potential of ML in improving diagnostic accuracy and enabling early intervention, ultimately improving patient outcomes.

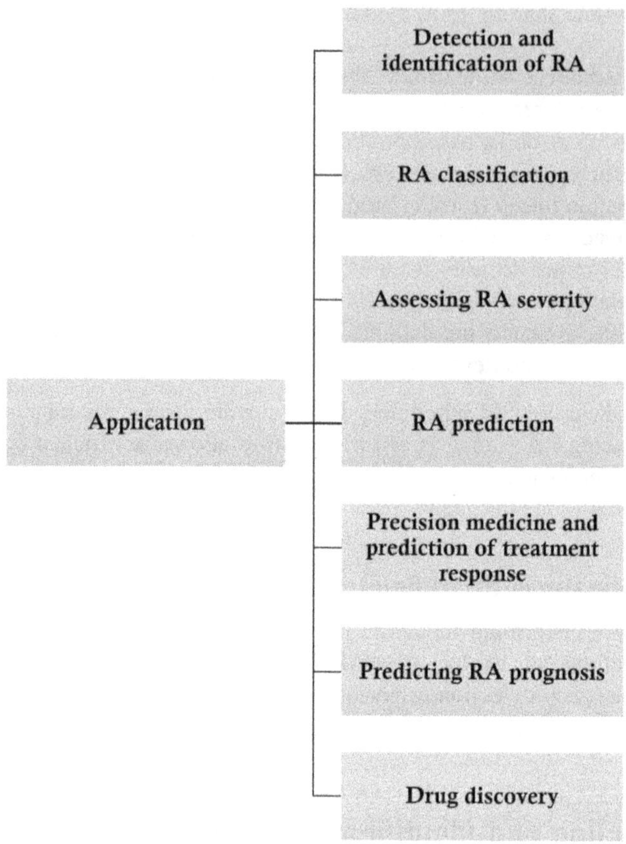

FIGURE 7.2

Overview of artificial intelligence applications in rheumatoid arthritis (RA) management.

One study compared identifying RA patients using ML extraction from EHRs versus traditional chart review (Maarseveen et al., 2021). The ML model performed well, offering a fast, efficient, and comparable alternative to traditional methods. The ML model, trained on 17,662 patients, performed well (sensitivity = 0.85, specificity = 0.99, positive predictive value = 0.86, and negative predictive value = 0.99). Among 1127 patients with both ML and criteria-based data, ML identified 373 RA cases, with 80% overlap with criteria-based selections. Baseline characteristics were comparable between ML and criteria-based selections. Some patients were exclusively identified by ML or criteria. Different ML cutoffs affected precision and sensitivity. The method offers fast, efficient, and similar patient selection to criteria-based review.

Other studies explored using ML with primary care data to predict RA diagnoses later confirmed in secondary care (Tarakci et al., 2023; Zhou et al., 2016). These studies demonstrated the efficacy of ML in analyzing EHR data for identifying

RA, offering the potential for faster diagnoses and improved resource allocation. Both the fuzzy expert system and ML approaches achieved impressive accuracy, particularly when incorporating patient history. Another study focused on identifying patients with RA through ML applied to primary care data. The study used data from the Farr Institute of Health Informatics Research, which links primary care records with secondary care rheumatology data, encompassing patients from two health boards in Wales, United Kingdom. Using standardized diagnostic codes known as reading codes, the researchers identified diagnoses, medications, and care processes within the dataset, in which 17% of patients received the diagnosis of RA in secondary care. A random forest method was employed to classify patients as either having RA or not based on the read codes. This approach assessed the relative frequency of each code as a risk factor without considering interactions between them. Subsequently, the model ranked the importance of individual predictors and selected the best ones for further analysis. To generate transparent classification rules, the researchers used a decision tree algorithm, which automatically generated if-then rules based on the selected predictors. Although such algorithms can be challenging to interpret owing to their black box nature, they facilitated the creation of clear classification rules. The performance of the model was evaluated on two datasets: one focused on Cardiff and another encompassed all individuals aged 16 and above within the primary care system from 2009 to 2012. Assessment metrics such as sensitivity, specificity, positive predictive value, and the prevalence of RA over time were employed to gauge the model's effectiveness. Overall, these studies underscored the potential of leveraging ML to analyze primary care EHR data for identifying complex medical conditions such as RA. The developed model exhibited performance comparable to existing expert-driven methods, suggesting its promise as a reliable and potentially more efficient alternative for RA phenotyping in primary care settings.

Research also explored deep learning for diagnosing RA from hand radiographs (Üreten et al., 2020). A CNN model achieved promising accuracy, showcasing the potential of deep learning in image-based RA detection. The study used a dataset composed of 135 radiographs, 61 of which were normal and 74 which had RA. The CNN model was trained on this dataset and achieved an accuracy of 73.3%. The CNN architecture consisted of six groups of convolutional layers, each containing a convolutional layer, batch normalization, and a Rectified Linear Unit (ReLU) activation function. These layers extracted increasingly complex features from the input images. This was followed by max-pooling layers for dimensionality reduction and further feature extraction. Finally, fully connected layers and a softmax layer performed the final classification task, determining the presence or absence of RA. To mitigate overfitting and improve generalizability, the researchers implemented online data augmentation techniques. This involved randomly translating and rotating the hand radiograph images within a small range, essentially creating variations of existing data without requiring additional images. This led to learning more robust features that were less sensitive to minor variations in image appearance, leading to improved performance.

Another study, by Qin et al. (2022), employed ML to identify blood biomarkers for diagnosing RA-ILD. This study examined ML to identify potential biomarkers for more specific RA diagnoses. These researchers analyzed 153 RA patients, 75 with RA-ILD and 78 without. Initial analysis revealed key differences between the groups: age, white blood cell count, inflammatory markers, and specific biomarkers such as KL-6, D-dimer, and tumor markers. To classify RA-ILD patients further, the study employed ML techniques (LASSO, random forest, and Partial Least Squares regression). These models achieved high accuracy (area under the curve [AUC] between 0.84 and 0.95). Notably, they identified age, KL-6, D-dimer, and CA19-9 as important biomarkers. Receiver operating characteristic (ROC) curve analysis confirmed that these biomarkers could effectively distinguish between RA-ILD and non-ILD patients, both individually and combined. The combination of all four achieved the highest AUC (0.928). The study also explored correlations between biomarkers and disease activity. Interestingly, whereas D-dimer levels correlated with disease activity in all RA patients, KL-6 and CA19-9 levels did not correlate specifically in the RA-ILD group. This suggests a potential role for these markers in interstitial lung disease development rather than RA itself. Overall, this study highlights the usefulness of ML algorithms in identifying biomarkers for RA-ILD diagnosis. It also suggests potential markers such as KL-6, D-dimer, and CA19-9 for further investigation and potential clinical application.

Another study explores the combination of glycan analysis with ML algorithms to improve discrimination accuracy, particularly for seronegative RA patients (Chocholova et al., 2018). This research emphasizes the potential of combining different data sources and analysis techniques for enhanced diagnostic capabilities. Analyzing 100 serum samples from healthy individuals and RA patients, the researchers employed both standard immunoassays and lectin-based glycan analysis. To delve deeper and discover crucial RA markers, data mining using artificial neural networks (ANNs) was implemented to distinguish accurately between healthy and RA individuals. Notably, the study unveiled the significant contribution of glycan analysis, particularly when combined with traditional immunoassays, in enhancing discrimination accuracy, especially for seronegative RA (S− RA) patients. Focusing on seropositive RA (S+ RA) patients, the study demonstrated that incorporating glycan analysis with immunoassays yielded superior discrimination accuracy compared with relying solely on immunoassays. Specifically, a combination of lectin enzyme-linked lectin-binding assay (ELLBA) with immunoassays achieved an impressive 92.5% accuracy in distinguishing non-RA from S+ RA patients, surpassing the 83.3% achieved by immunoassays alone. In addition, ML algorithms, particularly ANN, proved to be valuable tools for data mining and pattern recognition, further bolstering the diagnostic capabilities of the combined approach. However, traditional immunoassays alone were ineffective in identifying S− RA patients. Remarkably, glycan analysis, specifically when performed with ELLBA using adsorbed protein A, successfully identified 43.8% of S− RA patients. This finding underscores the potential of glycan analysis as a complementary diagnostic tool for identifying RA patients who might be missed by standard methods. The study concludes the important role of ML algorithms in handling large datasets and enhancing the diagnostic precision of glycan analysis in RA diagnosis.

This section has explored the possibilities of integrating ML with diverse data sources for RA detection and identification. Ongoing research showcases the role of ML in this field.

7.6 Machine learning for rheumatoid arthritis classification

ML is making significant progress in various aspects of RA management, particularly in classification. A study by Ho et al. (2018) investigated the potential of combining thermal imaging with ML to classify RA. The researchers employed decision tree and support vector machine (SVM) algorithms, along with ensemble learning and rigorous 10-fold cross-validation for evaluation. They assessed model performance using metrics such as accuracy, sensitivity, specificity, and ROC curves derived from the confusion matrix.

Another study explored using ML to classify traditional Chinese medicine (TCM) syndromes in RA patients (Xie et al., 2020). Accurate syndrome classification is first step for effective TCM treatment. This research aimed to improve classification accuracy using ML methods. The study analyzed data from over 1700 RA patients, extracting 200 features and employing feature selection to identify the 42 most critical ones. They evaluated six ML models, including K-nearest neighbor, SVM, decision tree, random forest, ANN, and AdaBoost. Among these, ANN achieved the highest classification accuracy (88%) after feature selection. Despite the effectiveness, the study acknowledged limitations such as heterogeneous data and potential bias owing to simplifying features into binary categories. In addition, the retrospective nature of the data necessitates validation through prospective studies. Overall, these work highlight the potential of ML for classifying RA, emphasizing the importance of feature selection for better model performance.

7.7 Assessing rheumatoid arthritis severity

Inconsistency in subjective evaluations can be a barrier to optimal treatment planning for RA severity. Researchers are exploring ML approaches to provide more objective and automated methods for severity assessment, particularly using metacarpophalangeal joint ultrasound images.

One such study proposed an automated grading method for assessing RA severity in ultrasound images (Yang et al., 2020). This approach combines image processing and ML algorithms to:

- **Segment bone erosion and synovial thickening:** A novel Gaussian scale space method effectively identifies both regions of interest, outperforming traditional thresholding techniques.

- **Extract quantitative:** The analysis extracts both geometric features (area, thickness, and shape) and texture features derived from gray-level co-occurrence matrix (GLCM) analysis from the segmented regions.
- **Perform feature selection:** Minimal redundancy−maximal relevance identifies the most informative features for classification.
- **Train and evaluate SVM models:** Models are trained with various feature combinations (GLCM only, geometric only, and combined) and evaluated on unseen data.

This process, depicting the various stages from image segmentation and feature extraction to model training and evaluation, is visually represented in Fig. 7.3.

The results demonstrate the effectiveness of the proposed approach. It achieves accurate segmentation of both bone erosion and synovial thickening, outperforming standard methods. Combining geometric and texture features further improves classification accuracy, reaching 92.5%. This suggests that incorporating both types of information provides a more comprehensive representation of RA severity compared with relying solely on one. Importantly, the SVM model trained with both GLCM and geometric features achieved superior performance compared with using either alone.

Overall, this study presents a promising ML-based approach for the automated and objective assessment of RA severity in ultrasound images. This has the potential to reduce reliance on subjective evaluations and contribute to more personalized and effective treatment plans for RA patients. However, further validation with larger datasets and exploration of its applicability to other types of arthritis or joint diseases are warranted.

7.8 Prediction of rheumatoid arthritis

Early diagnosis and intervention are crucial for improving outcomes in RA. ML offers promising potential for predicting RA at its early stages. One such study explored various ML algorithms to achieve this goal (Gurumoorthy et al., 2018).

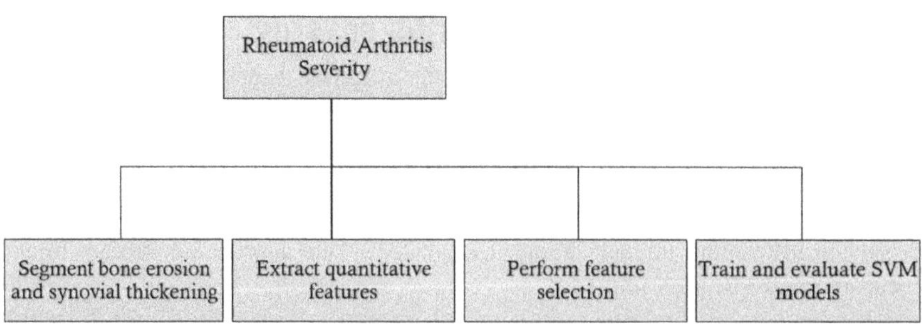

FIGURE 7.3

Assessing rheumatoid arthritis severity.
SVM, support vector machine.

The study gathered data from hospitals, including clinical characteristics and genetic factors from blood samples and general examinations. The researchers focused on patients with viral fever lasting over 6 weeks, who later developed RA, and compared them with patients with similar fever but no RA. The researchers then analyzed different ML algorithms such as AdaBoost, SVMs, and ANNs for their effectiveness in predicting RA.

Their proposed model, named the machine learning–based ensemble analytic approach (MLEAA), combined these algorithms in a two-step process. During the training phase, the researchers processed data using Hadoop's MapReduce framework to extract relevant features. These features were then fed into AdaBoost, SVM, and ANN algorithms during the prediction phase. A voting system combined the results from each algorithm to generate the final prediction. The authors claim that this multilearner ensemble approach for early arthritis prediction achieved better results than individual algorithms, suggesting its potential for early RA diagnosis.

7.9 Precision medicine and treatment response

Precision medicine, also referred to as personalized medicine, enhances health care by tailoring treatments to each patient's unique profile. This approach considers factors such as genetics, the environment, lifestyle, and specific biomarkers to guide health care decisions, diagnosis, and treatment plans. Recognizing the individual biological makeup allows precision medicine to maximize treatment effectiveness and minimize side effects. RA exemplifies the potential of precision medicine. Because of its varied indications and individual responses to treatments, RA often benefits from a personalized approach. In RA, precision medicine leverages diverse molecular markers, imaging techniques, and clinical assessments to optimize diagnosis, predict disease progression, and select the most effective therapies for individual patients.

ML significantly advances precision medicine for RA. ML algorithms analyze large patient data, including genetics, clinical records, imaging, and treatment outcomes. Leveraging ML, health care providers can design and develop models to predict RA prognosis, treatment response, and side effects. This enables better patient health care and more accurate personalized treatment plans. In addition, ML algorithms can aid in discovering new biomarkers and therapeutic targets. By analyzing complex biological data, hidden relationships among genes, molecular pathways, and disease mechanisms can be investigated to discover new drug targets and targeted therapies. Furthermore, ML enables the integration of diverse data sources such as omics data, EHRs, and patient-reported outcomes. This broad view of RA patients helps develop more precise treatment strategies (Aletaha, 2020).

Researchers explored the use of ML to predict long-term remission in RA patients who were receiving biologic drugs. Comparing various algorithms, they found the extreme gradient boosting algorithm to be most accurate in predicting remission using readily available patient data. This suggests that ML could be a valuable tool for doctors, enabling them to personalize treatment plans and improve patient outcomes (Venerito et al., 2022).

Methotrexate (MTX) remains the primary treatment for RA, but some patients receiving second- or third-line medications still experience treatment failure. Implementing precision medicine based on each patient's unique biological profile could potentially reduce this treatment's lack of responsiveness, benefiting both patients and the health care system. Data used to choose the most effective treatment plan can range from basic variables such as sex and age to complex information derived from proteomics and transcriptomics. ML algorithms have shown promising results in predicting response to MTX, achieving AUCs as high as 0.84 using only demographic and clinical data. Omics data further offer valuable insights for predicting treatment response. A random forest model using gut microbiome data achieved an AUC of 0.84 for identifying MTX responders. Notably, using omics data seems to be more beneficial for predicting responses to second- or third-line biological DMARDs compared with MTX. Moreover, imaging data can be incorporated into models predicting treatment response (Alimirzaei and Kieslich, 2023). However, studies employing AI to predict treatment response in RA face limitations including the use of diverse evaluation methods, a lack of comparisons between studies, and a predominant focus on MTX. Furthermore, few studies have evaluated treatment outcomes using nonbiological DMARDs. This poses a knowledge gap, because such research could identify patients who require safe and effective alternatives given the higher cost and accessibility limitations associated with bDMARDs (Maciejewski et al., 2021).

Despite promising results in ML-based prediction for RA treatments such as MTX, limitations hinder real-world application. These limitations include varied methodologies and a focus primarily on MTX response. In addition, exploring cost-effective alternatives for nonbiological DMARD response prediction becomes crucial owing to limitations in accessibility. The success of ML in predicting RA prognosis motivates the exploration of its potential in other diseases and treatments. A study by Hammam et al. (2023) exemplifies this by investigating ML to predict vision-threatening Behçet disease (VTBD) in patients. The researchers used data from a large patient cohort and compared various ML models with traditional analysis methods. They also aimed to identify risk factors for VTBD development. Their most accurate model, extreme gradient boosting (XGBoost), significantly outperformed traditional methods in predicting VTBD (area under the receiver operating characteristic curve of 0.85 vs. 0.64). This analysis demonstrated key risk factors such as high disease activity, history of smoking, and daily steroid use. To achieve these results, the researchers collected extensive data on demographics, clinical symptoms, disease duration, treatments, and lab markers. Disease activity was measured using a standardized scoring system, whereas ophthalmologists conducted detailed eye examinations. This research focuses on the ability of ML to handle complex relationships and predict the presence or absence of VTBD. The results of their experiments highlight the potential of ML, particularly XGBoost, in predicting VTBD. It identifies key risk factors and also suggests potential clinical applications for personalized treatment decisions and the prevention of vision loss. Further research and validation are needed, especially regarding observed sex differences in VTBD predictors.

Similarly, a study (Castro Corredor and Calvo Pascual, 2023) explored using ML models to predict when patients with rheumatic disease might discontinue biosimilar medicines. Considering data imbalances, the researchers employed different techniques to improve prediction accuracy. Their findings showed better performance in predicting discontinuation (F1-score: 0.63) compared with disease worsening (F1-score: 0.52). Age and switching of biosimilar drugs were identified as significant factors influencing both predictions. This study emphasizes the potential of ML to identify patients at risk of discontinuation, allowing earlier interventions and more informed treatment decisions. Another study aimed to predict clinical response to antitumor necrosis factor (anti-TNF) therapy using ML algorithms on multiomics data (Tao et al., 2021). By analyzing gene expression and DNA methylation profiles, the researchers identified different molecular signatures associated with reactions to different anti-TNF drugs. This led to the development of highly accurate ML models (85.9% and 88% for different drugs) for predicting treatment response. This approach is an example of personalized treatment plans by tailoring anti-TNF therapy based on individual patient profiles. Overall, this study indicates the potential of integrating multiomics data and ML algorithms to predict clinical response to anti-TNF therapy in RA patients.

These studies showcase the various applications of ML in predicting disease progression and treatment response in rheumatic disease. The capacity to identify patients at risk or predict individual responses to medications is promising in personalized medicine. However, challenges such as data heterogeneity and model validation require careful consideration to ensure robust and reliable implementation in clinical practice.

7.10 Predicting rheumatoid arthritis prognosis

Accurate measurement of disease activity is crucial for RA. These measurements are needed to determine treatment plans, assess response to therapy, and predict prognosis. This section explores how ML algorithms contribute to advances in predicting RA prognosis through various approaches. The Disease Activity Score in 28 Joints (DAS28) serves as a widely used severity measure. Studies have demonstrated that adaptive deep neural networks can outperform non—deep learning methods for DAS28 prediction, potentially leading to more precise treatment decisions (Hammam et al., 2023).

RA patients are often susceptible to various comorbidities, including cardiovascular, atherosclerotic, musculoskeletal, and neurological disorders. Carotid ultrasound, a noninvasive imaging technique, can assess atherosclerotic plaque buildup. In this context, ML algorithms can enhance cardiovascular risk stratification in RA patients, offering valuable insights for preventive interventions (Wei et al., 2020). Furthermore, ML classifiers incorporating cardiovascular risk factors have been shown to outperform classical risk scores in predicting cardiovascular events in this population.

The increasing adoption of wearable devices such as pulse oximeters and cardiac Holter monitors allows for continuous monitoring of disease activity and physical activity levels in RA patients. These devices, particularly those specifically designed for flare tracking, hold promise for improving patient care and predicting the future disease course (Smuck et al., 2021).

7.11 Drug discovery

Rheumatic diseases are chronic and require long-term treatment. To improve efficiency, drug discovery is expensive. Zhao and colleagues created ML models to propose treatments by inspecting the expression profiles of patients treated with proven drugs. Their results aligned with an external database, shifting research from assumption-based to known data, and overcoming challenges in handling large amounts of information (DiMasi et al., 2016).

Using established knowledge from patients who demonstrably responded to specific treatments, the model circumvents the need for extensive exploratory research based on theoretical assumptions. In addition, ML algorithms excel at handling large datasets efficiently, overcoming a significant challenge often encountered in traditional drug discovery pipelines. The findings from this study were validated by comparing the generated candidate drugs with an external database. This validation step bolsters confidence in the proposed approach and highlights its potential for identifying effective treatment options.

7.12 Machine learning limitations and challenges

Despite promising results, ML faces limitations hindering its clinical application in RA management. These include the need for large, high-quality datasets, the lack of standardization, and black box models that are difficult to explain. In addition, ethical concerns regarding data privacy and potential discrimination against patients identified as high-risk require careful consideration.

Looking forward, research focuses on overcoming these limitations. Standardizing data collection and developing self-optimizing algorithms can improve efficiency and reliability. Collaborations among data scientists, clinicians, and patients is crucial to build interpretable models and ensure ethical implementation. Exciting applications such as voice analysis and social media monitoring hold promise for personalized RA management and improved patient experience (Kedra et al., 2021).

Although challenges remain, ML presents significant potential to revolutionize RA management, requiring collaborative efforts to ensure its responsible and effective integration into clinical practice.

7.13 Future directions

This study identifies eight crucial areas for future research to harness the full potential of AI in managing RA and similar autoimmune disorders. These recommendations emphasize ethical considerations, model transparency, and robust validation. Future directions are depicted in Fig. 7.4. Key areas for advancement include:

- Ethical and Responsible Development: Adherence to established guidelines such as Checklist for Artificial Intelligence in Medical Imaging (CLAIM) and National Health Service frameworks ensures responsible AI development and use, minimizing potential harm.
- Transparency and Reproducibility: Sharing complete source code allows researchers to verify and improve AI models through external testing, fostering collaboration and trust.

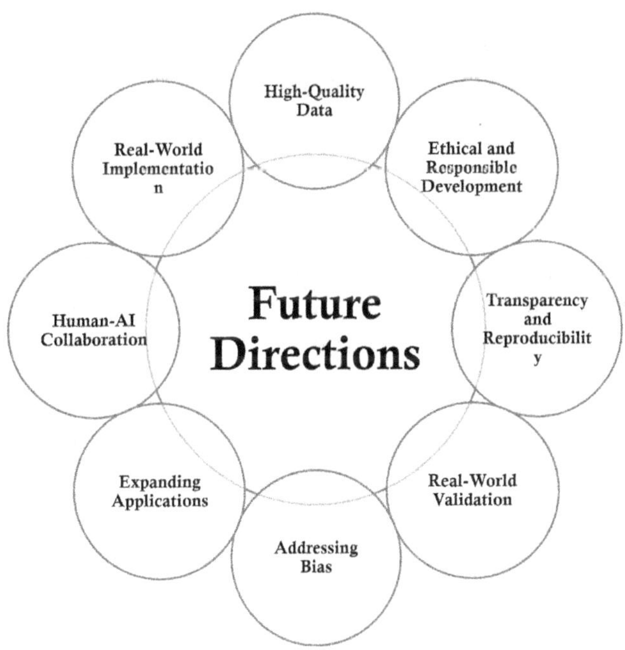

FIGURE 7.4

Future directions in machine learning research for rheumatoid arthritis.
AI, artificial intelligence.

- Real-World Validation: Conducting rigorous external validation on independent datasets is crucial to assess the true effectiveness of AI models in clinical settings.
- High-Quality Data: Using large-scale, diverse datasets encompassing various patient information (demographics, clinical, genomic, imaging, etc.) improves model accuracy and generalizability.
- Addressing Bias: Mitigating potential algorithm bias requires awareness and using datasets that adequately represent diverse populations to ensure fairness and inclusivity.
- Expanding Applications: Exploring AI to assess extraarticular manifestations beyond joints in RA patients can provide a more holistic view of the disease.
- Human-AI Collaboration: Evaluating the combined performance of AI and human experts compared with individual experts can offer more practical and effective solutions.
- Real-World Implementation: Designing prospective, multicenter trials is essential to evaluate the effectiveness of AI models in real-world clinical settings.

This study examined future research exploring AI applications in other autoimmune and inflammatory rheumatic diseases, potentially transforming diagnosis, treatment, and overall patient care.

The chapter concludes by underscoring the need for future research to develop interventions that address patient beliefs and fears, ultimately improving adherence to RA treatment regimens. This chapter investigates motivational interviewing approaches and targeted educational initiatives.

7.14 Conclusion

This chapter has examined the field of AI, particularly ML applications for RA classification, showing its potential to affect clinical practices significantly and enhance patient health. Delving deeper into this field reveals several key considerations. ML's unique ability to extract hidden patterns and insights from big datasets makes personalized medicine approaches easier than before. With this approach, treatment plans are tailored to individual patient profiles. Imagine a future in which genetic predispositions, disease activity biomarkers, and even social determinants of health inform treatment decisions. This data-driven approach holds the promise of optimized treatment efficacy, minimized side effects, and ultimately, enhanced patient experiences and quality of life.

ML's predictive capabilities extend far beyond treatment selection. Its potential lies in forecasting disease progression, anticipating flares, and identifying patients at high risk for complications. This early warning system empowers timely interventions, potentially preventing irreversible damage and improving long-term health outcomes. Although ML offers immense potential, ethical considerations demand careful attention. Protecting patient privacy, mitigating algorithmic bias, and ensuring transparency are paramount. Fostering collaboration among data scientists, clinicians, patients, and policymakers is crucial for responsible development, ethical implementation, and the continuous improvement of ML applications in RA management.

AI disclosure

During the preparation of this work the author(s) used https://app.grammarly.com/in to check grammar. After using this tool/service, the author(s) reviewed and edited the content as needed and take(s) full responsibility for the content of the publication.

References

Ahalya, R.K., Umapathy, S., Krishnan, P.T., Joseph Raj, A.N., 2022. Automated evaluation of rheumatoid arthritis from hand radiographs using machine learning and deep learning techniques. Proc. IME H J. Eng. Med. 236 (8), 1238–1249. https://doi.org/10.1177/09544119221109735.

Aletaha, D., 2020. Precision medicine and management of rheumatoid arthritis. J. Autoimmun. 110, 102405.

Alimirzaei, F., Kieslich, C.A., 2023. Machine learning models for predicting membranolytic anticancer peptides. Comput. Aided Chem. Eng. 52, 2691–2696. https://doi.org/10.1016/B978-0-443-15274-0.50428-5.

Borhani, M., Soofiani, N., Ebrahimi, E., Asadollah, S., 2014. First report on growth and reproduction of turcinoemacheilus bahaii (esmaeili, sayyadzadeh, özulug, geiger and freyhof. Zayandeh Roud River, Iran. Aust. Environ. Sci. 2.

Borhani, M., Soofiani, N.M., Ebrahimi, E.S.A., Asadollah, S., 2017. First report on growth and reproduction of Turcinoemacheilus Bahaii. Zayandeh Roud River, Iran. Aust. Environ. Sci. 2 (1), 1014.

Borhani, M., Kalbassi, M.R., Psenicka, M., Falahatkar, B., Crawford, A., 2022. Identification and isolation of sprmatogonial stem cells from beluga (Huso. J. Aquat. Physiol. Biotechnol. 10 (2), 1–24.

Castro Corredor, D., Calvo Pascual, L.Á., 2023. Imbalanced machine learning classification models for removal biosimilar. PLoS One 18 (11).

Chocholova, E., Bertok, T., Jane, E., Lorencova, L., Holazova, A., Belicka, L., Belicky, S., Mislovicova, D., Vikartovska, A., Imrich, R., Others, 2018. Glycomics meets artificial intelligence–potential of glycan analysis for. Clin. Chim. Acta 481, 49–55.

DiMasi, J.A., Grabowski, H.G., Hansen, R.W., 2016. Innovation in the pharmaceutical industry: new estimates of R&D costs. J. Health Econ. 47, 20–33.

Gurumoorthy, S., Rao, B.N.K., Gao, X.-Z., Shanmugam, S., Preethi, J., 2018. Design of rheumatoid arthritis predictor model using machine learning. Cognit. Sci. Artif. Intell.: Adv. Appl. 67–77.

Hammam, N., Bakhiet, A., El-Latif, E.A., El-Gazzar, I.I., Samy, N., Noor, R.A.A., El-Shebeiny, E., El-Najjar, A.R., Eesa, N.N., Salem, M.N., Others, 2023. Development of machine learning models for detection of vision threatening. BMC Med. Inf. Decis. Making 23 (1), 37.

Ho, S., Elamvazuthi, I., Lu, C.K., 2018. Classification of rheumatoid arthritis using machine learning algorithms. In: 2018 IEEE 4th International Symposium in Robotics and Manufacturing, pp. 1–6.

Kedra, J., Davergne, T., Braithwaite, B., Servy, H., Gossec, L., 2021. Machine learning approaches to improve disease management of patients with. Expet Rev. Clin. Immunol. 17 (12), 1311–1321.

Maarseveen, T.D., Maurits, M.P., Niemantsverdriet, E., van der Helm-van Mil, A.H.M., Huizinga, T.W.J., Knevel, R., 2021. Handwork vs machine: a comparison of rheumatoid arthritis patient. Arthritis Res. Ther. 23 (1), 174.

Maciejewski, M., Sands, C., Nair, N., Ling, S., Verstappen, S., Hyrich, K., Barton, A., Ziemek, D., Lewis, M.R., Plant, D., 2021. Prediction of response of methotrexate in patients with rheumatoid. Sci. Rep. 11 (1), 7266.

Momtazmanesh, S., Nowroozi, A., Rezaei, N., 2022. Artificial intelligence in rheumatoid arthritis: current status and future perspectives: a state-of-the-art review. Rheumatol. Ther. 9 (5), 1249–1304. https://doi.org/10.1007/s40744-022-00475-4.

Mohammadabadi, S.M.S., Zawad, S., Yan, F., Yang, L., 2023a. Speed up federated learning in heterogeneous environment: a dynamic tiering approach. arXiv preprint arXiv:2312.05642. https://doi.org/10.48550/arXiv.2312.05642.

Mohammadabadi, S.M.S., Liu, Y., Canafe, A., Yang, L., 2023b. Towards distributed learning of PMU data: a federated learning based event classification approach. In: 2023 IEEE Power & Energy Society General Meeting (PESGM), pp. 1–5. https://doi.org/10.1109/PESGM52003.2023.10252920.

Mohammadabadi, S.M.S., Yang, L., Yan, F., Zhang, J., 2024a. Communication-efficient training workload balancing for decentralized multi-agent learning. arXiv Preprint arXiv:2405.00839. https://doi.org/10.48550/arXiv.2405.00839.

Mohammadabadi, S.M.S., Entezami, M., Moghaddam, A.K., Orangian, M., Nejadshamsi, S., 2024b. Generative Artificial Intelligence for Distributed Learning to Enhance Smart Grid Communication. International Journal of Intelligent Networks 5 (2024), 267–274. https://doi.org/10.1016/j.ijin.2024.05.007. https://www.sciencedirect.com/science/article/pii/S2666603024000265.

Obayya, M., Alamgeer, M., S. Alzahrani, J., Alabdan, R., N. Al-Wesabi, F., Mohamed, A., Alsaid Hassan, M.I., 2022. Artificial intelligence driven biomedical image classification for robust rheumatoid arthritis classification. Biomedicines 10 (11), 2714. https://doi.org/10.3390/biomedicines10112714.

Orange, D.E., Agius, P., DiCarlo, E.F., Robine, N., Geiger, H., Szymonifka, J., McNamara, M., Cummings, R., Andersen, K.M., Mirza, S., Figgie, M., Ivashkiv, L.B., Pernis, A.B., Jiang, C.S., Frank, M.O., Darnell, R.B., Lingampali, N., Robinson, W.H., Gravallese, E., Bykerk, V.P., Goodman, S.M., Donlin, L.T., 2018. Identification of three rheumatoid arthritis disease subtypes by machine learning integration of synovial histologic features and RNA sequencing data. Arthritis Rheumatol. 70 (5), 690–701. https://doi.org/10.1002/art.40428.

Qin, Y., Wang, Y., Meng, F., 2022. Identification of biomarkers by machine learning classifiers to assist diagnose rheumatoid arthritis- associated interstitial lung disease. Arthritis Res. Ther. 24 (1), 1–12.

Sajjadi Mohammadabadi, S.M., 2024. Advancing communication efficiency in electric vehicle systems: a survey of generative ai and distributed machine learning strategies. Available at SSRN 4791891. https://doi.org/10.2139/ssrn.4791891.

Smuck, M., Odonkor, C.A., Wilt, J.K., Schmidt, N., Swiernik, M.A., 2021. The emerging clinical role of wearables: factors for successful. NPJ Digit. Med. 4 (1), 45.

Stoel, B.C., 2019. Artificial intelligence in detecting early RA. Semin. Arthritis Rheum. 49, S25–S28.

Sundaramurthy, S., Saravanabhavan, C., Kshirsagar, P., 2020. Prediction and classification of rheumatoid arthritis using ensemble machine learning approaches. In: 2020 International Conference on Decision Aid Sciences and Application, DASA 2020, pp. 17–21. https://doi.org/10.1109/DASA51403.2020.9317253.

Tao, W., Concepcion, A.N., Vianen, M., Marijnissen, A.C.A., Lafeber, F.P., Radstake, T.R., Pandit, A., 2021. Multiomics and machine learning accurately predict clinical response to. Arthritis Rheumatol. 73 (2), 212–222.

Tarakci, F., Ozkan, I.A., Yilmaz, S., Tezcan, D., 2023. Diagnosing rheumatoid arthritis disease using fuzzy expert system and. J. Intell. Fuzzy Syst. 44 (4), 5543–5557.

Üreten, K., Erbay, H., Maraş, H.H., 2020. Detection of rheumatoid arthritis from hand radiographs using a. Clin. Rheumatol. 39, 969–974.

Üreten, K., Maraş, H.H., 2022. Automated classification of rheumatoid arthritis, osteoarthritis, and normal hand radiographs with deep learning methods. J. Digit. Imag. 35 (2), 193–199. https://doi.org/10.1007/s10278-021-00564-w.

Venerito, V., Angelini, O., Fornaro, M., Cacciapaglia, F., Lopalco, G., Iannone, F., 2022. A machine learning approach for predicting sustained remission in. J. Clin. Rheumatol. 28 (2), e334–e339.

Vodencarevic, A., Tascilar, K., Hartmann, F., Reiser, M., Hueber, A.J., Haschka, J., Bayat, S., Meinderink, T., Knitza, J., Mendez, L., Hagen, M., Krönke, G., Rech, J., Manger, B., Kleyer, A., Zimmermann-Rittereiser, M., Schett, G., Simon, D., 2021. Advanced machine learning for predicting individual risk of flares in rheumatoid arthritis patients tapering biologic drugs. Arthritis Res. Ther. 23 (1). https://doi.org/10.1186/s13075-021-02439-5.

Wang, S., Hou, Y., Li, X., Meng, X., Zhang, Y., Wang, X., 2021. Practical implementation of artificial intelligence-based deep learning. Front. Pharmacol. 12, 765435.

Wei, T., Yang, B., Liu, H., Xin, F., Fu, L., 2020. Development and validation of a nomogram to predict coronary heart disease. Aging (Albany NY) 12 (4), 3190.

Xie, J., Li, Y., Wang, N., Xin, L., Fang, Y., Liu, J., 2020. Feature selection and syndrome classification for rheumatoid arthritis. Eur. J. Integr. Med. 34, 101059.

Yang, T., Zhu, H., Gao, X., Zhang, Y., Hui, Y., Wang, F., 2020. Grading of metacarpophalangeal rheumatoid arthritis on ultrasound images. IEEE Access 8, 67137–67146.

Zhou, S.-M., Fernandez-Gutierrez, F., Kennedy, J., Cooksey, R., Atkinson, M., Denaxas, S., Siebert, S., Dixon, W.G., O'Neill, T.W., Choy, E., Others, 2016. Defining disease phenotypes in primary care electronic health records by a. PLoS One 11 (5).

CHAPTER

8 Early detection approach for analysis of osteoarthritis using artificial intelligence and machine learning

Chander Prabha

Chitkara University Institute of Engineering and Technology, Chitkara University, Punjab, India

8.1 Introduction

The common joint disease of osteoarthritis (OA) is characterized by the breakdown of cartilage. Cartilage is a smooth and resilient tissue that cushions the ends of bones in joints. Degradation of cartilage leads to swelling, pain, and difficulty moving the joint (Ryd et al., 2015). In severe conditions, cartilage wears away leading to the rubbing of bones against each other, causing further pain and thus finally joint damage. The most trivial form of arthritis is OA which affects billions of people globally. Joint areas most affected in humans are the knees, hips, hands, and spine. The prime factor in this disease is aging, along with joint injuries, obesity, genetics, repetitive stress on joints, and bone deformities. Symptoms vary from mild to severe, initially starting with normal pain and leading to stiffness. This might be due to a period of inactivity, excessive use, pressure applied on some joints, a grating sensation during movement of joints, the loss of flexibility, and swelling.

OA management delineates symptoms to improve the functions of joints. Treatment of OA may include an amalgamation of modifications of lifestyle, medications, and physical therapy, and in severe cases, surgery. Changes in lifestyle such as staying active, maintaining a healthy weight, and avoiding tasks that put excessive stress on the joints can help manage symptoms of OA and naturally slow the progression of the disease. Antiinflammatory and pain relief medications can help manage symptoms. In severe cases, surgery in the form of joint replacement helps restore the functions of joints and relieves pain. OA is chronic; if it is not taken care, the damage caused by it cannot be reversed. However, due care and early diagnosis can help one lead an active, healthy life through a comprehensive treatment plan (Chen et al., 2017).

Identifying Osteoarthritis: A comprehensive physical examination is conducted to assess the degree of discomfort, muscular weakness, and flexibility in motion to diagnose OA. A complete medical record is obtained evaluating past injuries and

https://doi.org/10.1016/B978-0-443-32892-3.00008-7

genealogy. If a physical checkup reveals OA, additional diagnostic tests could be carried out, such as x-rays, magnetic resonance imaging (MRI) scans, blood tests to screen for other diseases, and an analysis of fluid that lubricates the joint (Flores et al., 2021).

Symptoms: A variety of symptoms lead to OA ranging from mild to severe, and they increase over time. The main indicators are shown in Table 8.1.

The severity of OA symptoms can differ greatly among individuals and may change, frequently intensifying with excessive use of the joint or during times of inactivity. Although OA is primarily linked to aging, its symptoms can affect people of all ages, particularly those with previous joint injuries or who are overweight. Such conditions can increase joint stress and possibly speed the deterioration of cartilage. For persons with these symptoms, it is essential to get advice from a doctor for a precise examination and then tailored treatment. Immediate action can aid in managing OA better, enhance joint mobility, and decelerate the progression of OA.

The identification of OA via artificial intelligence (AI) uses numerous technologies to analyze, detect, and predict OA's progression (Caliva et al., 2021). Deterioration of joint cartilage and the bone beneath results in pain and rigidity. The integration of machine learning (ML) methods in exploring OA with AI marks a stride toward progress. The ability of ML to work on huge data enhances comprehension and handling of OA. Specifically, ML and deep learning (DL), the branches of AI, have the potential to improve OA treatment by analyzing MRI scans (images), identifying biomarkers, and creating predictive models. This evaluation probes into the techniques and use of ML within an OA study and health care applications for customized health care solutions, with a commitment to content originality. Table 8.2 shows the contribution of AI in managing and handling OA.

Table 8.1 Osteoarthritis symptoms.

Symptom	Description
Joint pain	During movement, the affected joints may hurt.
Stiffness	Occurs after a period of inactivity and is most noticeable on waking up in the morning.
Tenderness	Light pressure near the joint might cause discomfort
Loss of flexibility	The joint may have a diminished capacity to achieve complete range of movement.
Grating sensation	An individual could experience a sensation of grinding when operating the joint, possibly accompanied by sounds of popping or crackling.
Bone spurs	Additional bone fragments resembling hard bumps may develop around the joint in question.
Swelling	This condition may result from inflammation of soft tissues surrounding the joint.

Table 8.2 Artificial intelligence contribution in osteoarthritis management.

Contribution	Description	Factor	Summary
AI imaging analysis in OA	A key use of AI in addressing OA involves examining medical imagery, including x-rays, magnetic resonance images, and ultrasounds. Deep learning algorithms, which fall under the broader category of ML, have shown notable success in detecting nuanced alterations in joint anatomy that might be difficult for the human eye to discern (Chang et al., 2021).	Automated diagnosis	AI models identify indicators of OA in imaging data. They have the potential to diagnose the condition earlier than conventional approaches. They enhance patient results by allowing treatment to commence more promptly.
		Severity assessment	The severity of OA is evaluated via AI algorithms because these are able to measure the extent of damaged joints via pattern recognition in imaging data.
Biomarker identification	Biomarkers are recognized by AI and are biological markers present in bodily fluids or blood (used to quantify indicators of the presence of OA) (Lazzarini et al., 2017).	Predictive modeling	ML models are trained on vast datasets to identify potential biomarkers pertaining to OA treatment responses, enhancing the prediction of OA disease and providing treatment approaches for individuals.
		Genetic factors	To identify mutations linked to a heightened risk of OA, AI technologies are used. These insights into genetic factors can enhance understanding of the disease.
Predictive models for disease progression	Health care professionals can use prediction models to create specific treatment strategies considering the unique risk factors and disease progression of	Personalized treatment plans	Health care professionals can use prediction models to create specific treatment strategies considering the unique risk factors and disease progression of individuals.

Continued

Table 8.2 Artificial intelligence contribution in osteoarthritis management.—*cont'd*

Contribution	Description	Factor	Summary
	individuals (Guan et al., 2022).	Outcome simulation	The results of various treatment alternatives are achieved by AI model simulation. Thus, resulting in making well-informed choices and being capable of estimating the responses of patients to numerous treatment strategies.
		Risk stratification	ML models assist individuals who are at a heightened risk of developing OA, facilitating the initiation of preventive actions.
ML in imaging analysis	Diagnosis and evaluation of OA depend on imaging techniques such as x-rays, magnetic resonance imaging, ultrasound, and ML, particularly through deep learning models such as convolutional neural networks.	Detection and diagnosis	ML algorithms can be trained to detect initial indications of OA, occasionally even before the appearance of clinical symptoms, enabling potentially earlier interventions
		Progression monitoring	Through the assessment of alterations in joint imagery over time, ML models can track the progression of the disease, providing valuable information about the speed at which OA is progressing and the efficacy of treatment approaches.
Personalized medicine	This method capitalizes on ML's prowess in parsing complex data and spotting patterns. It facilitates a more tailored and effective treatment regimen for individuals with OA (Liukkonen et al., 2017).	Customized treatment recommendations	The patient's distinctive profile is taken into account (such as genetic markers, lifestyle, and disease traits) by ML algorithms to recommend the most efficacious treatment.

Table 8.2 Artificial intelligence contribution in osteoarthritis management.—*cont'd*

Contribution	Description	Factor	Summary
		Biomarker discovery	New biomarkers that signal the presence of OA are discovered, enabling prompt diagnosis and the development of therapeutic approaches.

AI, *artificial intelligence;* ML, *machine learning;* OA, *osteoarthritis.*

8.2 Osteoarthritis detection methods using artificial intelligence and machine learning

To detect OA quickly with high accuracy, many AI and ML algorithms are used to sort medical data such as the radiological scans, biomarkers, and histories of patients. These methods are compared with traditional diagnostic methods that rely on interpretation by humans and may introduce bias. This bias is removed by AI and ML approaches that offer a path to efficient and consistent evaluations. Some important methods are:

- **Convolutional Neural Networks:** These methods are good at analyzing medical imaging (e.g., computed tomography scans, MRIs, and x-rays) for identifying joints for any structural change that leads to the confirmation of OA. The self-learning concept is adopted by these convolutional neural networks (CNNs) to identify patterns and traits related to OA (e.g., osteophyte production, subchondral sclerosis, and joint space constriction). These networks have excellent precision capability to gain knowledge from large-picture datasets, which allows them to perform better as time passes (Mahum et al., 2021).
- **Deep Learning for Symptom Prediction:** Beyond CNNs, DL models are skilled in predicting how OA will progress by examining data collected over time. These models can incorporate a wide range of inputs, including attributes of the patient, genetic data, symptom reports, and changes in imaging over time. Such models can delineate intricate links and predict the course of an illness, which helps with the creation of specific therapy regimens (Yeoh et al., 2021).
- **Natural Language Processing for Clinical Notes:** Informal medical records can contain facts that natural language processing (NLP) may retrieve, evaluate, and convert into organized data. Then, using the processed data, one may determine OA hazards, symptoms, and therapeutic outcomes. By using existing medical data, NLP reduces the need for extra testing and offers insights derived solely from standard clinical procedures.
- **ML for Biomarker Analysis:** To identify OA, ML algorithms can examine genetic or metabolic biomarkers taken from blood tests or synovial fluid, among

other biological specimens. Patterns that may escape human inspection can be found using this technology. It offers an alternative to surgery for the early identification and continued tracking of OA, maybe even before significant structural changes show up on imaging tests (Lazzarini et al., 2017; Hafezi-Nejad et al., 2017).

- **Predictive Analytics for Risk Assessment:** Through the use of a combination of patient data including genes, lifestyle factors, and prior injuries, predictive analytics can assess the risk of developing OA. For those at danger, this preventative strategy makes it possible to put safeguards into place. It customizes risk assessment, enabling targeted actions meant to prevent or delay the onset of OA (Teo et al., 2021).
- **Transfer Learning for Small Datasets:** Transfer modeling provides an alternative when there is a shortage of OA datasets by allowing machines trained on large datasets of similar tasks to be used for OA identification. Without requiring a huge quantity of OA-specific data, this strategy can significantly improve efficiency. It makes it easier to create models that work, especially when OA-specific information is limited (Yeoh et al., 2023).

These methods demonstrate the various ways that AI and ML can be used to enhance the identification and treatment of OA. To overcome current obstacles and realize the full potential of these advances in health care facilities, ongoing studies and developments are essential. Challenges and considerations are shown in Table 8.3.

8.3 Artificial intelligence—aided model

Fig. 8.1 shows an AI-aided model consisting of ML. The ML is further categorized into localization (followed by detection and segmentation), classification, and prediction of OA disease.

In the initial phase of the OA ML model, segmentation and object identification techniques are used to identify the knee joint. Whereas object segmentation is a finer localization technique that uses a mask to lie on a spot of interest with an accurate shape created on the perimeter of the item, object detection uses a rectangular

Table 8.3 Challenges and considerations.

Challenge	Description
Data privacy and security	Ensuring the privacy and security of sensitive patient data is paramount.
Ethical considerations	Biases in datasets need to be addressed to prevent unequal treatment recommendations.
Interpretability	Models need to be developed that provide transparent decision-making processes to gain trust from health care professionals.

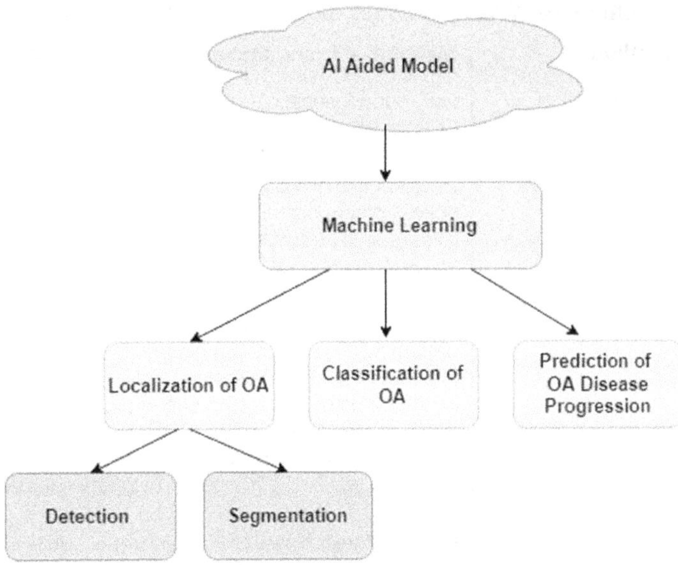

FIGURE 8.1

Artificial intelligence (AI)-aided model. *OA*, osteoarthritis.

boundary box to locate a region of interest. Object localization is a crucial stage that facilitates subsequent ML processes by assisting in the extraction of appropriate image sections and removing unnecessary image components. In classification, the primary diagnostic challenge is determining the severity of knee OA. Most studies (Abedin et al., 2019; Chen et al., 2019; Norman et al., 2019; Tomas et al., 2020) constructed the categorization model using the Kellgren-Lawrence (KL) grading method. A small amount of research examined the categorization of osteoarthritic meniscus and cartilage tissue (Pedoia et al., 2019), or osteoarthritic knee (Hirvasniemi et al., 2021). One of the biggest challenges in managing knee OA illness is the prognosis, or the ability to forecast future events in the disease. A linear process was formerly used to simulate knee OA disease; however, a few researchers have questioned this assumption (Roemer et al., 2011; Vos et al., 2012). Numerous long-term investigations were conducted to simulate advances in knee OA (Collins et al., 2021). Table 8.4 shows some prominent techniques of ML for OA.

8.4 Artificial intelligence preventive solution for osteoarthritis

AI provides novel ways to avoid OA, a form of arthritis that can cause pain, stiffness, and limited movement. Scientists and medical professionals can anticipate the start of OA, detect risk factors promptly, and put individualized preventative methods into

Table 8.4 Machine learning techniques for osteoarthritis.

Method		Machine learning techniques
Localization	Detection	Fully convolutional neural network (Abedin et al., 2019; Chen et al., 2019), random forest regression voting (Chen et al., 2019), YOLOv2 network (Chen et al., 2019)
	Segmentation	Decision tree classifier (Gornale et al., 2020, 2021), active contour algorithm (Gornale et al., 2020), U-Net (Chang et al., 2021; Pedoia et al., 2019; Norman et al., 2018; Yong et al., 2021; Huang et al., 2022), Res-U-Net (Cheung et al., 2021)
Classification		Convolutional neural network (CNN) with Local binary pattern (LBP) and histogram oriented of gradients (HOG) (Mahum et al., 2021), ResNet, Visual Geometry Group (VGG) (Chen et al., 2019; Yong et al., 2021), Random forest model (Abedin et al., 2019; Pedoia et al., 2019), 3D-CNN (Pedoia et al., 2019), deep Siamese CNN (Tiulpin et al., 2018), *K*-nearest neighbor (Gornale et al., 2020)
Prediction		Random forest classifier (Lazzarini et al., 2017), DenseNet CNN (Tolpadi et al., 2020), Logistic regression classifier (Tolpadi et al., 2020; Bowes et al., 2020), deep CNN (Tiulpin et al., 2019; Pierson et al., 2021), DeepSurv (Jamshidi et al., 2021)

practice using AI and ML technology. Table 8.5 presents a few AI-powered OA preventative measures.

Implementation Challenges: AI has a great deal of potential for OA, but putting it into practice will present difficulties, such as protecting data privacy, resolving the unequal possession of technology, and requiring interdisciplinary cooperation to create practical, empirically supported AI solutions (Kokkotis et al., 2020). Early identification, tailored therapies, and ongoing monitoring made possible by AI can completely transform the avoidance of OA. Yet, the moral and inclusive application of such AI-driven remedies, their incorporation into medical facilities, and their confirmation through research studies will determine their efficacy.

8.5 Machine learning preventive solution for osteoarthritis

ML is a category of AI. It offers potent instruments for the early diagnosis and treatment of OA (a prevalent joint condition). ML systems may find insights and trends within complicated datasets that experienced analysts might not see right away. This potential can be used to create OA prevention plans that emphasize decreasing risk factors along with prompt intervention. Table 8.6 presents ML preventive strategies and approaches (Binvignat et al., 2022).

Table 8.5 Artificial intelligence-powered osteoarthritis preventative measures.

OA preventative measures	Description	Benefits
Early risk detection	To identify people who are at an elevated risk of having OA, AI models can examine enormous amounts of data that include gene information, lifestyle characteristics (such as dietary habits, physical activity, and body weight), and exposure to the environment.	Early identification allows for the implementation of targeted interventions to reduce risk factors before OA develops.
Personalized lifestyle recommendations	By unique risk identities, technologies can offer tailored lifestyle and fitness advice consisting of nutritional recommendations to lower irritation, customized weight management plans, and workouts meant to bolster the muscles surrounding joints.	Customized plans help individuals adopt healthier habits that are effective in lowering the risk of OA
Predictive modeling for joint health	AI algorithms can forecast a person's future joint wellness by evaluating historical health data and suggesting individuals who might profit from early therapy.	Enables proactive management of joint health, potentially delaying or preventing the onset of OA symptoms
Wearable technology integration	Data from stress on joints, wearables that track spinal alignment, and physical activity are analyzed via AI that provides instantaneous feedback and advice for patients to avoid activities that could put people at risk for OA.	Promotes adjusting routines to lessen the strain on joints.
Nutritional guidance and supplements	AI models examine dietary data and medical research to suggest diets or supplements. These suggestions have preventive impacts on the condition of OA, such as those high in omega-3 fatty acid intake.	Promotes the condition of joints as a whole and may lower the chance of having OA.
Rehabilitation and physiotherapy optimization	AI helps optimize rehabilitation regimens for patients with OA and makes sure these activities are carried out correctly depending on progression to avert the worsening of OA.	Personalizes rehabilitation to individual needs, enhancing recovery and minimizing OA risk.

Continued

Table 8.5 Artificial intelligence-powered osteoarthritis preventative measures.—*cont'd*

OA preventative measures	Description	Benefits
Community and support systems	The use of AI-driven platforms helps people at risk of OA be connected to professional guidance, forums, and individualized coaching resources to stay motivated and stick to preventative strategies.	Enhances engagement and compliance with prevention strategies through community support.

AI, *artificial intelligence;* OA, *osteoarthritis.*

Table 8.6 Machine learning preventive solutions.

Factor	Approach	Preventive strategy
Identification of high-risk individuals	ML models can identify individuals more likely to sustain OA by analyzing routine, biological, and historical information using supervised learning algorithms. This entails assessing variables such as age, sex, weight, genetic susceptibilities, and prior joint injuries.	Targeted preventative actions, including lifestyle changes and initial OA symptom monitoring, are made possible by prompt detection.
Analysis of biomechanical data	Wearable technology data that monitor alignment, posture, and mobility are examined by ML models. These models can uncover modest kinematic indicators of risk for OA by pointing to trends that could result in stress on joints and degeneration.	It is possible to create individualized fitness and rehabilitation plans to address structural abnormalities and lessen joint stress.
Lifestyle and environmental factor analysis	ML can find associations among lifestyle and environmental variables and the onset of OA by analyzing these elements using methods based on unsupervised learning. Diet, the degree of physical exercise, and exposure to specific ecological risk factors are examples.	Supplies individualized guidance on food, exercise, and other lifestyle adjustments to reduce recognized risk factors.
Predictive modeling for early intervention	Models are created via ML techniques to forecast the course of OA depending on initial signs and associated risks. These models can predict a patient's early occurrence of OA progression.	To stop the progression of the condition, early treatments can be started, such as physical therapy or medicine.

Table 8.6 Machine learning preventive solutions.—*cont'd*

Factor	Approach	Preventive strategy
Optimizing physical therapy with reinforcement learning	By ML techniques (reinforcement learning [RL]), the physical therapy regimens can be made more effective. RL makes ongoing adjustments in the model based on the patient's success.	Personalized physical therapy regimens that change with the patient as they continue to meet their demands, improving joint health and possibly postponing the beginning of OA.
Dietary recommendations	To find dietary needs that affect the risk of OA, ML may evaluate dietary data along with individual health records. The effect of specific nutrition on the health of joints is covered by this approach.	Provide tailored dietary advice to promote joint health and lessen irritation, which may minimize the incidence of OA.

ML, *machine learning;* OA, *osteoarthritis.*

Implementation Challenges: It is important to examine data protection carefully as well as ethical use of AI to guarantee that these developments are safe for numerous individuals when adopting ML-driven preventive approaches. Furthermore, validation must be via practical application and clinical studies to verify safety and efficacy. Thus, ML presents a feasible path toward creating efficient approaches to OA prevention.

8.6 Early detection of osteoarthritis using artificial intelligence and machine learning

To identify OA at an initial stage, a multistep process is used involving the proper analysis of scans, clinical facts, and patient feedback in conjunction with AI and ML. A proactive approach seeks to slow the progression of OA. It enhances patient results and lessens the financial cost burden of treating patients at the advanced stages of the illness (Kwon et al., 2020). The method involves creating unique, innovative AI models made especially for analyzing and identifying OA. Fig. 8.2 presents a method for OA detection. Figs. 8.3–8.5 present the data collection and preprocessing, feature engineering and selection, and model development and training, respectively.

- **Evaluation and Iteration:** These are is done via measures including sensitivity (true positive rate), precision, specificity (true negative rate), and area under the curve for receiver operating characteristics analysis. They are used to assess the performance of the model. Iteration can be used for parameter modification to alter the methods for better performance results.
- **Real-world Testing Implementation:** Testing is done clinically to help medical professionals in the early identification of OA and arriving at decisions. Patient feedback is critical to improve the results and alter the model for clinical trials.
- **Patient privacy and Ethical Considerations:** Always use the AI model ethically. It must be impartial, comprehensible, and transparent. Focusing on the moral implications along with protection of patients are an important ethical concern. Thus ensuring data privacy of patients.

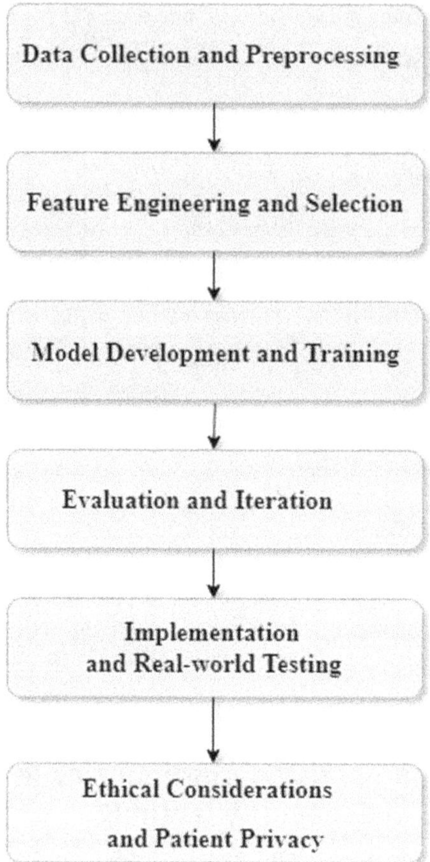

FIGURE 8.2

Method for early detection of osteoarthritis.

Using the method described in Fig. 8.2, it is feasible to build a strong AI and ML framework for the early identification of OA that is exclusive and caters to the particular demands of patients, including distinctive features for analyzing OA.

8.7 Case studies of osteoarthritis using artificial intelligence and machine learning

This section presents some hypothetical examples of technological advances used to enhance the diagnosis, treatment, and visual appearance of OA imagery. These case studies show how ML and AI have the potential to improve treatment of OA in the field of medicine.

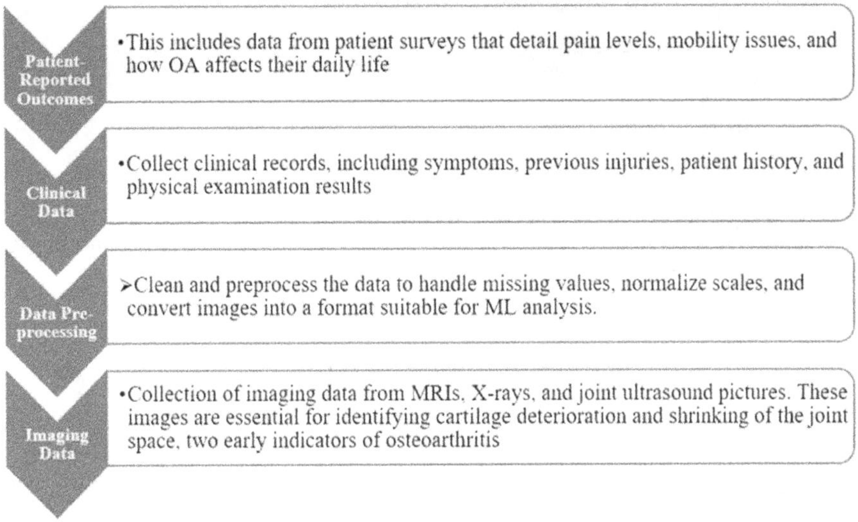

FIGURE 8.3

Data collection and preprocessing. *ML*, machine learning; *MRI*, magnetic resonance imaging; *OA*, osteoarthritis.

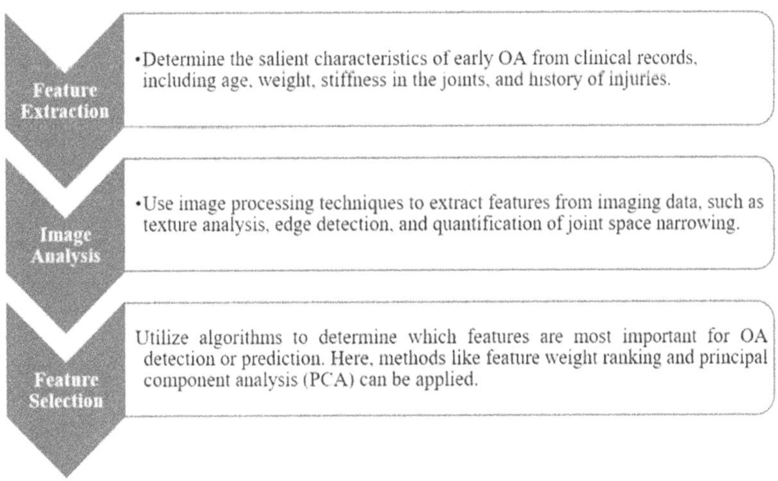

FIGURE 8.4

Feature engineering and selection. *OA*, osteoarthritis.

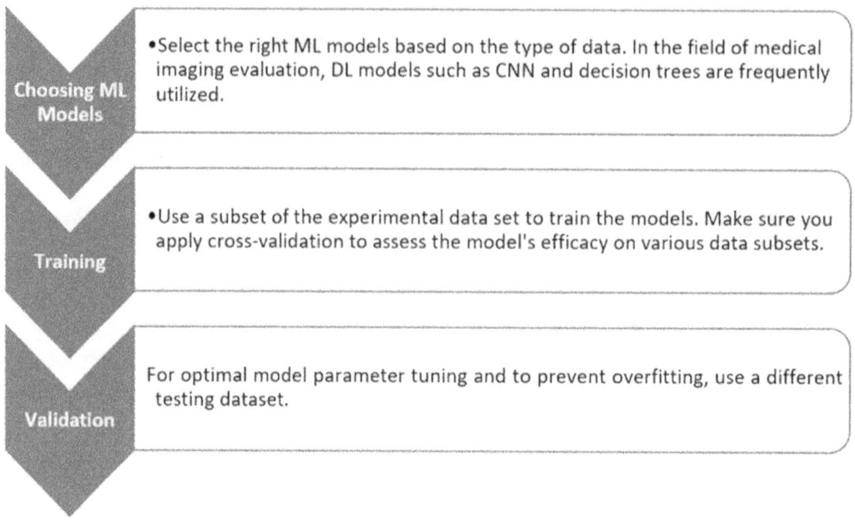

FIGURE 8.5

Model development and training. *CNN*, convolutional neural network; *DL*, deep learning; *ML*, machine learning.

8.7.1 Case study 1: Early detection of osteoarthritis through imaging analysis

An algorithm based on DL was created (Jakaite et al., 2021) to identify OA early in knee x-ray images. The aim was to detect OA in the early stages of notable symptoms for early treatments. Their AI model gives a higher accuracy rate than traditional diagnostic methods. OA patients identified in the early stage by the AI model showed slowed disease progression in follow-up studies. The approach consists of data collection, model development, and validation (Fig. 8.6).

8.7.2 Case study 2: Use of machine learning algorithms in creating personalized treatment plans

Clinical professionals at OA treatment clinics used integrated ML algorithms in their practice to create customized treatment therapies based on the patients' history (Kasula, 2021). The outcome resulted in improved pain management and flexibility of joints compared with standardized procedures. The approach is shown in Fig. 8.7.

8.7.3 Case study 3: With artificial intelligence predicting osteoarthritis progression

Using AI, a study was proposed (Moustakidis et al., 2023) that aimed to predict the progression of OA in patients, facilitating early and preventive treatment strategies. The model accurately predicted patients at risk of fast OA progression. Early

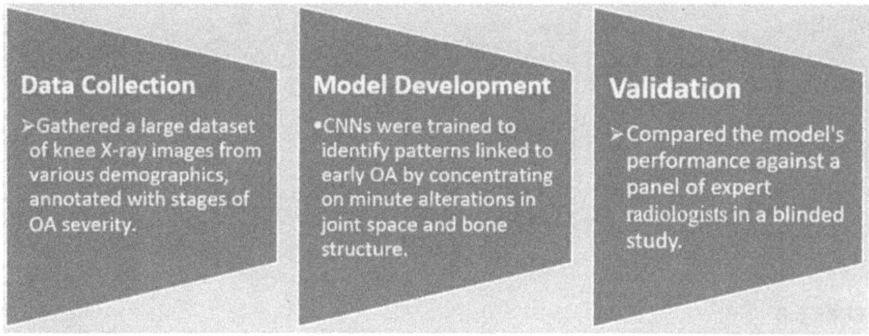

FIGURE 8.6

Image analysis approach steps. *CNN*, convolutional neural network; *OA*, osteoarthritis.

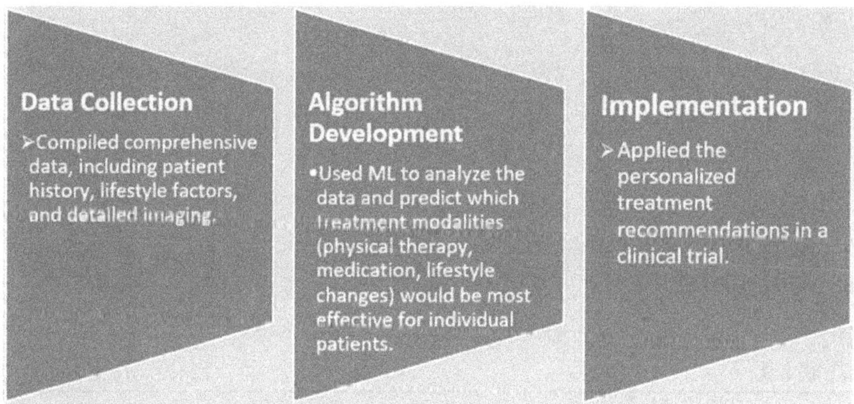

FIGURE 8.7

Machine learning (ML) algorithm approach.

intervention led to improved quality of life by significantly lowering the rate of OA-related surgeries. The approach used in AI is shown in Fig. 8.8.

8.7.4 Case study 4: Enhancing patient monitoring through machine learning and wearable devices

A tech startup (Tariq, 2024) collaborated with orthopedic specialists to invent a wearable device to monitor joint health in real time using ML to analyze patient data for early symptoms of OA deterioration. Patients who used wearable devices with ML analysis integration experienced fewer OA flare-ups. They reported higher satisfaction with the OA treatment, attributing it to the personalized and proactive approach (Fig. 8.9).

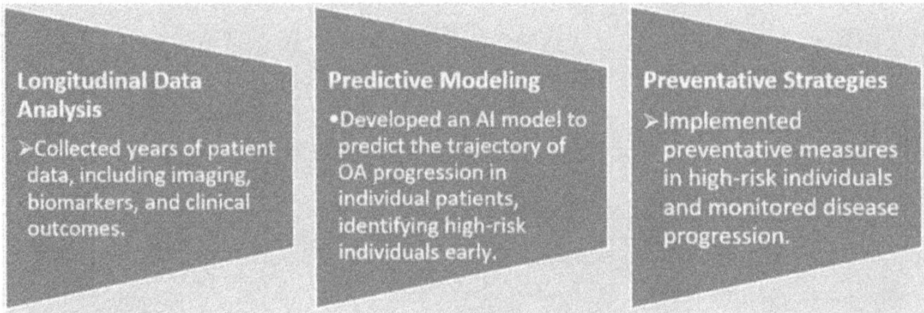

FIGURE 8.8

Artificial intelligence (AI) approach used. *OA*, osteoarthritis.

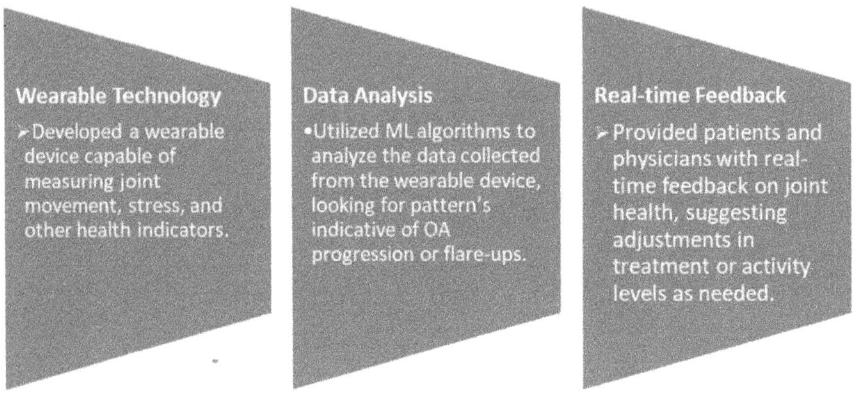

FIGURE 8.9

Proactive approach via machine learning (ML) in wearable technology. *OA*, osteoarthritis.

8.8 Challenges and future directions

The use of AI in OA identification leads to many benefits. It also has numerous challenges including issues regarding data privacy, the need for expansive and varied datasets to train AI models adequately, and the models' adaptability across different demographics. Furthermore, integrating AI technologies into clinical environments requires validation through clinical studies and regulatory clearances, essential measures to confirm their safety and efficacy in practical settings (Tariq, 2024).

In the future, further advances in AI for OA will concentrate on enhancing precision in AI predictions. This would help build AI models for early detection and personalize treatment options. Advances in using AI to manage and treat cartilage arthritis will be greatly aided by interdisciplinary cooperation among AI researchers, researchers in biology, physicians, and patients. This will guarantee that the solutions created are up to date and precisely aligned with the requirements of patients (McCabe et al., 2022; Verma et al.).

Finally, AI has the potential to revolutionize the understanding, diagnosis, and treatment of OA. Using AI's powers to analyze images, biomarker identification, and modeling predictions, medical professionals may provide patients who have this crippling illness with more accurate, customized, and effective care. AI's promise for OA care highlights a major development in achieving treatment results and standards of living (Behl et al., 2021; Kumar et al., 2024).

References

Abedin, J., Antony, J., McGuinness, K., et al., 2019. Predicting knee osteoarthritis severity: comparative modeling based on patient's data and plain X-ray images. Sci. Rep. 9 (1), 5761.

Behl, T., et al., 2021. Multifaceted role of matrix metalloproteinases in neurodegenerative diseases: pathophysiological and therapeutic perspectives. Int. J. Mol. Sci. 22 (3), 1413.

Binvignat, M., et al., 2022. Use of machine learning in osteoarthritis research: a systematic literature review. RMD Open 8 (1) e001998.

Bowes, M.A., Kacena, K., Alabas, O.A., et al., 2020. Machine learning, MRI bone shape and important clinical outcomes in osteoarthritis: data from the Osteoarthritis Initiative. Ann. Rheum. Dis. 80 (4), 502–504.

Caliva, F., Namiri, N.K., Dubreuil, M., Pedoia, V., Ozhinsky, E., Majumdar, S., 2021. Studying osteoarthritis with artificial intelligence applied to magnetic resonance imaging. Nat. Rev. Rheumatol. 18 (2), 112–121.

Chang, G.H., Park, L.K., Le, N.A., et al., 2021. Subchondral bone length in knee osteoarthritis: a deep learning-derived imaging measure and its association with radiographic and clinical outcomes. Arthritis Rheumatol. 73 (12), 2240–2248.

Chen, Di, et al., 2017. Osteoarthritis: toward a comprehensive understanding of pathological mechanism. Bone Res. 5, 16044. https://doi.org/10.1038/boneres.2016.44.

Chen, P., Gao, L., Shi, X., Allen, K., Yang, L., 2019. Fully automatic knee osteoarthritis severity grading using deep neural networks with a novel ordinal loss. Comput. Med. Imag. Graph. 75, 84–92.

Cheung, J.C.-W., Tam, A.Y.-C., Chan, L.-C., Chan, P.-K., Wen, C., 2021. Superiority of multiple-joint space width over minimum-joint space width approach in the machine learning for radiographic severity and knee osteoarthritis progression. Biology 10 (11).

Collins, J.E., Neogi, T., Losina, E., September 2021. Trajectories of structural disease progression in knee osteoarthritis. Arthritis Care Res. 73 (9), 1354–1362.

Flores, A.M., Demsas, F., Leeper, N.J., Ross, E.G., 2021. Leveraging machine learning and artificial intelligence to Improve Peripheral Artery disease detection, treatment, and outcomes. Circ. Res. 128 (12), 1833–1850.

Gornale, S.S., Patravali, P.U., Hiremath, P.S., 2020. Automatic detection and classi<cation of knee osteoarthritis using Hu's invariant moments. Front Robot AI 7, 591827.

Gornale, S.S., Patravali, P.U., Hiremath, P.S., 2021. Identification of region of interest for assessment of knee osteoarthritis in radiographic images. Int. J. Med. Eng. Inf. 13 (1), 64–74.

Guan, B.C., Liu, F., Haj Mizaian, A., et al., February 2022. Deep learning approach to predict pain progression in knee osteoarthritis. Skeletal Radiol. 51 (2), 363–373.

Hafezi-Nejad, N., Guermazi, A., Roemer, F.W., et al., 2017. Prediction of medial tibiofemoral compartment joint space loss progression using volumetric cartilage measurements: data from the FNIH OA biomarkers consortium. Eur. Radiol. 27 (2), 464–473.

Hirvasniemi, J., Klein, S., Bierma-Zeinstra, S., Vernooij, M.W., Schiphof, D., Oei, E.H.G., 2021. A machine learning approach to distinguish between knees without and with osteoarthritis using MRI-based radiomic features from tibial bone. Eur. Radiol. 31 (11), 8513–8521.

Huang, C., Xu, Z., Shen, Z., et al., 2022. DADP: dynamic abnormality detection and progression for longitudinal knee magnetic resonance images from the osteoarthritis initiative. Med. Image Anal. 77, 102343.

Jakaite, L., Schetinin, V., Hladůvka, J., Minaev, S., Ambia, A., Krzanowski, W., 2021. Deep learning for early detection of pathological changes in X-ray bone microstructures: case of osteoarthritis. Sci. Rep. 11 (1), 1–9.

Jamshidi, A., Pelletier, J.P., Labbe, A., Abram, F., Martel-Pelletier, J., Droit, A., October 2021. Machine learning-based individualized survival prediction model for total knee replacement in osteoarthritis: data from the osteoarthritis initiative. Arthritis Care Res. 73 (10), 1518–1527.

Kasula, B.Y., 2021. Machine learning in healthcare: revolutionizing disease diagnosis and treatment. IJRCTD 3 (3).

Kokkotis, C., Moustakidis, S., Papageorgiou, E., Giakas, G., Tsaopoulos, D.E., 2020. Machine learning in knee osteoarthritis: a review. Osteoarthr. Cartil. Open 2 (3), 100069.

Kumar, R., Gupta, M., Abraham, A., 2024. A critical analysis on vertebra identification and cobb angle estimation using deep learning for scoliosis detection. IEEE Access 12, 11170–11184. https://doi.org/10.1109/ACCESS.2024.3353794.

Kwon, S.B., Ku, Y., Han, H.-S., Lee, M.C., Kim, H.C., Ro, D.H., 2020. A machine learning-based diagnostic model associated with knee osteoarthritis severity. Sci. Rep. 10 (1), 1–8.

Lazzarini, N., Runhaar, J., Bay-Jensen, A.C., et al., 2017. A machine learning approach for the identification of new biomarkers for knee osteoarthritis development in overweight and obese women. Osteoarthritis Cartilage 25 (12), 2014–2021.

Liukkonen, M.K., Mononen, M.E., Klets, O., Arokoski, J.P., Saarakkala, S., Korhonen, R.K., 2017. Simulation of subject specific progression of knee osteoarthritis and comparison to experimental follow-up data: data from the osteoarthritis initiative. Sci. Rep. 7 (1), 9177.

Mahum, R., Rehman, S.U., Meraj, T., et al., 2021. A novel hybrid approach based on deep CNN features to detect knee osteoarthritis. Sensors 21 (18), 6189.

McCabe, P.G., Lisboa, P., Baltzopoulos, B., Olier, I., 2022. Externally validated models for first diagnosis and risk of progression of knee osteoarthritis. PLoS One 17 (7) e0270652.

Moustakidis, S., Kokkotis, C., Tsaopoulos, D., 2023. Patient-specific modeling of pain progression: a use case on knee osteoarthritis patients using machine learning algorithms. In: Digital Human Modeling and Medicine. Elsevier, pp. 805–828.

Norman, B., Pedoia, V., Majumdar, S., 2018. Use of 2D U-Net convolutional neural networks for automated cartilage and meniscus segmentation of knee MR imaging data to Determine Relaxometry and Morphometry. Radiology 288 (1), 177–185.

Norman, B., Pedoia, V., Noworolski, A., Link, T.M., Majumdar, S., June 2019. Applying Densely connected convolutional neural networks for staging osteoarthritis severity from plain radiographs. J. Digit. Imag. 32 (3), 471–477.

Pedoia, V., Norman, B., Mehany, S.N., Bucknor, M.D., Link, T.M., Majumdar, S., 2019. 3D convolutional neural networks for detection and severity staging of meniscus and PFJ cartilage morphological degenerative changes in osteoarthritis and anterior cruciate ligament subjects. J. Magn. Reson. Imag. 49 (2), 400–410.

Pierson, E., Cutler, D.M., Leskovec, J., Mullainathan, S., Obermeyer, Z., January 2021. An algorithmic approach to reducing unexplained pain disparities in underserved populations. Nat. Med. 27 (1), 136+.

Roemer, F.W., Crema, M.D., Trattnig, S., Guermazi, A., 2011. Advances in imaging of osteoarthritis and cartilage. Radiology 260 (2), 332–354.

Ryd, L., et al., 2015. Pre-osteoarthritis: definition and diagnosis of an Elusive clinical entity. Cartilage 6 (3), 156–165.

Tariq, M.U., 2024. Advanced wearable medical devices and their role in transformative remote health monitoring. In: Advances in Healthcare Information Systems and Administration. IGI Global, Hershey, PA, pp. 308–326.

Teo, K., Yong, C.W., Chuah, J.H., Murphy, B.P., Lai, K.W., 2021. Early detection of readmission risk for decision support based on clinical notes. J. Med. Imaging Health Inform. 11 (2), 529–534.

Tiulpin, A., Thevenot, J., Rahtu, E., Lehenkari, P., Saarakkala, S., 2018. Automatic knee osteoarthritis diagnosis from plain radiographs: a deep learning-based approach. Sci. Rep. 8 (1), 1727.

Tiulpin, A., Klein, S., Bierma-Zeinstra, S.M.A., et al., 2019. Multimodal machine learning-based knee osteoarthritis progression prediction from plain radiographs and clinical data. Sci. Rep. 9 (1), 20038.

Tolpadi, A.A., Lee, J.J., Pedoia, V., Majumdar, S., 2020. Deep learning predicts total knee replacement from magnetic resonance images. Sci. Rep. 10 (1), 6371.

Tomas, K.A., Kidzinski, L., Halilaj, E., et al., 2020. Automated classification of radiographic knee osteoarthritis severity using deep neural networks. Radiol. Artif. Intell. 2 (2), e190065.

Verma K., et al., Latest Tools for Data Mining and Machine Learning. Open Access.

Vos, T., et al., 2012. Years lived with disability (YLDs) for 1160 sequelae of 289 diseases and injuries 1990–2010: a systematic analysis for the Global Burden of Disease Study 2010. Lancet 380 (9859), 2163–2196.

Yeoh, P.S.Q., et al., 2021. Emergence of deep learning in knee osteoarthritis diagnosis. Comput. Intell. Neurosci. 2021, 4931437.

Yeoh, P.S.Q., Lai, K.W., Goh, S.L., Hasikin, K., Wu, X., Li, P., 2023. Transfer learning-assisted 3D deep learning models for knee osteoarthritis detection: data from the osteoarthritis initiative. Front. Bioeng. Biotechnol. 11.

Yong, C.W., Lai, K.W., Murphy, B.P., Hum, Y.C., 2021. Comparative study of encoder-decoder-based convolutional neural networks in cartilage Delineation from knee magnetic resonance images. Curr Med Imaging 17 (8), 981–987.

CHAPTER

Preanalysis of ankylosing spondylitis using machine learning

9

Chander Prabha[1], Md Mehedi Hassan[2]

[1]*Chitkara University Institute of Engineering and Technology, Chitkara University, Punjab, India;*
[2]*Computer Science and Engineering Discipline, Khulna University, Khulna, Bangladesh*

9.1 Introduction

One kind of arthritic condition that inflames the spine's ligaments and joints is called ankylosing spondylitis (AS) and is also known as axial spondyloarthritis. Peripheral joints such as the hips, ankles, and knees may also be affected. The spine's ligaments and joints normally permit us to bend and move. Stiffness may result from inflammation in the spine's joints and tissues if you have AS. In extreme situations, this could lead to the spine's vertebrae (bones) fusing. The result of a fused vertebrae might be an immobile and stiff spine. If the ribs are affected with AS, it may be difficult to breathe deeply. Fig. 9.1 shows a healthy spine and an AS-affected spine. Axial spondyloarthritis has two types. If AS is diagnosed via X-ray, it is called

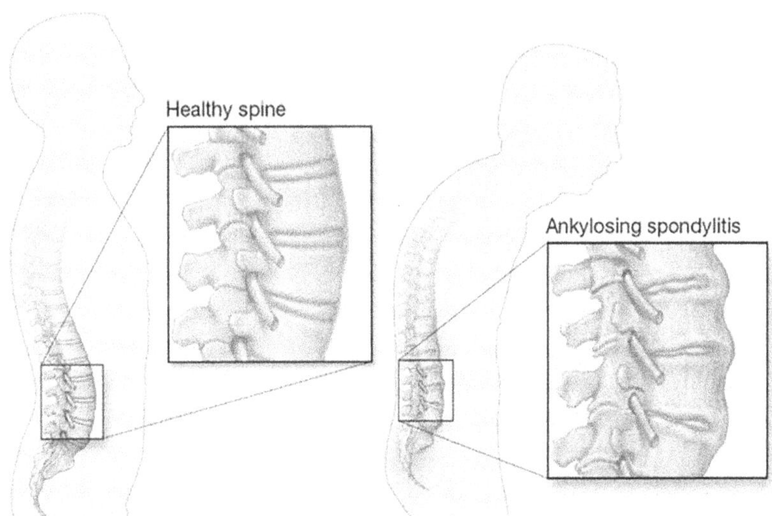

FIGURE 9.1

A healthy spine and ankylosing spondylitis–affected spine.

Diagnosing Musculoskeletal Conditions using Artificial Intelligence and Machine Learning to Aid Interpretation of Clinical Imaging
https://doi.org/10.1016/B978-0-443-32892-3.00009-9

AS. However, when it cannot be diagnosed via x-ray it is called nonradiographic axial spondyloarthritis. In this case, it is diagnosed through symptoms, blood tests, and other imaging tests (Inman, 2021).

A common symptom of AS is sporadic, temporary bouts of back discomfort and stiffness. Others, however, experience excruciating, continuous pain and a loss of spine flexibility. Furthermore, other symptoms could appear based on whatever other body parts are affected by the illness. AS patients may experience uveitis (eye disease), psoriasis (skin disease), or inflammatory bowel (gut disease). Frequent lower spine and buttock pain and stiffness, which develop gradually, are perhaps the most prevalent early signs of AS. Initially, pain may be experienced on one side exclusively or on alternating sides. Typically, the pain is not specific; instead, it is dull and widespread. It is generally harder in the morning and at night, although a hot bath or some gentle exercise will help with this discomfort and stiffness. In addition, a slight fever, decreased appetite, and feeling uncomfortable are possible in the early stages of AS. AS causes inflammation rather than physiologic back pain. Both sides commonly experience persistent (chronic) discomfort that lasts for at least 3 months.

Pain and stiffness may progress up the spine and into the neck over the course of months or years. It is also possible for pain and soreness to radiate to the heels, hips, thighs, shoulder blades, and ribs. The most affected areas in the body of patients with AS are shown in Fig. 9.2.

Although there is no known cure for AS, there are numerous ways to manage the symptoms. Physical activity, and mental and/or occupational treatment to enhance posture and movement, and medicine to assist reduce inflammation, control pain, and decrease the disease's progression are examples of suggested interventions.

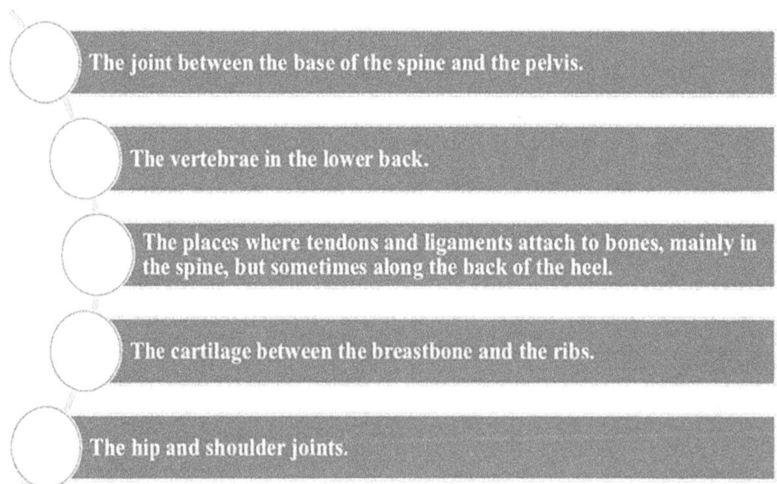

FIGURE 9.2

Affected areas in patient's body with ankylosing spondylitis.

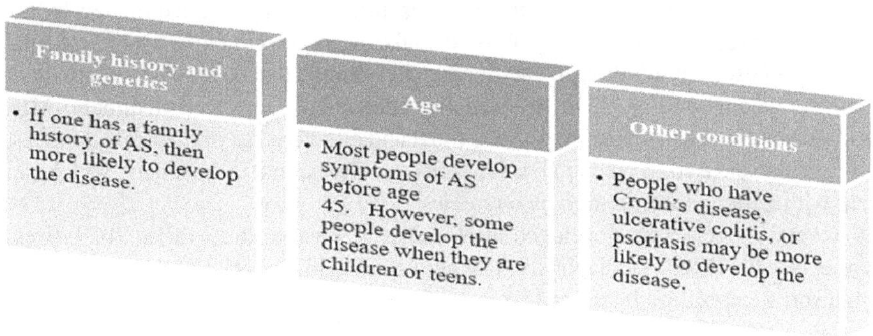

FIGURE 9.3

Factors responsible for ankylosing spondylitis (AS).

Most persons with AS can lead fulfilling lives if they receive treatment (Linden et al., 1984). Fig. 9.3 shows factors that increase the risk of developing AS.

The most typical sign of AS is discomfort and stiffness in the hips and/or lower spine. The symptoms could spread to other parts of the body or vertebrae over time. Individuals may find that pain is worse in the middle of the night or after extended sitting because it usually gets worse during times of rest or idleness. Movement and physical activity often reduce discomfort. AS indications differ among individuals. Although some people experience occasional minor pain, others will experience serious persistent pain. Whether severe or mild, AS symptoms may aggravate during flares and get better amid remissions (Sieper and Poddubnyy, 2017).

The rest of this chapter is organized as follows. Section 10.2 presents related work by various researchers. Section 10.3 discusses the concept of preanalysis using machine learning (ML) of AS. Section 10.4 presents challenges faced while performing a preanalysis. Strategies used to tackle those challenges are presented in Section 10.5. Finally, Section 10.6 concludes the chapter.

9.2 Related work

The identification and management of AS are difficult, particularly in underdeveloped nations where there is a shortage of personnel and diagnostic equipment. Innovative methods such as ML can help with early detection of AS and improve therapy tailored to patients' prognosis. The goal of ML, a subfield of artificial intelligence, is to replicate human-like learning with high accuracy to make decisions that do not require human intervention. These models are fed large volumes of data. Using statistical methods, systems are trained to generate predictions, classify data, or uncover previously unknown but significant details and correlations from datasets containing both labeled and unlabeled raw data. ML technology has a bright future in aiding health care because it makes it feasible to link routinely gathered patient information, allowing for ongoing monitoring of the patient's route via the health care system. This method demonstrates the usefulness of ML-trained models by using smartphone-captured photographs to make accurate conclusions in real-time smart diagnoses. In addition to facilitating the prompt detection of populations at risk and recommending them to specialist centers for improved treatment, ML-assisted solutions can accomplish the optimal and effective use of medical resources. ML can also be used to predict how AS patients will respond to treatment, which can help with patient-specific therapy strategies.

Several models were developed by Walsh et al. (Walsh et al., 2018, 2020; Bressem et al., 2021) for AS diagnosis. The algorithm achieved 91.1% accuracy in distinguishing sacroiliitis by using text documents from the electronic medical record (Walsh et al., 2019). In addition, different techniques applied to identical data demonstrated areas under the curve (AUCs) ranging from 0.86 to 0.96, confirming the presence of axial spondyloarthropathy (Lee et al., 2020). Deodhar et al. (2020) created a model using clinical and pharmaceutical claims data to detect AS, achieving a positive likelihood ratio of 6.24%.

Joo et al. (2020) used ML to forecast radiologic progression. They employed a training set consisting of 253 cases and a test set consisting of 173 cases. The various models achieved an equilibrium accuracy of over 65% in the test set, with the random forest model achieving the greatest accuracy of 69.3%. Furthermore, the generalized linear model and support vector machine demonstrated superior performance, achieving an AUC of 0.78. Table 9.1 presents numerous methods through which researchers worked on AS.

9.3 Preanalysis

AS is a chronic inflammatory disease primarily affecting the sacroiliac joints and spine. Early diagnosis is important to manage symptoms and prevent severe complications. ML offers a powerful approach to improve diagnosis and management by identifying patterns and predicting disease progression from various types of data. This preanalysis involves key steps and considerations in using ML for AS (Ahammad et al., 2024). Early diagnosis can be done from genetic data, imaging data (e.g., magnetic resonance imaging [MRI], x-rays), and clinical records. An ML model will be developed to assist in diagnosing AS, predicting the progression of the disease and identifying major risk factors associated with AS. The detailed step-by-step approach to conducting a preanalysis of AS follows.

Data Collection: The various data sources are clinical data (medical history, patient demographics, symptoms, treatment records, etc.), imaging data (computed tomography scans, MRI, x-rays of the spine and sacroiliac joints), laboratory data (C-reactive protein and erythrocyte sedimentation rate levels), and genetic data (relevant genetic markers and human leukocyte antigen-B27 status). Fig. 9.4 shows the preprocessing of various types of data.

Exploratory Data Analysis: This is done via statistical analysis, visualization, and checking class imbalances if any (Fig. 9.5).

Table 9.1 Methods that researchers used.

References	Results and other factors			
Castro-Zunti et al. (2020)	Method RF	Accuracy 96%	Recall 98%	AUC 0.97
	Method Modified Inception V3	Accuracy 99%	Recall 98%	AUC 0.97
	Inception V3 has 8.4% sensitivity and 9.5% specificity.			
	Modified Inception V3 classifies 1 sacroiliac joint in 58.5 ms			
	Dataset: 681 grayscale images of sacroiliac joints.			
Zhu et al. (2022)	Used least absolute shrinkage and selection operator (LASSO), RF, and support vector machine recursive feature elimination			
	Dataset: 348 patients with AS; training 258 patients with AS and 247 patients without AS; validation 90 patients with AS and 113 patients without AS			
	Seven factors considered: mean platelet volume, erythrocyte sedimentation rate, albumin, aspartate aminotransferase, red blood cell count, and creatinine			
	Results: during training *C* value: 0.878, AUC: 0.8779; during validation *C* value: 0.823 and AUC: 0.8232			
Lee et al. (2021)	Dataset: 611 AS patients			
	Methods: RF, LR			
	RF accuracy: 72.6%, LR accuracy: 68.9%			
	RF AUC: 0.638, LR AUC: 0.565			
	RF F1 score 0.841, LR F1 score 0.803			
	RF precision-recall 0.808, LR precision-recall 0.754			
Walsh et al. (2017)	Dataset: electronic medical record (1123 AS patients)			
	Methods: RF, LR, extreme gradient boosting			
	Best with RF with mean accuracy: 73.73% and AUC 0.79			

AS, *ankylosing spondylitis;* AUC, *area under the curve;* LR, *logistic regression;* RF, *random forest.*

Model Selection and Development: For preanalysis, various supervised and unsupervised ML models and deep learning models can be used with the data. Further various techniques for classification, clustering, and so on are used to analyze and predict the output. The numerous techniques involved are shown in Fig. 9.6.

Model Training and Validation: Data are split into training, test, and validation sets. Furthermore, cross-validation is used to fine-tune the hyperparameters. The output is predicted in terms of evaluation metrics: accuracy, recall, precision, F1-score, and receiver operating characteristic-AUC in case of classification, and mean absolute error, R^2 score, and mean squared error for regression. The model is interpreted using Local Interpretable Model-agnostic Explanations (LIME) or SHapley Additive exPlanations (SHAP).

Monitoring and Deployment: Integration of the model is done into a clinical decision support system. After integration, the model is continuously monitored

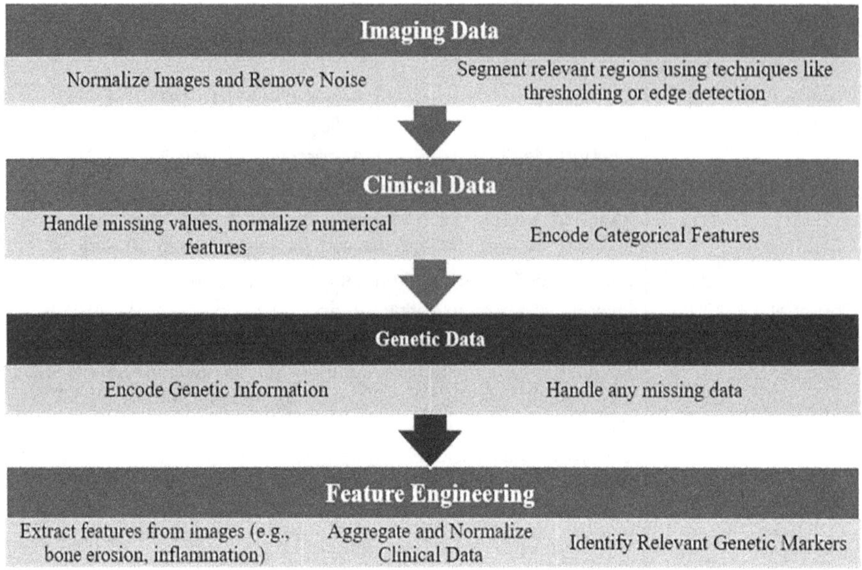

FIGURE 9.4

Data preprocessing.

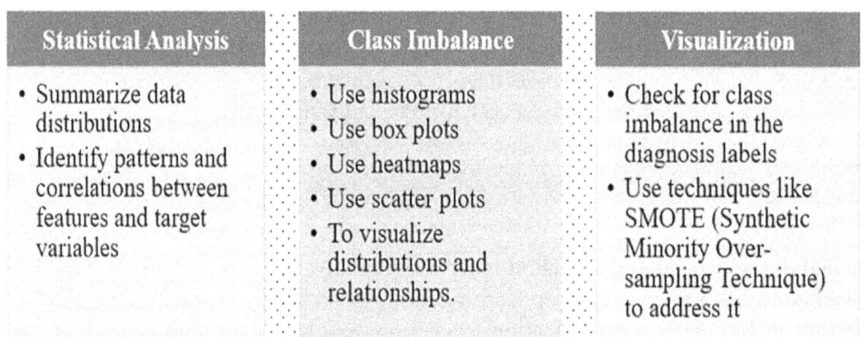

FIGURE 9.5

Exploratory data analysis.

for performance, retraining new data and ensuring compliance related to health regulations.

Early screening and treatment could be revolutionized by using ML for the preanalysis of AS. Using rigorous data gathering, powerful modeling, and cautious deployment, ML might offer notable perspectives and assistance to health care practitioners, enhancing patients' results. Table 9.2 presents some of articles that used ML for the preanalysis of AS.

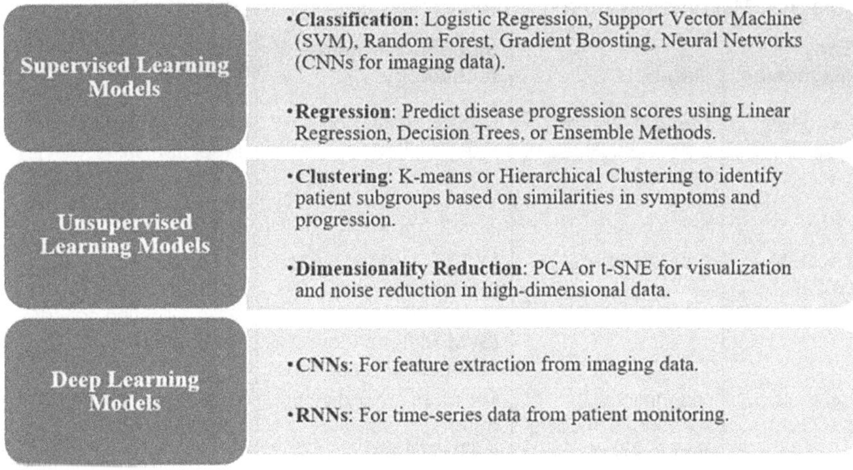

FIGURE 9.6

Techniques used in model selection and development. CNNs, Convolution Neural Network's; PCA, Principal Component Analysis; RNNs, Recurrent Neural Networks; t-SNE, t-Stochastic Neighbor Embedding.

9.4 Challenges faced

Preanalysis of AS using ML poses numerous challenges that need to be addressed to ensure the development of reliable and accurate models. This section presents some key challenges (Motta et al., 2020; Ranganathan et al., 2017).

9.4.1 Data availability and quality

Limited Data Availability: AS is a relatively rare condition, which makes it difficult to collect large, high-quality datasets.

Heterogeneous Data: Data might be inconsistent in format and quality owing to collection from multiple sources (e.g., different hospitals and labs).

Incomplete Records: The records of patients might be incomplete. Some may be missing critical information (viz. genetic data, detailed medical history, full imaging studies, etc.).

Issues in Imaging Data: Imaging data could be noisy and of varying quality. This is because data are gathered from multiple machines and because of their settings used during acquisition.

Class Imbalance: In many datasets, groups of AS-affected patients might be much smaller compared with healthier ones, leading to imbalanced classes that can bias the model.

Table 9.2 Machine learning used in articles for preanalysis.

Early diagnosing	Dataset	Machine learning technology	Results
Deodhar et al. (2020)	Medical records from medical claim database	Logistic regression model optimized	Model A/B positive predictive value (PPV): 6.24% simplified logistic regression model PPV: 2.55%
Zhu et al. (2022)	Lab test indicators	LASSO, support vector machine recursive feature elimination (SVM-RFE), random forest, and nomograms	Training cohort area under the curve (AUC): 0.87 validation cohort AUC: 0.82
Ye et al. (2022)	Magnetic resonance imaging and medical records	Minimum redundancy - maximum relevance (mRMR), LASSO, multivariable logistic regression analysis, and nomograms	Rad-score AUC: 0.82; Clinical radiomics nomogram model AUC: 0.9
Wen et al. (2022)	GEO	SVM-RFE, LASSO, RF, and nomograms	AUC >0.84
Han et al. (2022)	GEO	Weighted gene co-expression network analysis (WGCNA) and SVM-RFE	AUC >0.7
Zhao et al. (2020)	Electronic health records	Logistic regression, multimodal automated phenotyping, and LASSO	AUC: 0.93

9.4.2 Feature engineering and selection

The process of determining and obtaining pertinent information from clinical, genomic, and imaging data can be intricate and necessitate specialized knowledge of the field. Model training and computing requirements might be complicated by high-dimensional data, particularly from genetic and imaging sources (Rahbar et al., 2018).

9.4.3 Interpretability of model

Advanced models such as deep neural networks can provide high levels of accuracy, but they often pose challenges in terms of interpretability. This is important in medical use and ensures compliance with regulations. It is also crucial to gain the trust of physicians.

9.4.4 Ethics and bias

Privacy Concerns: Patient data are sensitive and their handling requires strict adherence to privacy laws and regulations (e.g., Health Insurance Portability and Accountability Act of 1996 in the United States) to ensure data are anonymized and secure.

Data Bias: Biases in the training data (e.g., underrepresentation of certain demographics) can lead to biased predictions, which can exacerbate health disparities.

9.4.5 Regulatory compliance

Validation: Extensive validation and verification are required to ensure the model's reliability and safety before deployment in a clinical environment.

Health care Regulations: Compliance with medical regulations and obtaining necessary approvals for the use of ML models in clinical settings can be complex and time-consuming (Ramiro et al., 2023).

9.4.6 Interdisciplinary collaboration

Collaboration: Effective collaboration among data scientists, clinicians, radiologists, and geneticists is essential for successful model development and validation, but coordinating across disciplines can be challenging (Benavent et al., 2023).

9.4.7 Computational resources

Training complex models, especially those involving large imaging datasets or genetic data, requires significant computational power and resources, which can be a limiting factor (Kumar et al., 2024).

9.4.8 Continuous learning and adaptation

Model Updates: Medical knowledge and practices evolve, necessitating continuous updates to the model to maintain its relevance and accuracy. This requires ongoing data collection and model retraining (Benavent and Navarro-Compán, 2024; Sharma et al., 2023).

9.5 Strategies to address challenges

This section deals with various strategies to address the challenges mentioned (Fig. 9.7).

One can employ techniques such as data augmentation to enlarge the dataset and synthetic data production artificially to tackle the issue of limited data availability (Prabha et al., 2022; Sharma and Prabha, 2021). To address missing data, standardize diverse data sources, develop and apply reliable preprocessing pipelines, and

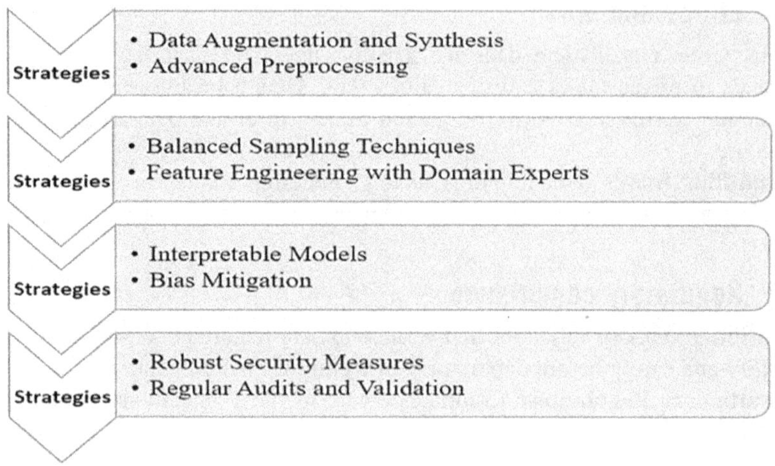

FIGURE 9.7

Strategies for addressing challenges.

enhance the overall accuracy of imaging data. Use methods such as the Synthetic Minority Oversampling Technique to achieve class balance in the dataset. Engage in collaboration with medical professionals to discern and construct significant characteristics from unprocessed data (Benavent and Navarro-Compán, 2024). Give priority to using comprehensible models and techniques such as SHAP and LIME to provide explanations for model forecasts. Employ bias identification and mitigation techniques to guarantee equitable and impartial predictions from the model. Implement stringent security measures for the preservation and use of data to adhere to privacy rules. Perform routine audits and validating operations to ensure continuous adherence to legal requirements and uphold model precision (Braun, 2018).

Tackling these issues in an organized way can greatly enhance the efficiency of ML models in the preliminary analysis of AS, improving early detection, medical treatment, and clinical outcomes (Verma et al., 2019).

9.6 Conclusion

For the efficient management of AS, ML can help with early identification and patient-tailored treatment. ML preanalysis for AS necessitates a systematic method for gathering data, preparing it, performing feature engineering, developing models, and conducting validation. Through the use of many kinds of data and sophisticated analytical methods, ML has the potential to enhance the early detection of diseases, forecast how they will progress, and increase the care of patients. Benefits are particularly greater in areas lacking in human specialists and diagnostic resources. Usually overburdened health care systems can receive essential support in a timely and cost-effective manner through the application of ML-trained models for AS diagnostic and therapy.

References

Ahammad, S.H., et al., 2024. Advanced model based machine learning technique for early stage prediction of ankylosing spondylitis under timely analysis with featured textures. Multimed. Tools Appl. 1–21.

Benavent, D., Capelusnik, D., Ramiro, S., Navarro-Compán, V., 2023. Looking through the window of opportunity in axial spondyloarthritis? Joint Bone Spine 90 (6), 105603.

Benavent, D., Navarro-Compán, V., 2024. Exploring the latest advances in axial spondyloarthritis management. Nat. Rev. Rheumatol. 20 (2), 79–80.

Braun, J., 2018. Axial spondyloarthritis including ankylosing spondylitis. Rheumatology 57 (Suppl. 1_6), vi1–vi3.

Bressem, K.K., et al., 2021. Deep learning for detection of radiographic sacroiliitis: achieving expert-level performance. Arthritis Res. Ther. 23 (1).

Castro-Zunti, R., Park, E.H., Choi, Y., Jin, G.Y., Ko, S.-B., 2020. Early detection of ankylosing spondylitis using texture features and statistical machine learning, and deep learning, with some patient age analysis. Comput. Med. Imag. Graph. 82, 101718.

Deodhar, A., et al., 2020. Use of machine learning techniques in the development and refinement of a predictive model for early diagnosis of ankylosing spondylitis. Clin. Rheumatol. 39 (4), 975–982.

Han, Y., et al., 2022. Identification of diagnostic mRNA biomarkers in whole blood for ankylosing spondylitis using WGCNA and machine learning feature selection. Front. Immunol. 13, 956027.

Inman, R.D., 2021. Axial spondyloarthritis: current advances, future challenges. J. Rheum. Dis. 28 (2), 55–59.

Joo, Y.B., Baek, I.-W., Park, Y.-J., Park, K.-S., Kim, K.-J., 2020. Machine learning–based prediction of radiographic progression in patients with axial spondyloarthritis. Clin. Rheumatol. 39 (4), 983–991.

Kumar, R., Gupta, M., Abraham, A., 2024. A critical analysis on vertebra identification and cobb angle estimation using deep learning for scoliosis detection. IEEE Access 12, 11170–11184. https://doi.org/10.1109/ACCESS.2024.3353794.

Lee, T.-H., et al., 2020. Conventional disease-modifying antirheumatic drugs therapy may not slow spinal radiographic progression in ankylosing spondylitis: results from an 18-year longitudinal dataset. Ther. Adv. Musculoskelet. Dis. 12.

Lee, S., et al., 2021. Machine learning-based prediction model for responses of bDMARDs in patients with rheumatoid arthritis and ankylosing spondylitis. Arthritis Res. Ther. 23 (1).

Linden, S.V.D., Valkenburg, H.A., Cats, A., 1984. Evaluation of diagnostic criteria for ankylosing spondylitis. Arthritis Rheum. 27 (4), 361–368.

Motta, F., Carena, M.C., Selmi, C., Vecellio, M., 2020. MicroRNAs in ankylosing spondylitis: function, potential and challenges. J. Transl. Autoimmun. 3, 100050.

Prabha, C., Singh, J., Agarwal, S., Verma, A., Sharma, N., 2022. Introduction to computational intelligence in healthcare. In: Computational Intelligence in Healthcare. CRC Press, Boca Raton, pp. 1–15.

Rahbar, M.H., et al., 2018. Harmonization, data management, and statistical issues related to prospective multicenter studies in Ankylosing spondylitis (AS): experience from the Prospective Study of Ankylosing Spondylitis (PSOAS) cohort. Contemp. Clin. Trials Commun 11, 127–135.

Ramiro, S., et al., 2023. ASAS-EULAR recommendations for the management of axial spondyloarthritis: 2022 update. Ann. Rheum. Dis. 82 (1), 19–34.

Ranganathan, V., Gracey, E., Brown, M.A., Inman, R.D., Haroon, N., 2017. Pathogenesis of ankylosing spondylitis — recent advances and future directions. Nat. Rev. Rheumatol. 13 (6), 359–367.

Sharma, N., Prabha, C., 2021. Computing paradigms: an overview. In: 2021 Asian Conference on Innovation in Technology (ASIANCON), pp. 1–6.

Sharma, S., Malik, M., Prabha, C., Al-Rasheed, A., Alduailij, M., Almakdi, S., 2023. Robust image watermarking using LWT and stochastic gradient firefly algorithm. Comput. Mater. Contin. 75 (1), 393–407.

Sieper, J., Poddubnyy, D., 2017. Axial spondyloarthritis. Lancet 390 (10089), 73–84.

Verma, K., et al., 2019. Latest Tools for Data Mining and Machine Learning. *Open Access.*

Walsh, J.A., et al., 2017. Identifying axial spondyloarthritis in electronic medical records of US veterans. Arthritis Care Res. 69 (9), 1414–1420.

Walsh, J.A., et al., 2018. Cohort identification of axial spondyloarthritis in a large healthcare dataset: current and future methods. BMC Musculoskelet. Disord. 19 (1).

Walsh, J.A., Rozycki, M., Yi, E., Park, Y., 2019. Application of machine learning in the diagnosis of axial spondyloarthritis. Curr. Opin. Rheumatol. 31 (4), 362–367.

Walsh, J.A., et al., 2020. Identification of axial spondyloarthritis patients in a large dataset: the development and validation of novel methods. J. Rheumatol. 47 (1), 42–49.

Wen, J., Wan, L., Dong, X., 2022. Novel peripheral blood diagnostic biomarkers screened by machine learning algorithms in ankylosing spondylitis. Front. Genet. 13, 1032010.

Ye, L., et al., 2022. A predictive clinical-radiomics nomogram for diagnosing of axial spondyloarthritis using MRI and clinical risk factors. Rheumatology 61 (4), 1440–1447.

Zhao, S.S., et al., 2020. Incorporating natural language processing to improve classification of axial spondyloarthritis using electronic health records. Rheumatology 59 (5), 1059–1065.

Zhu, J., et al., 2022. Development and validation of a machine learning-based nomogram for prediction of ankylosing spondylitis. Rheumatol. Ther 9 (5), 1377–1397.

CHAPTER

10 Analysis and identification of gout flares using machine learning

Akey Sungheetha[1], Rajesh Sharma R[1], John Blake[2], Chinnaiyan R[3], Ketema Adere[4] and Bahiru Shifaw[4]

[1]*Computer Science and Engineering, Alliance University, Bengaluru, Karnataka, India;* [2]*University of Aizu, Aizuwakamatsu, Japan;* [3]*Lingaya's Vidyapeeth University, Faridabad, Haryana, India;* [4]*Computer Science and Engineering, Adama Science and Technology University, Adama, Ethiopia*

10.1 Introduction

Gout is a common and crippling ailment that increases the risk of cardiovascular disease (De Vera et al., 2010; Seminog and Goldacre, 2013). The roles of patients with gout in an Egyptian college are categorized and measured to investigate heart disease by examining both clinical and demographic data. The use of natural language processing (NLP) and machine learning (ML) has significantly improved the effectiveness and accuracy of disease management, garnering increased attention in the health care industry. ML has been used to identify gout flares (GFs) using electronic clinical notes (Zheng et al., 2014) and to classify the leukemia and uric acid profiles of patients with gout similar to the grey wolf optimization model (Sharma et al., 2019). As a result, health care professionals are better equipped to treat and prevent flares with these advanced technologies. This chapter explores the assessment of different metabolites using ML, linked to both frequent and infrequent gout flares. The cases discussed involve inflammatory arthritis connected to metabolic syndrome, diabetes, obesity, hypertension, and heart disease. By developing a statistical model using ML techniques, researchers can gain insights into the fundamental metabolic processes that cause gout flares, potentially leading to new diagnostic and treatment strategies for gout. The terminologies used are similar to those for characterizing asthma and chronic obstructive pulmonary disease, which, like gout, vary based on individual characteristics (Parry et al., 2018).

10.1.1 How natural language processing and machine learning are used to forecast gout flares

This section provides an overview of how ML and NLP enable effective information analysis through sophisticated algorithms and models, processing vast volumes of data to generate classifications and predictions. These technologies also support

Diagnosing Musculoskeletal Conditions using Artificial Intelligence and Machine Learning to Aid Interpretation of Clinical Imaging
https://doi.org/10.1016/B978-0-443-32892-3.00010-5

the identification of potential health variables, proactively recognizing diseases in a manner akin to archived multiobjective simulated annealing, which optimizes and classifies data to ensure optimal results (Wang, 2022; Sungheetha et al., 2017). Intelligent techniques are also required for effective classification (Rajesh Sharma and Sungheetha, 2017). Data collected through Internet of Things devices create reliable predictions of results (Chauhan et al., 2023). This chapter explores the increased risk of cardiovascular disease, challenges in screening patients with gout, and advances in ML-driven diagnostic models that enhance screening and diagnosis. Retrospective screening was conducted on clinical data from adult patients with gout admitted to Jiangxi Provincial People's Hospital between January 1, 2016 and November 1, 2020. Patients with the diagnosis of gout or gout concurrent with heart disease were included, using the 2015 Gout Diagnostic Criteria (Neogi et al., 2015) and considering conditions such as coronary revascularization, unstable angina, and myocardial infarction. The chapter concludes with significant findings and their implications recommendations for future research on GF management and prediction.

10.1.2 Problems of detecting gout owing to lack of diagnosis codes

There are considerable difficulties in correctly diagnosing gout flares owing to the lack of particular diagnostic codes, which also negatively affects treatment. We can improve the identification and management of gout flares and thus improve the standard of care for patients with this crippling type of arthritis by tackling this issue through campaigning, studies, and cooperation within the medical community. These two approaches are used in the combined sampling approach to solve the imbalance issue. Following a combined sampling procedure, it was observed that the updated training set contained 123 patients with gout and 121 with gout and Coronary Heart Disease (CHD), compared with 41 individuals with gout and 203 with gout and CHD in the initial training set. Health care providers can prevent additional damage to joints and improve long-term results for patients by promptly detecting gout flares based on primary observations of symptoms by nurses in the hospital emergency room. The application of ML and NLP techniques to identify and predict gout flares using electronic medical records (EMRs) is examined in this chapter. Nevertheless, there are few studies on using ML models to predict CHD in patients with gout. Using statistical analysis or ML algorithms, the goal of this study was to develop a reliable diagnostic model for screening CHD in patients with gout, based on basic clinical factors. Historically, conducting reliable database studies on gout has been difficult owing to the absence of specific diagnostic codes for flares (Bantelay et al., 2023).

10.1.3 The current medical code system and its limitations for gout flares

The existing medical coding system lacks sufficient diagnostic codes for gout flares, which complicates the process of diagnosing this condition. Gout, a type of

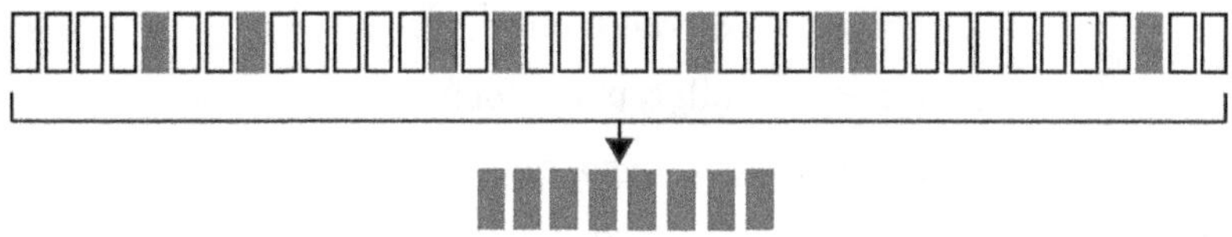

FIGURE 10.1

Clinical factors distilled into final feature inputs.

inflammatory arthritis, manifests with acute, severe discomfort along with redness and swelling in the affected area, often caused by the deposition of uric acid crystals in joints and surrounding tissues.

From the raw data, 38 clinical factors were gathered; however, after the variable selection process, as shown in Fig. 10.1 only eight clinical factors were ultimately used as feature inputs over eight ML models: support vector machine (SVM), context gradient booster equipment (Light), gradient boosting decision tree, artificial neural network, logistic regression, supervised decision tree, random forest, and extreme gradient boosting (XGBoost).

Gout flares are short, sharp episodes that have the potential to impair a person's quality of life seriously. However, researchers and medical professionals can more accurately and precisely identify GFs automatically by applying computer-based implementation techniques such as ML and NLP algorithms to digital clinical notes. Diagnosis coding that is imprecise can cause several problems. First, because medical professionals might not be able to identify and distinguish gout flares positively from other forms of joint pain, this could lead to underdiagnosis or misunderstanding of the condition. According to our findings, random forest with XGBoost models performed better in terms of memory and accuracy, whereas stepwise logistical regression and SVM produced higher-quality area under the receiver operating characteristics curve (AUC) values. This type of misdiagnosis could therefore cause people with gout flares to wait longer for the proper course of action and management techniques, exacerbating the symptoms.

Moreover, high-risk variables that help predict CHD in patients with gout were identified by interpretable ML analysis. The development of predictive models for hospital risk factors linked to gout flares in comorbid patients is also examined in this chapter. The lack of accurate and uniform coding for GFs hinders the collection of reliable epidemiological data, monitoring of disease trends, and conduct of insightful research studies. The experimental findings mentioned earlier demonstrated that the ML-based predictive models demonstrated exceptional performance. Identifying high-risk factors linked to CHD in patients with gout is also a crucial question in clinical practice. A multifaceted approach is needed to address the difficulties in diagnosing gout flares caused by insufficient diagnostic codes. A snippet from the International Classification of Diseases, Tenth Revision, Clinical Modification (ICD-10-CM) diagnosis code M10 for gout is shown in Fig. 10.2.

- M1A.49X0 ...... without tophus (tophi)
- M1A.49X1 ...... with tophus (tophi)
- M1A.9 Chronic gout, unspecified
  - M1A.9XX0 ...... without tophus (tophi)
  - M1A.9XX1 ...... with tophus (tophi)
- **M10 Gout**
  - M10.0 Idiopathic gout
    - M10.00 ...... unspecified site
    - M10.01 Idiopathic gout, shoulder
      - M10.011 Idiopathic gout, right shoulder
      - M10.012 Idiopathic gout, left shoulder
      - M10.019 Idiopathic gout, unspecified shoulder

FIGURE 10.2

Extract from ICD-10-CM diagnostic code M10 for gout.

The ICD-10-CM diagnostic code is hierarchical code with multiple subdivisions for each disease. The full code is available at https://www.icd10data.com/ICD10CM/Codes/M00-M99/M05-M14/M10-/M10.

To capture the distinct clinical features of these episodes adequately, health care organizations and legislators may need to push for the creation of special diagnostic codes for gout flares within the ICD system (Hao et al., 2014; Babichev et al., 2003; Beneke et al., 1997; Stahl, 2020; Abbott et al., 2019). Researchers can create predictive models to assist health care providers in better managing and treating patients with gout in hospital settings by identifying critical factors that lead to flares of the disease while the patient is in the hospital, similar to Convolutional Neural Network (CNN) model developments (Bantelay et al., 2023). Accurately diagnosing and treating gout flares can be challenging, though, because of the absence of specific diagnostic codes. As a result, the treatment of patients and their outcomes would be enhanced by the ability to identify, record, and track GFs more precisely in medical settings.

10.1.4 Significance of early identification for focused interventions and sustained joint health

The importance of early detection cannot be overstated, especially in maintaining joint function and enabling targeted interventions over time. Without proper management, gout can progress to chronic tophaceous gout, necessitating early diagnosis and treatment. Early identification of problems allows for prompt and appropriate interventions, significantly influencing the prognosis and overall outcomes for individuals with joint-related conditions.

Over 200,000 visits to the hospital emergency department (ED) are attributed to gout annually, accounting for 0.2% of all visits. Enhancing patients with continuity of care for gout is crucial, particularly after an ED visit. It is often the case that GFs handled in the ED do not receive the best follow-up care, so strategies for early identification and referral of GF patients throughout an ED visit are needed (J.D. Osborne et al., 2020a). Experimental results may use statistical techniques, such as Analysis of Variance (ANOVA) to improve performance (Sungheetha et al., 2019). Although patients are encouraged to maintain activity, the underestimation of treatment time for those who remain seated for extended periods is notable. Individualized care might benefit more from treatment using artificial intelligence models (Schaller et al., 2016; Sungheetha et al., 2023). It is challenging to assess one's activity level accurately because people often overestimate their activity when asked to quantify it, compared with objective measurements (Schaller et al., 2016).

Early detection can also assist people in avoiding the need for future, more expensive and invasive interventions. Patients' general quality of life may be enhanced, and the cost of health care associated with more involved interventions may be decreased. Addressing joint-related issues early can prevent the necessity for surgery or other invasive procedures, which entail higher risks and longer recovery times. Recommendations for altering critical health behaviors such as physical activity often include self-monitoring, a widely recognized and effective method for behavior modification (Mercer et al., 2016; Eisele et al., 2019).

Moreover, long-term maintenance of optimal joint health and mobility can be facilitated by early detection. The use of activity trackers can help sustain long-term joint function from early detection of the disease (Spencer et al., 2006). Health care providers can develop individualized treatment plans that target the root causes of joint-related problems by early joint identification. Long-term patient outcomes can be improved by doing this, because it can help people maintain autonomy and their standard of living as they age. Ultimately, this preventive measure can enhance patients' general level of life by halting additional joint damage and averting the need to undergo additional invasive procedures. During flare episodes, patients experience intensified pain and symptoms along with increased inflammation, which can be assessed through physical examinations, such as by observing an increase in joint swelling. Accurate assessment of these flares in clinical practice is crucial to understand the state of the disease and the effectiveness of treatment, ensuring the best possible outcomes for joint health (Hewlett et al., 2012; Gossec et al., 2016). Therefore, to guarantee the best possible results for joint health, people ought to seek immediate medical care when they experience joint-related symptoms.

10.2 Automatic recognition of gout attacks

The introduction of voice-based responsive automated devices simplified the process of obtaining patient reports remotely (Bennett et al., 2016). Patients with gout can respond verbally or by interacting with the device's touchscreen as part of treatment (Lee et al., 2003; Daftary et al., 2017). Researchers are looking into automated techniques to identify gout flares to enhance diagnostic accuracy. Autonomous identification systems employ sophisticated algorithms and ML methods to scrutinize patient information and identify patterns or tendencies suggestive of gout episodes. Despite good agreement in field tests between physician chart abstraction and automated electronic health record (EHR) abstraction, site investigators have faced challenges in pinpointing past gout flares. By integrating data from EMRs, patient-reported symptoms, and other relevant sources, these systems enable health care providers to take prompt and accurate action during suspected gout flares. When cases were found, there was no false-positive rate.

Using wearable technology to track behavioral and physiological indicators linked to gout attacks is a single instance of an automatic diagnosis system for GFs. Real-time information regarding changes that might indicate a gout flare can be obtained from these devices, which can measure parameters such as skin conductivity, joint temperature, and physical activity levels. Five percent of patients from the three clinical sites ($n = 96$ cases) and 5% from the final quarter of 2015 in Rheumatology Informatics System for Effectiveness (RISE) ($n = 515$ cases) met the measure quality specification, which calls for using Urate-Lowering Therapy (ULT) in cases of tophaceous or active gout. Using ML techniques to analyze these data, medical professionals are able to foresee and prevent gout flares, facilitating prompt treatment and intervention. Testing data indicate that this measure does not mistake numerator cases, even though it might miss certain denominator cases (in which a patient is supposed to be receiving ULT). As a result, although it might not catch all possible denominator cases, it still points to a significant performance gap (FitzGerald et al., 2018).

An additional method for automatically identifying GFs is to analyze text data from reports from patients and EMRs using NLP algorithms. After ULT is started or changed, Serum Uric Acid (SUA) surveillance is conducted, and subject to Quality Management System (QMS) approval, incorporated in RISE. Serum urate levels should be checked 6 months after starting or adjusting ULT in a patient with gout since they are essential for the best possible management of ULT. Patients who have recently started ULT or had a change in ULT are the focus of M4. The SUA needs to be checked 6 months after starting or changing a dose to comply with treat-to-target guidelines. Clinical sites expressed concern that it was challenging to abstract the precise date associated with a change in ULT dose, so further study is needed. Patterns and trends indicating an impending gout flares can be found through NLP algorithms by extracting and evaluating key phrases, symptoms, or contextual information associated with gout flares. Afterward, these data can be used to notify medical professionals and initiate the necessary actions to stop or lessen the effects of the flare. This assessment expands upon an earlier assessment idea put forth by different clinical specialists (Mikuls et al., 2004). The results were promising. M4 received a high global rating of 8.6 (on a scale of 1−9) and the QMS gave approval for field testing, RISE implementation, and National Quality Forum (NQF) submission.

The potential transformation in the way in which gout is identified and managed by automated detection systems for flares cannot be overstated. These systems can improve patient outcomes, reduce the impact of gout flares, and elevate the standard of care provided to patients with gout by offering real-time insights into patient data and enabling proactive interventions. Moreover, by preventing unnecessary emergency room visits, hospital stays, and long-term complications associated with gout, these systems can reduce the overall cost of gout-related health care. To maximize these technologies and make them broadly accessible to physicians and patients, more study and advancement in this field are required.

10.2.1 Using machine learning and natural language processing to identify gout flares in automated clinical notes

Gout, a common inflammatory arthritis condition, is characterized by abrupt and intense flares of joint pain, swelling, and redness. It is brought on by uric acid crystals deposited in the bones, which causes discomfort and inflammation. The use of clinical decision-support instruments in EMR systems has the potential to enhance the identification, management, and outcomes of gout and other conditions in high-volume clinical settings, including EDs (Alsolamy et al., 2014; Nguyen et al., 2014). For gout to be effectively managed and treated, flares must be promptly and accurately diagnosed. Gout flares can, however, be difficult to diagnose because they can resemble other forms of arthritic or joint disorders.

Assessments of algorithms using NLP techniques focused on identifying acute GFs based on the chief symptoms noted in nursing ED records when patients present with joint problems. Health care professionals can record patient encounters and gather crucial clinical data with the help of automated clinical notes. Health care professionals have traditionally written clinical notes by hand, which can be labor-intensive and error-prone. Findings showed that when it came to the real-time recognition of GFs within the ED, the term gout in the main symptom in itself demonstrated a strong positive value for prediction (Osborne et al., 2020a,b). On the other hand, the creation and analysis of clinical notes can be automated owing to technological advances, especially in the areas of ML with NLP. Our aim was to create a robust automatic GF alert system based on EMRs that could be implemented in Epic and Cerner. We wanted to make sure the system could withstand expert scrutiny of medical records, which is the reference standard in this field. With the help of enormous datasets of clinical notes, algorithms that use ML can be trained to identify patterns and connections among the notes' various features. Based on input data, including the patient's history, symptoms, or physical examination results, these algorithms can be used to generate their clinical notes.

By merging information from the list of medical problems of a patient, our GF detection and alert system improved upon our prior rules-based NLP GF detection method to detect GFs in the ED. By removing and evaluating relevant details from unstructured text data, NLP techniques can improve the processing of medical notes. After our GF alert is successfully developed and validated, we plan to carry out a prospective study that focuses on gout care in the ED. The ED treats acute gout in a large number of patients, especially those from underserved racial and ethnic groups. This makes the ED a novel and applicable setting for gout research.

A large dataset containing medical notes from patients experiencing flares can be analyzed by ML algorithms to identify patterns and correlations among various factors, including age, sex, multiple medical conditions, and medications taken, linked to a higher risk of flares. Health care providers can receive real-time alerts from this predictive model through integration with EHR systems, enhancing the effectiveness and precision of gout diagnosis and management. It is anticipated that this approach may be successfully and semiautomatically scaled by analyzing large amounts of dictated data and employing supervised learning methods to guide the selection and development of rules. By leveraging technology to analyze and obtain information from clinical notes, health care professionals can make more informed decisions and provide more personalized treatment to patients experiencing gout flares. Although the results were good, they were not as excellent as those found in the text. Artificial intelligence detection methods also improve the efficiency of the results (Olana et al., 2021).

10.2.2 Developing an analytical paradigm with machine learning algorithms

Computational intelligence has become a potent instrument for deciphering intricate datasets and drawing insightful conclusions. Without explicit programming, ML algorithms such as SVM, neural networks, and decision tree can identify patterns and relationships in data. Ascorbic acid, fructose, lactose, uric acid, and mannose were the only sugars in which the biosensor could detect glucose with precision. Changes in the blood's uric acid level, which is between 140 and 430 M, can result in stones in the kidneys, and gout (Chen et al., 2014). With this ability, new avenues for developing analytical paradigms that offer more in-depth comprehension and predictive power across a range of fields have become possible. We must first specify the problem we wish to solve as well as the data we can access to use ML algorithms to develop an analytical paradigm. Biosensors are a useful tool for monitoring blood uric acid levels, which is a crucial task (Castillo-Ortega et al., 2002).

Serious kidney-related diseases can arise from variations in the levels of urea, another metabolic by-product produced by the kidneys (Sekhar et al., 2017; Patil et al., 2023). Next, we must choose a suitable ML algorithm, or set of algorithms, to model relationships in the data and generate precise predictions. This might entail optimizing performance by augmenting the hyperparameters of several algorithms and testing them with a trial of distributed ML models. The amount of urea was tracked using ZnO nanorods (Patil et al., 2023). To measure cholesterol, a variety of ZnO nanotubes were placed on a silicon/silver substrate (Ahmad et al., 2014). For the customer attrition example, we might decide to predict churn probability based on client characteristics by combining random forest and logistic regression algorithms. The sensor's sensitivity, measured at 0.5 nM limit of detection (LOD), was approximately 79.40 AmM1 cm^2. Comparably, nanorods made of ZnO in a field-effect transistor, a composite of CuO and ZnO, and ZnO with multiwalled carbon nanotube-covered Pt/Au electrodes, have been effectively used to detect cholesterol with good sensitivity and a low LOD (Ahmad et al., 2013; Wang et al., 2012; Batra et al., 2015). After our ML model has been trained on the data, we can assess its effectiveness using metrics such as the F1 score, accuracy, precision, and recall. This will assist us in evaluating the predictive power of the model and pinpointing areas in need of development. Fang et al. (2016) successfully used a biosensor with electrodes made of gold to detect the presence in creatine kinase, an enzyme generated during damage to muscles of the brain and other body muscles. After that, we can apply the model in an actual environment and use it to forecast fresh data. Finding hidden trends and conclusions in the data that may not be visible through conventional analysis techniques is a main benefit of using algorithms for ML to develop an analytical paradigm. Sensor-based devices allow ML algorithms to adapt and acquire new data to overcome traditional analytical methods. Although more study in this field is required to confirm these results and apply them to clinical settings, metabolomics appears to have promising applications in the treatment of gout.

10.3 Early identification of severe gout

During consultation, physicians frequently diagnose conditions that have the potential to worsen over time, cause early death, or both. Early stage diseases are usually more stable and have a higher clinical cure rate than late-stage diseases, which progress more quickly and have a lower clinical cure rate. Patients with this excruciating form of arthritis must be given the diagnosis of acute gout as soon as possible to manage the condition effectively and avoid complications. Phosphorous acid crystals accumulate in the joints as a result of gout, which can cause acute, intense flares of pain and inflammation, including redness. Patients typically experience numerous conditions rather than only one illness if they are identified and treated in the clinical setting at the earliest stages of the disease's development. Thus, the most important issue for clinical physicians to address before staging is how to diagnose a particular illness correctly. A person's quality of life can be greatly affected by these flares, which can be crippling. Thus, for prompt treatment and symptom relief, early detection of symptoms of gout and prompt diagnosis are essential.

Understanding the common symptoms of acute gout is the first step toward early detection of the illness. This chapter decomposes the actual clinical evaluation process of diagnosing the stages of gout into two steps: diagnosis and staging. As such, the set of medical records used for the diagnosis of staging gout can be further narrowed to include only patients who have received a diagnosis of gout. The most typical symptom of gout is acute, sudden pain in any number of joints, most commonly the elbow, big toe, ankle, or knee. Warmth, redness, and swelling in the afflicted joint may be present in addition to pain. Fever and chills are also experienced by certain patients during episodes of gout. When providing a diagnosis for patients with joint pain, medical professionals should ask about these symptoms, particularly if the patient has risk factors for gout such as high blood pressure, being overweight, having kidney disease, or having relatives with a history of the condition. This method increases the accuracy of the diagnosis and minimizes the impact of undiagnosed patient medical record data on the task of staging.

10.4 Models for predicting risk of hospitalization from gout flare

Health care providers can prevent hospitalizations and other negative outcomes by identifying patients at higher risk of experiencing a GF. This section discusses various predictive models for assessing the risk of hospitalization owing to gout flares.

10.4.1 Development of hospitalization risk prediction models for patients with gout

Building prediction models using information from EHRs along with additional clinical sources is a method of predicting the likelihood that patients with gout will require hospitalization. By identifying patients who might benefit from more frequent monitoring and focused interventions to avoid hospitalizations, these models can assist clinicians and health care providers. One such model used logistic regression to create a predictive contrast-associated acute kidney injury (CA-AKI) risk model with an AUC of 0.767 (confidence interval,: 0.764−0.770) on extraction and 0.761 on verification for any stage of acute kidney injury. This model's positive and negative predictive values were 19.2% and 96.1%, respectively. The AUC serves as a measure of the model's ability to discriminate between those who will and will not require hospitalization.

Researchers usually start by choosing pertinent predictor or risk factors that have been demonstrated to be linked to increased hospitalization risk in this group of people, to create a prediction framework for the risk of hospitalization in patients with gout. Clinical features (such as comorbidities and disease severity), medication use, biomarkers, and patterns of health care use are a few examples of these predictors. The validation cohort's AUC value for the risk model predicting CA-AKI phases 2 and 3 was 0.818, with positive and negative predictive outcomes of 13% and 98.5%, respectively. By combining these predictors, researchers can create a predictive model that estimates a person's risk of hospitalization using ML and statistical methods. The prediction model can be created and validated using methods such as logistical regression, random forest, SVM, and neural networks. With high-risk patients, clinicians can use the framework to prioritize resources and actions such as medication optimization, lifestyle changes, and routine follow-up appointments.

10.4.2 Participation of patients with gout in the classification process

The methods of classification procedures for patients with gout depend on the disease severity and assessment measures. These data are essential for assigning patients to various risk groups and choosing the best course of treatment. Data collected from the beginning of 2013 to April 2014 from 25 centers across 16 countries on four continents involved 983 individuals (509 cases and 472 controls). Evaluation of risk factors for gout can also involve the patient. Lifestyle habits, including food, alcohol intake, and physical activity levels, can affect the risk of developing gout. Patients can share this information with you. Those who actively participate in this evaluation will be better able to make educated decisions regarding their health and understand how their way of life may affect the condition. Of these, 702 (71.4%) actively participated in managing the condition, which can significantly reduce the frequency and severity of gout attacks.

10.5 Conclusion

The research discussed in this chapter emphasizes how crucial it is to use ML and NLP methods to detect and anticipate gout flares accurately. Through the use of differential metabolites and clinical note analysis, health care providers can create early detection strategies for acute gout, which can result in prompt interventions and better long-term joint function. We also show that GF ED detection is achievable with clinical criteria, but deployment may be difficult owing to vendor restrictions. Furthermore, the investigation of hospitalization risk factor prediction models linked to gout flares highlights the need for improved diagnostic tools in the medical field. In the future, we aim to assess the effectiveness of the NLP ED algorithm in facilitating efficient referral pathways to outpatient care and clinical research studies, which could significantly improve patient outcomes. Prospects for advancing the screening and treatment of gout, particularly with comorbid conditions such as cardiovascular disease, are promising with the development of an automated learning-based diagnostic model. This research contributes to the advancement of personalized medicine within rheumatology by improving patient care.

10.5.1 Summary of main conclusions and their implications for further study

Our research has yielded an unparalleled understanding of the metabolic underpinnings of the frequency of GF, showcasing unique metabolomic profiles for individuals experiencing frequent and infrequent flares. These results show that metabolomic histories can distinguish among various clinical presentations of gout.

10.5.2 Supplementary suggestions for managing and predicting gout flares

The goal of future research should be to find new biomarkers to track the progression of the disease and forecast gout flares. Current treatments such as colchicine, nonsteroidal antiiflammatory drugs (NSAIDs), and oral corticosteroids are largely evidence-based, unlike combined therapies and intraarticular corticosteroid injections, which still require further trials for validation. For patients with frequent, poorly controlled flares, consideration of the interleukin-1 blockade is advised despite its costs and potential for infection, especially in those who cannot tolerate other medications, such as corticosteroids, NSAIDs, or colchicine.

Examining how lifestyle choices might help prevent gout flares: The development of flares of gout has been linked to lifestyle factors such as diet, weight, and exercise. To develop treatments that may assist in preventing these attacks, more research is required to ascertain how these variables related to lifestyle affect the likelihood of gout flares. To provide quick treatment for flares, the task force emphasizes the importance of patient education and the pill-in-the-pocket approach. This is because the workforce believes that individuals must take an active role in managing the disease.

Evaluating the influence of comorbidities on the risk of gout flares: Obesity, elevated blood pressure, and kidney disease are just a few illnesses to which gout is frequently linked. Better management of the condition may result from an understanding of how these comorbidities affect the likelihood of gout flares and influence gout treatment. Over the coming years, the European League Against Rheumatism (EULAR) and American College of Rheumatology (ACR) recommendations will surely need to be updated. In fact, we expect to have access to fresh information on new or developing medications, particularly uricosurics.

Additionally, although genetics are known to influence gout development significantly, their role in flare prognosis remains unclear. Identifying genetic markers related to gout flares and exploring their interactions with other risk factors are vital areas for future research.

Personalized treatment strategies are crucial for managing the flares of gout because the disease is heterogeneous and patients experience flares at varying frequencies and intensities. The development of individualized treatment plans that consider individual variations in disease presentation, complications, and genetic factors should be the main goal of future research.

Additional research is required to advance our knowledge of GF management and forecasting. We can provide better care for patients with gout and lessen the load of this crippling condition by finding indicators for GF estimation, researching the influence of comorbidities and lifestyle factors on GF hazards, looking into the function of genealogy in GF forecasting, and creating individualized treatment plans.

References

Abbott, T.M.C., Abdalla, F.B., Avila, S., Banerji, M., Baxter, E., Bechtol, K., Becker, M.R., Bertin, E., Blazek, J., Bridle, S.L., Brooks, D., Brout, D., Burke, D.L., Campos, A., Carnero Rosell, A., Carrasco Kind, M., Carretero, J., Castander, F.J., Cawthon, R., Chang, C., Chen, A., Crocce, M., Cunha, C.E., Da Costa, L.N., Davis, C., De Vicente, J., Derose, J., Desai, S., Di Valentino, E., Diehl, H.T., Dietrich, J.P., Dodelson, S., Doel, P., Drlica-Wagner, A., Eifler, T.F., Elvin-Poole, J., Evrard, A.E., Fernandez, E., Ferté, A., Flaugher, B., Fosalba, P., Frieman, J., García-Bellido, J., Gaztanaga, E., Gerdes, D.W., Giannantonio, T., Gruen, D., Gruendl, R.A., Gschwend, J., Gutierrez, G., Hartley, W.G., Hollowood, D.L., Honscheid, K., Hoyle, B., Huterer, D., Jain, B., Jeltema, T., Johnson, M.W.G., Johnson, M.D., Kim, A.G., Krause, E., Kuehn, K., Kuropatkin, N., Lahav, O., Lee, S., Lemos, P., Leonard, C.D., Li, T.S., Liddle, A.R., Lima, M., Lin, H., Maia, M.A.G., Marshall, J.L., Martini, P., Menanteau, F., Miller, C.J., Miquel, R., Miranda, V., Mohr, J.J., Muir, J., Nichol, R.C., Nord, B., Ogando, R.L.C., Plazas, A.A., Raveri, M., Rollins, R.P., Romer, A.K., Roodman, A., Rosenfeld, R., Samuroff, S., Sanchez, E., Scarpine, V., Schindler, R., Schubnell, M., Scolnic, D., Secco, L.F., Serrano, S., Sevilla-Noarbe, I., Smith, M., Soares-Santos, M., Sobreira, F., Suchyta, E., Swanson, M.E.C., Tarle, G., Thomas, D., Troxel, M.A., Vikram, V., Walker, A.R., Weaverdyck, N., Wechsler, R.H., Weller, J., Yanny, B., Zhang, Y., Zuntz, J., 2019. Dark Energy Survey year 1 results: constraints on extended cosmological models from galaxy clustering and weak lensing. Phys. Rev. D 99 (12). https://doi.org/10.1103/PhysRevD.99.123505.

Ahmad, R., Tripathy, N., Kim, S.H., Umar, A., Al-Hajry, A., Hahn, Y.B., 2014. High performance cholesterol sensor based on ZnO nanotubes grown on Si/Ag electrodes. Electrochem. Commun. 38, 4–7. https://doi.org/10.1016/j.elecom.2013.10.028.

Ahmad, R., Tripathy, N., Hahn, Y.B., 2013. High-performance cholesterol sensor based on the solution-gated field effect transistor fabricated with ZnO nanorods. Biosens. Bioelectron. 45 (1), 281–286. https://doi.org/10.1016/j.bios.2013.01.021.

Alsolamy, S., Al Salamah, M., Al Thagafi, M., Al-Dorzi, H.M., Marini, A.M., Aljerian, N., Al-Enezi, F., Al-Hunaidi, F., Mahmoud, A.M., Alamry, A., Arabi, Y.M., 2014. Diagnostic accuracy of a screening electronic alert tool for severe sepsis and septic shock in the emergency department. BMC Med. Inf. Decis. Making 14 (1). https://doi.org/10.1186/s12911-014-0105-7.

Babichev, S.A., Ries, J., Lvovsky, A.I., 2003. Quantum scissors: teleportation of single-mode optical states by means of a nonlocal single photon. Europhys. Lett. 64 (1), 1–7. https://doi.org/10.1209/epl/i2003-00504-y.

Bantelay, L.M., Abebe, M., Sharma Rajendran, R., Sungheetha, A., Sengottaiyan, N., 2023. Heuristic pneumonia and tuberculosis detection in X-ray images using convolutional neural networks. In: 2023 Annual International Conference on Emerging Research Areas: International Conference on Intelligent Systems, AICERA/ICIS 2023. https://doi.org/10.1109/AICERA/ICIS59538.2023.10420329.

Batra, N., Tomar, M., Gupta, V., 2015. ZnO−CuO composite matrix based reagentless biosensor for detection of total cholesterol. Biosens. Bioelectron. 67, 263−271. https://doi.org/10.1016/j.bios.2014.08.029.

Beneke, M., Buchalla, G., Dunietz, I., 1997. Mixing-induced CP asymmetries in inclusive B decays. Phys. Lett. B 393 (1−2), 132−142. https://doi.org/10.1016/s0370-2693(96)01648-6.

Bennett, A.V., Keenoy, K., Shouery, M., Basch, E., Temple, L.K., 2016. Evaluation of mode equivalence of the MSKCC bowel function instrument, LASA quality of life, and subjective significance questionnaire items administered by web, interactive voice response system (IVRS), and paper. Qual. Life Res. 25 (5), 1123−1130. https://doi.org/10.1007/s11136-015-1162-9.

Castillo-Ortega, M.M., Rodriguez, D.E., Encinas, J.C., Plascencia, M., Méndez-Velarde, F.A., Olayo, R., 2002. Conductometric uric acid and urea biosensor prepared from electroconductive polyaniline−poly(n-butyl methacrylate) composites. Sensor. Actuator. B Chem. 85 (1−2), 19−25. https://doi.org/10.1016/s0925-4005(02)00045-x.

Chauhan, B.K., Banupriya, M., Venkatagurunatham, K., Sungheetha, A., Kumari, C.H.D.V.P., Karthick, L., 2023. Internet of Things routing protocol with mobility awareness and energy efficiency. In: 2023 3rd International Conference on Advance Computing and Innovative Technologies in Engineering, ICACITE 2023, pp. 437−441. https://doi.org/10.1109/ICACITE57410.2023.10183117.

Chen, X., Wu, G., Cai, Z., Oyama, M., Chen, X., 2014. Advances in enzyme-free electrochemical sensors for hydrogen peroxide, glucose, and uric acid. Microchim. Acta 181 (7−8), 689−705. https://doi.org/10.1007/s00604-013-1098-0.

Daftary, A., Hirsch-Moverman, Y., Kassie, G.M., Melaku, Z., Gadisa, T., Saito, S., Howard, A.A., 2017. A qualitative evaluation of the acceptability of an interactive voice response system to enhance adherence to isoniazid preventive therapy among people living with HIV in Ethiopia. AIDS Behav. 21 (11), 3057−3067. https://doi.org/10.1007/s10461-016-1432-8.

De Vera, M.A., Rahman, M.M., Bhole, V., Kopec, J.A., Choi, H.K., 2010. Independent impact of gout on the risk of acute myocardial infarction among elderly women: a population-based study. Ann. Rheum. Dis. 69 (6), 1162−1164. https://doi.org/10.1136/ard.2009.122770.

Eisele, A., Schagg, D., Krämer, L.V., Bengel, J., Göhner, W., 2019. Behaviour change techniques applied in interventions to enhance physical activity adherence in patients with chronic musculoskeletal conditions: a systematic review and meta-analysis. Patient Educ. Counsel. 102 (1), 25−36. https://doi.org/10.1016/j.pec.2018.09.018.

Fang, C.S., Oh, K.H., Oh, A., Lee, K., Park, S., Kim, S., Park, J.K., Yang, H., 2016. An ultrasensitive and incubation-free electrochemical immunosensor using a gold-nanocatalyst label mediating outer-sphere-reaction-philic and inner-sphere-reaction-philic species. Chem. Commun. 52 (34), 5884−5887. https://doi.org/10.1039/c6cc00353b.

FitzGerald, J.D., Mikuls, T.R., Neogi, T., Singh, J.A., Robbins, M., Khanna, P.P., Turner, A.S., Myslinski, R., Suter, L.G., 2018. Development of the American college of rheumatology electronic clinical quality measures for gout. Arthritis Care Res. 70 (5), 659−671. https://doi.org/10.1002/acr.23500.

Gossec, L., Portier, A., Landewé, R., Etcheto, A., Navarro-Compán, V., Kroon, F., van der Heijde, D., Dougados, M., 2016. Preliminary definitions of 'flare' in axial spondyloarthritis, based on pain, BASDAI and ASDAS-CRP: an ASAS initiative. Ann. Rheum. Dis. 75 (6), 991–996. https://doi.org/10.1136/annrheumdis-2015-208593.

Hao, Z., AghaKouchak, A., Nakhjiri, N., Farahmand, A., 2014. Global integrated drought monitoring and prediction system. Sci. Data 1 (1). https://doi.org/10.1038/sdata.2014.1.

Hewlett, S., Sanderson, T., May, J., Alten, R., Bingham, C.O., Cross, M., March, L., Pohl, C., Woodworth, T., Bartlett, S.J., 2012. 'I'm hurting, I want to kill myself': rheumatoid arthritis flare is more than a high joint count-an international patient perspective on flare where medical help is sought. Rheumatology 51 (1), 69–76. https://doi.org/10.1093/rheumatology/keq455.

Lee, H., Friedman, M.E., Cukor, P., Ahern, D., 2003. Interactive voice response system (IVRS) in health care services. Nurs. Outlook 51 (6), 277–283. https://doi.org/10.1016/S0029-6554(03)00161-1.

Mercer, K., Li, M., Giangregorio, L., Burns, C., Grindrod, K., 2016. Behavior change techniques present in wearable activity trackers: a critical analysis. JMIR mHealth uHealth 4 (2). https://doi.org/10.2196/mhealth.4461.

Mikuls, T.R., MacLean, C.H., Olivieri, J., Patino, F., Allison, J.J., Farrar, J.T., Bilker, W.B., Saag, K.G., 2004. Quality of care indicators for gout management. Arthritis Rheum. 50 (3), 937–943. https://doi.org/10.1002/art.20102.

Neogi, T., Jansen, T.L.T.A., Dalbeth, N., Fransen, J., Schumacher, H.R., Berendsen, D., Brown, M., Choi, H., Edwards, N.L., Janssens, H.J.E.M., Lioté, F., Naden, R.P., Nuki, G., Ogdie, A., Perez-Ruiz, F., Saag, K., Singh, J.A., Sundy, J.S., Tausche, A.K., Vaquez-Mellado, J., Yarows, S.A., Taylor, W.J., 2015. 2015 gout classification criteria: an American college of rheumatology/European league against rheumatism collaborative initiative. Arthritis Rheumatol. 67 (10), 2557–2568. https://doi.org/10.1002/art.39254.

Nguyen, S.Q., Mwakalindile, E., Booth, J.S., Hogan, V., Morgan, J., Prickett, C.T., Donnelly, J.P., Wang, H.E., 2014. Automated electronic medical record sepsis detection in the emergency department. PeerJ 2 (1). https://doi.org/10.7717/peerj.343.

Olana, M.D., Rajesh Sharma, R., Sungheetha, A., Chung, Y.K., 2021. In: Applying Deep Learning Approach for Wheat Rust Disease Detection Using MosNet Classification Technique, vol 173. Springer Science and Business Media LLC, pp. 551–565. https://doi.org/10.1007/978-981-33-4305-4_41.

Osborne, J.D., Leary, O., Mudano, T., Booth, A., Rosas, J., Peramsetty, G., et al., 2020. Gout Emergency Department Chief Complaint Corpora (version 1.0). PhysioNet.

Osborne, J.D., Booth, J.S., O'Leary, T., Mudano, A., Rosas, G., Foster, P.J., Saag, K.G., Danila, M.I., 2020. Identification of gout flares in chief complaint text using natural language processing. In: AMIA … Annual Symposium proceedings. AMIA Symposium, pp. 973–982.

Parry, E.L., Thomas, M.J., Peat, G., 2018. Defining acute flares in knee osteoarthritis: a systematic review. BMJ Open 8 (7). https://doi.org/10.1136/bmjopen-2017-019804.

Patil, S., Sungheetha, A., Bansode, G.S., Kalaivaani, P.T., Kandaswamy, V.A., Jagannathan, S.K., 2023. Design and behavioral analysis of students during examinations using distributed machine learning. In: 2023 IEEE International Conference on Research Methodologies in Knowledge Management, Artificial Intelligence and Telecommunication Engineering, RMKMATE 2023. https://doi.org/10.1109/RMKMATE59243.2023.10369281.

Rajesh Sharma, R., Sungheetha, A., 2017. Segmentation and classification techniques of medical images using innovated hybridized techniques - a study. In: Proceedings of 2017 11th International Conference on Intelligent Systems and Control, ISCO 2017, pp. 192–196. https://doi.org/10.1109/ISCO.2017.7855979.

Schaller, A., Rudolf, K., Dejonghe, L., Grieben, C., Froboese, I., 2016. Influencing factors on the overestimation of self-reported physical activity: a cross-sectional analysis of low back pain patients and healthy controls. BioMed Res. Int. 2016, 1–11. https://doi.org/10.1155/2016/1497213.

Sekhar, P., Moore, Z., Aravamudhan, S., Khosla, A., 2017. A new low-temperature electrochemical hydrocarbon and NOx sensor. Sensors 17 (12). https://doi.org/10.3390/s17122759.

Seminog, O.O., Goldacre, M.J., 2013. Gout as a risk factor for myocardial infarction and stroke in England: evidence from record linkage studies. Rheumatology 52 (12), 2251–2259. https://doi.org/10.1093/rheumatology/ket293.

Sharma, R., Marikkannu, P., Sungheetha, A., 2019. Three-dimensional MRI brain tumour classification using hybrid ant colony optimisation and grey wolf optimiser with proximal support vector machine. Int. J. Biomed. Eng. Technol. 29 (1). https://doi.org/10.1504/ijbet.2019.096879.

Spencer, L., Adams, T.B., Malone, S., Roy, L., Yost, E., 2006. Applying the transtheoretical model to exercise: a systematic and comprehensive review of the literature. Health Promot. Pract. 7 (4), 428–443. https://doi.org/10.1177/1524839905278900.

Stahl, B., 2020. DeepSIP: Deep Learning of Supernova Ia Parameters, 0.42. Astrophysics Source Code Library.

Sungheetha, A., Mssujitha, R., Arthi, V., Sharma, R.R., 2017. Data analysis of multiobjective density based spatial clustering schemes in gene selection process for cancer diagnosis. In: Proceedings of 2017 4th International Conference on Electronics and Communication Systems, ICECS 2017, pp. 134–137. https://doi.org/10.1109/ECS.2017.8067854.

Sungheetha, A., Bharathi, B., Ganesan, D., Karthikeyan, T., Bindu Madhavi, N., Chairma Lakshmi, K.R., 2023. E-commerce business model analysis and success in urban areas using AI-distributed machine learning. In: 2023 IEEE International Conference on Research Methodologies in Knowledge Management, Artificial Intelligence and Telecommunication Engineering, RMKMATE 2023. https://doi.org/10.1109/RMKMATE59243.2023.10369718.

Sungheetha, A., Rajesh Sharma, R., Nuradis, J., 2019. Implication centered learning mechanism for exploring analysis of variance by means of linear regression in artificial neural networks. In: Proceedings of the 3rd International Conference on I-SMAC IoT in Social, Mobile, Analytics and Cloud, I-SMAC 2019, pp. 748–752. https://doi.org/10.1109/I-SMAC47947.2019.9032475.

Wang, C., Tan, X., Chen, S., Yuan, R., Hu, F., Yuan, D., Xiang, Y., 2012. Highly-sensitive cholesterol biosensor based on platinum–gold hybrid functionalized ZnO nanorods. Talanta 94, 263–270. https://doi.org/10.1016/j.talanta.2012.03.037.

Wang, J., 2022. Prediction of postoperative recovery in patients with acoustic neuroma using machine learning and SMOTE-ENN techniques. Math. Biosci. Eng. 19 (10), 10407–10423. https://doi.org/10.3934/mbe.2022487.

Zheng, C., Rashid, N., Wu, Y.L., Koblick, R., Lin, A.T., Levy, G.D., Cheetham, T.C., 2014. Using Natural Language Processing and machine learning to identify gout flares from electronic clinical notes. Arthritis Care Res. 66 (11), 1740–1748. https://doi.org/10.1002/acr.22324.

CHAPTER

Impact of amyloidosis on bones and its relationship to dementia

11

Krishnendu Ghosh[1], Ricardo Fitas[2], Eshita Dey[3] and Munmun Patra[3]

[1]*Department of Computer Science & Engineering, Indian Institute of Information Technology Dharwad, Dharwad, Karnataka, India;* [2]*Department of Mechanical Engineering, Technical University of Darmstadt, Hesse, Germany;* [3]*Department of Artificial Intelligence & Machine Learning, Techno International New Town, Kolkata, West Bengal, India*

11.1 Amyloidosis: A brief overview

Amyloidosis encompasses a spectrum of intricate and uncommon disorders characterized by the abnormal buildup of amyloid proteins, referred to as amyloid fibrils, within different organs and tissues across the body. This pathologic condition arises from misfolding proteins typically soluble into resistant, insoluble fibrils. As a result, affected organs stop working correctly. This phenomenon frequently accompanies various diseases. Each specific condition is distinguished by the aggregation of a distinct protein or peptide. Alzheimer's disease (AD) and type 2 diabetes stand out as prominent examples of amyloid disease. In 1838 Matthias Schleiden conceived the term "amyloid" from the Greek word for "starch-like." In the context of conceptualizing amyloidosis, pathologist Rudolph Virchow further highlighted tissue characteristics observed during autopsies that were previously referred to as "wax infiltration" or "Lardakus disease." Whether presenting as a primary ailment or as a secondary complication of other disorders, amyloidosisexhibits a diverse array of clinical manifestations, thereby complicating its diagnosis and management.

Commonly, amyloidosis is categorized into two distinct classes: systemic amyloidosis, which affects the body more than a second, and localized amyloidosis. Approximately 36 distinct types of amyloidosis arise from the misfolding of specific proteins. Among these 36 proteins, 14 are implicated in systemic amyloidosis, 19 are associated with localized amyloidosis, and three are classified as either. This hierarchical classification is listed in Table 11.1.

11.1.1 Systemic amyloidosis

Systemic forms of amyloidosis encompass primary amyloidosis, associated not only with multiple myeloma, but also with other plasma cell dyscrasias, reactive or secondary amyloidosis, and familial forms (Adav et al., 2016). Whereas systemic

Diagnosing Musculoskeletal Conditions using Artificial Intelligence and Machine Learning to Aid Interpretation of Clinical Imaging
https://doi.org/10.1016/B978-0-443-32892-3.00011-7

Table 11.1 Misfolded proteins for systematic and localized amyloidosis.

Systematic			Localized			
Light chain	TRwt	ATTR	AH	ABri	ADan[a]	Aβ
AA	ALECT2	AGel	AαSyn	ATau	APrP	ACal
AApoAI	AApoCII	ALys	AIAPP	AANF	APro	AIns
AApoAII	AApoCIII	Aβ2M	ASPC[b]	ACor	AMed	ALac
AApoAIV	AFyb		AOAAP	ASemI	AEnf	

[a] *ADan is the product of the same gene as ABri.*
[b] *Not proven by amino acid sequence analysis.*

amyloidosis can potentially affect the entire body, it typically targets specific tissues and organs. In contrast, localized amyloidosis includes limited forms affecting various organs, focal amyloid deposits, and senile amyloidosis. Notably, no limited forms of amyloidosis are known to be associated with or progress to systemic involvement (Blondel et al., 2008). Amyloidosis can manifest in individuals aged 40 years and greater. This is particularly common with individuals who have idiopathic peripheral neuropathy, nephrotic syndrome, congestive heart failure, or unexplained hepatomegaly (Gertz et al., 1999). Common systemic amyloidosis has four types: (i) light chain (LC) (AL), (ii) inflammation (AA), (iii) dialysis-related ($A\beta_2M$), and (iv) hereditary (ATTR).

1. **Light Chain (AL)**: Patients with the diagnosis of AL amyloidosis typically exhibit weakly positive pyrophosphate or 3,3-diphosphono-1,2-propanodicarboxylic acid (PYP) scans. Therefore, when a positive PYP scan is present, it is imperative to conduct a free LC assay in blood and perform immunofixation tests in serum and urine to rule out AL amyloidosis definitively. Notably, the presence of macroglossia, although not pathognomonic for amyloidosis, warrants consideration, with potential differentials including hypothyroidism, tongue cancer, and acromegaly. Macroglossia becomes more indicative of amyloidosis once the diagnosis is confirmed, particularly related to AL amyloidosis. In addition, nail dystrophy may precede the diagnosis of amyloidosis by several years. Amyloid deposition in joints can lead to symptoms such as joint pain and stiffness, with the "shoulder pad sign" on physical examination representing a classic indicator of amyloidosis (Vaxman and Gertz, 2020).
2. **Inflammation (AA)**: This type of amyloidosis is related to chronic inflammatory conditions. Its symptoms are renal manifestations such as proteinuria (progression to nephrotic syndrome) and decreased renal function (progressing to kidney failure), which are frequently observed . Autonomic neuropathy, hepatomegaly, and cardiomyopathy are less frequent manifestations. Some common symptoms are fatigue and weight loss (Desport et al., 2012).

3. **Dialysis-related (Aβ_2M)**: Clinical manifestations are arthropathies such as tenosynovitis, shoulder pain, carpal tunnel syndrome, periarticular cysts, pathologic fractures, and destructive spondyloarthropathy. This disease is seen in patients who have undergone long-term hemodialysis (Shenoy et al., 2018). Synovial tissue biopsy is the method for detecting amyloid. Kidney transplant reverses the disease. Aβ_2M amyloidosis is a disabling disease that should be recognized and treated (Desport et al., 2012).
4. **Hereditary ATTR**: The primary clinical features typically include (familial) peripheral and autonomic neuropathy, although manifestations such as cardiomyopathy, renal failure, and ocular involvement (vitreous opacities) may also manifest. Severe cardiomyopathy may serve as an initial indicator of specific transthyretin (TTR) mutations. In addition, normal (wild-type) TTR can act as a precursor protein in advanced age, leading to senile systemic amyloidosis characterized by gradually progressive cardiomyopathy. Other potential manifestations encompass gastrointestinal complications and carpal tunnel syndrome (Desport et al., 2012).

11.1.2 Localized amyloidosis

Localized amyloidosis is where amyloid deposits are confined to specific organs or tissues without systemic involvement. In certain individuals, AL amyloidosis presents as a localized occurrence linked to the aggregation of monoclonal LCs in proximity to their production by focal plasma cell clones. Organs or tissues primarily affected in localized AL amyloidosis encompass not only the urinary tract but also the bladder, tracheobronchial system, larynx, lungs, ocular globe, and skin. Typical manifestations include swelling, diminished mobility, pain, functional impairments, and irregular heartbeat (Desport et al., 2012). The symptoms experienced by individuals with localized amyloidosis vary depending on the affected sites, as itemized below.

1. **Bladder**: Symptoms may include urinary frequency, urgency, or difficulty urinating.
2. **Heart**: Shortness of breath, orthopnea, paroxysmal nocturnal dyspnea, jugular venous distention, peripheral edema, or arrhythmia.
3. **Kidney**: Peripheral edema.
4. **Nerve**: Length-dependent symmetric peripheral neuropathy, autonomic neuropathy (erectile dysfunction, postural hypertension, voiding dysfunction, early satiety, or irregular bowel habits).
5. **Soft tissue**: Macroglossia, hoarseness, obstructive sleep apnea, dysarthria, enlarged salivary glands, dry mouth, arthropathy, jaw claudication, periorbital or other bruising, or carpal tunnel syndrome.
6. **Gastrointestinal tract**: Irregular bowel habits, early satiety, weight loss, gastroparesis, dysphagia, or gastrointestinal bleeding.

7. **Tracheobronchial system**: Symptoms may include coughing, wheezing, breath shortness, or difficulty breathing.
8. **Liver**: Hepatomegaly, unintentional loss of weight, early satiety, bleeding disorders.
9. **Spleen**: Rarely spontaneous splenic rapture, early satiety, or abdominal distension.
10. **Lung**: Shortness of breath or dry cough.

Amyloidosis can occur in different body organ systems, such as the kidneys, cardiovascular, respiratory, alimentary, liver, spleen, and nervous systems. Moreover, amyloidosis has varying effects for individuals from different age groups. Table 11.2 specifies the effect of amyloidosis for different ages and genders.

11.1.3 Diagnosing localized amyloidosis

Diagnosing localized amyloidosis requires a systematic approach encompassing clinical evaluation, imaging studies, and histopathologic confirmation. The algorithm shown in Fig. 11.1 outlines steps for diagnosing localized amyloidosis.

11.1.3.1 Kidney

If the kidney is affected by amyloidosis, that will be the main cause of death. In the glomerulus, amyloid is first deposited in the mesangium. Amyloid fibrils are seen around interstitial capillaries and adjacent fibroblasts in the area damaged earlier. The basement membrane is thickened and contains amyloid fibrils. Amyloid fibrils can also be seen within the tubular lumen and urinary sediment. Earlier, it was considered that the development of intracytoplasmic kidney-related amyloid fibrils

Table 11.2 Varying effects of amyloidosis for different age and gender groups.

	Distribution (%)					
	Group 1			Group 2		
Age group	**Male**	**Female**	**Total**	**Male**	**Female**	**Total**
<40	1	4	2	3	4	3
40–49	10	18	13	3	7	5
50–59	31	33	32	35	37	36
60–69	38	33	36	38	33	36
70–79	12	10	11	21	19	20
≥80	8	2	6	0	0	0
Number of patients	83	49	132	34	27	61
% of group	63	37	100	56	44	100
Mean age, year	62.5	57.5	60.6	61.2	61.0	61.1

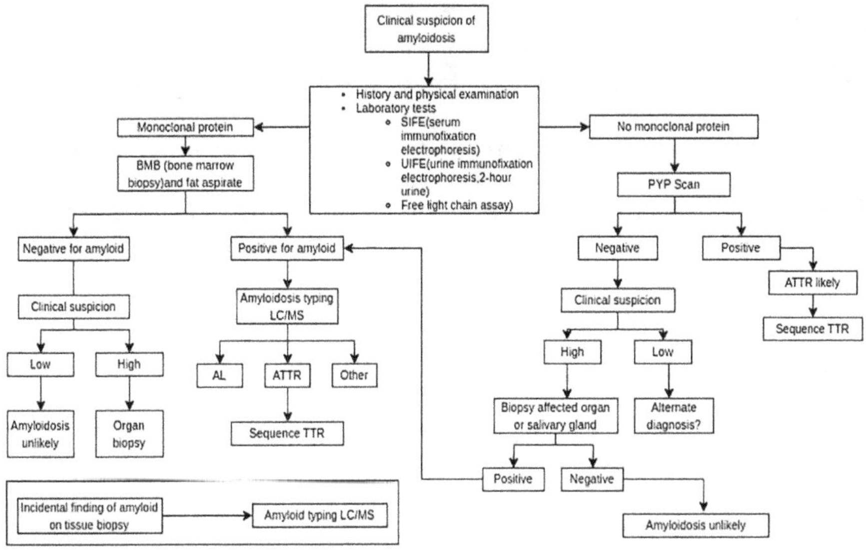

FIGURE 11.1

Steps for diagnosing localized amyloidosis.
BMB, bone marrow biopsy; *SIFE,* serum immunofixation electrophoresis; *UIFE,* urine immunofixation electrophoresis.

was an extracellular phenomenon. According to some studies, the more amyloid deposits there are in the glomerulus, the more severe the renal disease will be.

11.1.3.2 Cardiovascular system

Amyloidosis frequently affects the heart. Congestive heart failure (CHF), a chronic condition, is the predominant manifestation. Here, the heart fails to pump blood to meet the body's demands adequately. In patients with unexplained CHF, amyloidosis should be considered, particularly if the condition proves resistant to treatment, because cardiac failure in amyloidosis tends to worsen progressively and resist conventional therapies. Cardiac arrhythmias are prevalent, especially in individuals receiving digitalis, because deposition of amyloid in the sinus node can exacerbate arrhythmias, potentially leading to sudden cardiac death. Electrocardiogram abnormalities, notably changes in voltage, are common, with findings sometimes mimicking myocardial infarction. Amyloidosis may also mimic valvular heart disease, and angina pectoris and myocardial infarction can occur owing to amyloid deposition in coronary arteries. In addition, reports exist of myocardial rupture in cases of amyloidosis. Intermittent claudication, observed in some amyloidosis patients, is believed to result from amyloid involvement in arterioles, leading to fixed, nonexpansile vessels.

11.1.3.3 Respiratory tract

Amyloidosis in the respiratory tract is common but usually produces no symptoms. Based on experience, dyspnea is a rare symptom that results from extensive amyloid-related infiltration of the lung in patients with primary amyloidosis. Amyloid deposition may displace the alveolar capillaries and not significantly interfere with gas diffusion. Impairment of gas diffusion by amyloid by its infiltration of the alveolar capillaries has been reported in a case of myeloma with generalized amyloidosis.

11.1.3.4 Alimentary tract

Gastrointestinal involvement is a prevalent feature of systemic amyloidosis, often presenting with a broad spectrum of symptoms. Some patients may remain asymptomatic but others may experience widespread dysfunction within the gastrointestinal tract. Macroglossia, although it is not a common finding and is found in a minority of cases, can manifest prominently in the oral cavity. Despite its potential challenges, partial tongue resection is not routinely recommended. Recurrent hemorrhagic bullae may develop in the oral cavity. Swelling in the floor of the mouth may indicate primary amyloidosis. In contrast, focal amyloidosis can manifest as a mass on the soft palate, later spreading to involve cervical and axillary lymph nodes. Amyloidosis of the esophagus can lead to bleeding and perforation, with amyloid deposits found in the walls of gut vessels and the submucosa of the gastrointestinal tract. Prepyloric obstruction and partial obstruction of the small bowel from amyloid deposits have been documented, including instances of mass formation at the ligament of Treitz. Stomach amyloidosis may present with the absence of rugal patterns and reduced motility, whereas decreased motor activity is a common feature in the small bowel affected by amyloidosis.

11.1.3.5 Liver and spleen

Amyloidosis commonly affects the liver, and hepatomegaly is an initial indication and potential presenting symptom. Hepatomegaly may be substantial in size, although other physical signs of primary liver disease are typically uncommon. Despite considerable amyloid deposition, alterations in hepatic function are generally regarded as minimal. However, approximately 50% of patients who underwent liver function tests exhibited elevated serum alkaline phosphatase, serum glutamic-oxaloacetic transaminase (SGOT), and BSP levels. Portal hypertension resulting in esophageal bleeding caused by amyloid-related deposition has been documented, and spontaneous splenic rupture may also occur in cases of amyloidosis.

11.1.3.6 Nervous system

Peripheral neuropathy usually begins with paresthesia or pain, particularly in the legs, followed by weakness. It may also involve the autonomic nervous system with consequent diarrhea, lack of sweating, impotence, and orthostatic hypotension.

Sometimes, retention or the inability to control urine or feces excretion may be a problem (Kyle and Bayed, 1975).

11.2 Amyloidosis and bones

This section delves into the intricate interplay among amyloidosis, disorders characterized by the abnormal accumulation of misfolded proteins, and bone health. Amyloidosis can profoundly affect skeletal integrity by depositing amyloid fibrils within bone tissue, leading to complications such as osteoporosis, fractures, and joint abnormalities. Understanding the mechanisms underlying amyloid-induced bone pathology is essential for effective diagnosis and management strategies to preserve bone health in individuals affected by amyloidosis.

11.2.1 Microscopic appearance of amyloid

The amyloid material in primary amyloidosis appears amorphous and more or less homogeneous compared with amyloid in secondary amyloidosis. Amyloid deposits, particularly in the alveolar septa of the lungs, bones, and kidneys, can lead to calcification (Symmers, 1956).

11.2.2 Bones, joints, and tendons

Chronic renal failure patients undergoing long-term regular hemodialysis are at risk of developing amyloidosis, characterized by the excess extractable plasma protein β-microglobulin deposition in tendons, joints, and bones. Over time, various muscle disorders may manifest, as well as complications such as trigger finger, flexor tendon contracture, spontaneous tendon rupture, and pathologic fractures owing to amyloid bone cysts (Kurer et al., 1991). Amyloidosis is observed in 5%−15% of multiple myeloma cases, often presenting as radiolytic bone lesions, mainly solitary amyloidomas, and the skull and spine are common sites of involvement (McCarthy, 2020). Symptoms tend to manifest earlier in older patients (Kurer et al., 1991).

Extensive amyloidosis within the bone marrow can result in the replacement of hemopoietic tissue. Radiologically, bone amyloidoma presents as a proliferative lytic lesion, in which stippled radiodensity is potentially observable. Within the marrow space, synovial membrane, and adjacent soft tissues, waxy eosinophilic Congo red–positive deposits are typically found (McCarthy, 2020). The most severe amyloidosis manifestations often occur in long bones, where hemopoietic tissue predominates. Sometimes, the marrow cavity may become filled with amyloid, potentially extending into the Haversian canal system, leading to bone atrophy and dyspathologic fractures. Solitary amyloidoma of the bone typically progresses to diffuse amyloidosis or multiple myeloma (McCarthy, 2020). However, in many cases, amyloid deposition is primarily limited to the blood vessel walls within the bone (Symmers, 1956).

11.3 Prediction of amyloidosis using artificial intelligence and machine learning techniques

Amyloidosis, which is characterized by the uncontrolled deposition of amyloid proteins, poses a challenge in early diagnosis owing to its overlapping clinical presentation with more prevalent hypertrophic diseases. Furthermore, classifying amyloidosis into distinct groups based on the constituent proteins of amyloid deposits underscores the importance of accurate differentiation for appropriate therapeutic intervention. Diagnostic evaluation typically involves identifying clinical features, electrocardiographic abnormalities, and imaging findings suggestive of amyloidosis, often necessitating histologic confirmation of amyloid deposits. Automated diagnostic algorithms offer a promising avenue for overcoming early diagnosis hurdles, leveraging machine learning (ML) to extract pertinent information from raw data autonomously, reducing reliance on operator expertise. This review critically assesses various diagnostic modalities and the utility of artificial intelligence computational techniques in enhancing amyloidosis detection (Allegra et al., 2023). Table 11.3 provides an overview of the impact of amyloidosis on major organs.

11.3.1 Algorithm to develop a model to predict amyloidosis

The algorithm's learning and training techniques are grounded in the clinical records of hospitalizations, serving as the foundational data source. It encompasses each admission and discharge episode, using treatment-related information generated for patients as a data vector for aggregation. The substantial volume of clinical history entails high-dimensional data, contributing to the dataset's complexity. In addition, the absence of a definitive diagnosis results in a significant class imbalance, underscoring the challenge posed by the limited quality of disease prevalence

Table 11.3 Different effects of amyloidosis on different major organs.

Amyloid type	Precursor protein	Major organ involvement		
		Heart	Kidney	Liver
AL	Immunoglobulin light chain	>50%[a]	≥50%	10%–30%
ATTRv	Mutated transthyretin	≥50%[b]	Rare	Rare
ATTRwt	Wild-type transthyretin	≥50%[c]	Rare	Rare
ApoAI	Mutated apolipoprotein AI	≥10%[d]	≥10%	≥50%
AA	Serum amyloid A protein	≥10%	≥50%	≥10%
ALECT2	Leukocyte chemotactic factor 2	Rare	≥50%	≥10%

[a] *Usually absent, can be intense.*
[b] *Usually intense, can be absent in some variants.*
[c] *Usually intense.*
[d] *Present.*

(Agibetov et al., 2020). For a comprehensive understanding, the algorithm in Box 11.1 outlines the detailed steps of this process.

Box 11.1 Algorithm for developing a model to predict amyloidosis

Step 1: Construct the baseline linear model based on K logistic regression.
Step 2: Use a machine learning algorithm based on gradient tree boosting to improve linearity.
Step 3: Develop prediction models to perform nonlinear predictions.
Step 4: Compare the performance of all diagnostic algorithms.
Step 5: Then all the prediction models will be built into a training group so they may prove effective in patients who have amyloidosis and amyloidosis-unrelated heart failure.
Step 6: The performance of all prediction models needs to be evaluated separately.

11.3.2 Working procedure

The model to be built to detect amyloidosis follow some steps briefly described next.

11.3.2.1 Features extraction

For feature selection, sociodemographic variables, clinical progression indicators, and additional parameters were systematically recorded by medical and nursing personnel within the structured framework of electronic clinical records. These were composed of clinical data derived from assessments, supplementary examinations, diagnoses, and treatment recommendations. However, capturing the entirety of clinical evolution throughout the hospitalization process necessitated the extraction of unstructured data from clinical narratives contained within the free-text fields of electronic medical records authored by health care practitioners. Despite the inherent lack of structure in these narrative fields, this qualitative analysis may prove pivotal in uncovering subtle indicators to identify potential cases of undiagnosed amyloidosis, although with the challenge of requiring greater standardization for robust analysis.

11.3.2.2 Data processing

Data analysis commences with identifying diagnosed amyloidosis cases within the extracted medical records. The procedure outlined in Fig. 11.2 was employed to compile the patient characteristic vector. This figure shows an overview of the entire training dataset construction process. As depicted in the upper left corner of Fig. 11.2 (step 1), the clinical records were loaded into a data dictionary, facilitating efficient memory-based manipulation.

The data underwent an initial preprocessing stage (step 2) before new fields were generated and transformation pipelines were executed. During cleaning, columns or rows containing whitespace, special characters, null values, and constant values were removed. Also, values were standardized to join data tables from different sources and a new one-phase primary key field was created. Subsequently, new fields were added following the data combination (step 3), focusing on specific features associated with amyloidosis as identified by experts. Within step 4, numerous fields

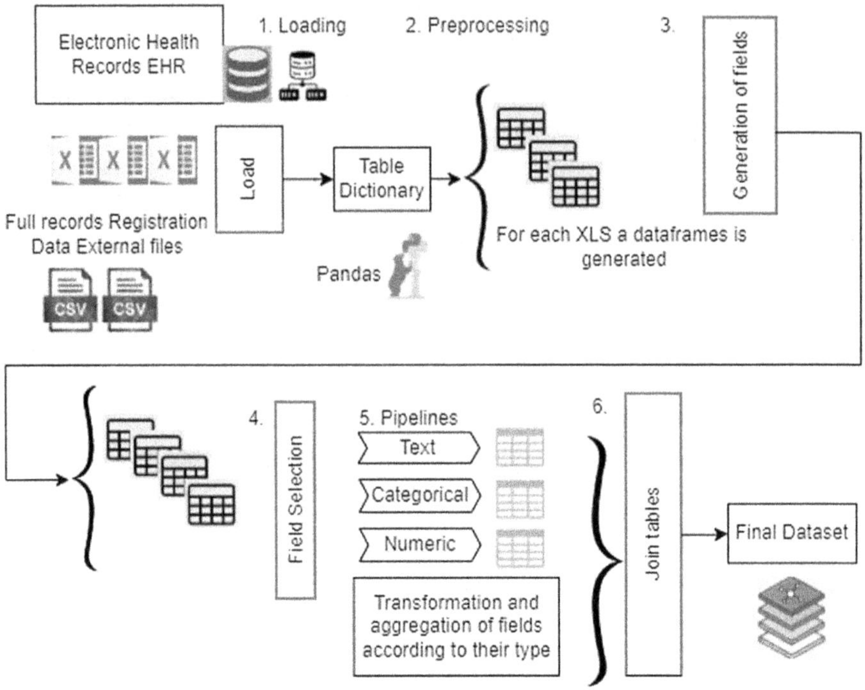

FIGURE 11.2

Steps for data processing.
CSV, comma seperated value; *XLS*, excel spreadsheet.

with low-quality data, memory depletion, and redundant operations were discarded. Step 5, depicted in Fig. 11.2, involved separately processing each variable type (textual, categorical, and numerical) to address their unique characteristics. Pipelines as programming artifacts defined a reproducible process that could be parallelized for efficient execution with new data. On completion of pipeline execution, the training variables were obtained. Finally, the table was joined using the episode key in a table/trainable matrix (step 6).

11.3.2.3 Validation techniques

In the validation technique, cross-validation is employed to train and test the model, in which a portion of the dataset is used for training and the remaining portion for testing. For instance, with the number of folds set to k = 5, each fold influences training on 80% of the data, whereas validation takes place on the remaining 20%. Stratification ensures the proportion of 0 and 1s remains consistent in both validation and training sets. Through this process, the algorithm learns from all available data, with varying patients used for training each time, making it particularly valuable when an objective determination is unavailable. Cross-validation aids in addressing issues associated with sparse data. During the selection phase of the model, the aim is to ascertain whether other complex models such as random forests or neural networks exhibit signs of overfitting. Notably, these models may encounter scenarios that need help identifying true positives in multiple cross-validation folds.

11.3.2.4 Algorithm design for training

Following an iterative process, transformations and operations of the steps of the algorithm proposed are outlined as: (1) imputation of missing values using the mean; (2) selection of the variables exhibiting variance greater than zero; (3) variables are normalized so that principal component analysis (PCA) may be implemented; (4) application of PCA to cover 99% of variance, effectively reducing the dimensionality of the feature space; (5) standardization of variables; (6) implementation of undersampling using the centroid of each cluster; (7) application of slight oversampling using the synthetic minority oversampling technique; and (8) training using logistic regression.

11.3.2.5 Metrics for evaluating results

Metrics such as accuracy, sensitivity, and specificity are pertinent to evaluate the efficacy of the proposed training and testing. To evaluate the model's learning capability under imbalanced class conditions, it was essential to consider the area under the receiver operating characteristic curve (Ghosh et al., 2022). Disease prevalence affects accuracy, specificity, and sensitivity, and the latter is particularly important for diseases with less prevalence. Conversely, F1, a metric measuring test accuracy, is highly sensitive to improvements or deteriorations in results and enables the assessment of success in positive classes, proving valuable in scenarios of imbalanced learning. When the algorithm is finished, it undergoes training, initially with standard samples and subsequently with patient samples (García-García et al., 2021).

11.3.3 Methods for detecting amyloidosis

There are various methods for detecting amyloidosis. In this section, we discuss two of them.

11.3.3.1 Naive Bayesian classification

A Naïve Bayesian classifier uses amyloidogenic and nonamyloidogenic derivatives for each microbe. Essentially, the probability, denoted as p, is determined for a mutation occurring at position x for both amyloidogenic (p_{AM}) and nonamyloidogenic (p_{NAM}) strains of the same organism. The raw values of p_{AM} and p_{NAM} may potentially be zero; hence, to mitigate this, the Laplace correction method is employed. This involves adding Eq. (11.1) to the numerator and Eq. (11.2) to the denominator. Complementary values, q_{AM} and q_{NAM}, indicating residue retention, are calculated as $1 - p_{AM}$ or $1 - p_{NAM}$, respectively.

These probabilities are used to determine amyloidogenic as well as nonamyloidogenic propensities so that a test sequence is generated from the same organism as a training set. Based on the assumption that there are mutations of s at positions that are defined by set M, the amyloidogenic potential is calculated as AM:

$$p_{\mathrm{AM}} = \prod_{x=1}^{n,n \notin M} q_{\mathrm{AM}_x} \times \prod_{x=1}^{n,n \notin M} p_{\mathrm{AM}_x} \tag{11.1}$$

and the probability of nonamyloidogenic is computed as:

$$p_{\mathrm{NAM}} = \prod_{x=1}^{n,n \notin M} q_{\mathrm{NAM}_x} \times \prod_{x=1}^{n,n \notin M} p_{\mathrm{NAM}_x} \tag{11.2}$$

Here, x represents the particular position, where mutation can cause amyloidosis or not (as illustrated in Fig. 11.1). If the amyloidogenic score (AM) exceeds the nonamyloidogenic score (NAM), the sequence is categorized as amyloidogenic; it is categorized as nonamyloidogenic otherwise. The classifier accuracy is verified against the test and the training sets used. Given the restricted number of sequences available, validation is conducted on a preliminary basis. This involves leave-one-out cross-validation that is performed for all amyloidogenic and nonamyloidogenic derivatives alongside a onetime test validation holdout.

11.3.3.2 Decision tree

The development of a weighted decision tree offers quantitative evaluations of mutations' individual and collective impacts based on their location (e.g., Complementarity Determining Region [CDR]/Framework Region [FR]), exposure levels, and alterations in sheet formation propensity. Initially, we create distinct mutation matrices for residues buried within the CDR, buried residues within the FR, exposed residues within the CDR, and alignment of exposed residues within the FR for amyloidogenic and nonamyloidogenic derivatives.

The algorithm defines *exposed residues* as those accounting for 25% or more of the accessible surface residues. Exposure data for each alignment is generated using the structural germline sequence homologs. These data are visually represented to facilitate analysis, followed by postprocessing involving nonamyloidogenic subtraction from amyloidogenic matrix images. An image in which relative intensity reflects the dominance of specific mutations is the result. In addition, a binary matrix contains mutations exclusive to the amyloid formers. The matrix is organized based on an ascending trend of β-sheet formation (Table 11.4), with original residues in rows and columns replacing the remainder. Mutations to the diagonal right indicate an increased tendency toward sheet formation, whereas those to the diagonal left signify decreasing trends. These observed trends in the matrices are then employed as weights attached to tree branches. The objective is to determine whether it is possible to generalize the pathways taken by both amyloid and nonamyloid formers or if they depend on the microbe. This approach identifies paths that optimize the amyloidogenic and nonamyloidogenic derivatives separation for each bacterium (David et al., 2010).

Table 11.4 β-Sheet Amino Acids Forming Propensities.

Amino acid	ΔG (kcal/mol)	Amino acid	ΔG (kcal/mol)	Amino acid	ΔG (kcal/mol)
Thr	1.1	Val	0.82	Cys	0.52
Ile	1.0	Met	0.72	Leu	0.51
Tyr	0.96	Ser	0.70	Arg	0.45
Phe	0.86	Trp	0.54	Lys	0.27
Gln	0.23	Asn	−0.08	Gly	−1.2
Glu	0.01	Asp	−0.94	Pro	< −3
Ala	0.00	His	−0.02		

11.4 Artificial intelligence and machine learning–based approach to amyloidosis detection model

Regression analysis using the least absolute shrinkage and selection operator was conducted based on ML models, and their diagnostic efficacy was assessed. The integration of traditional magnetic resonance (MR) metrics with radiomics texture features in a combined model exhibited superior discriminatory performance compared with the traditional MR metrics model alone, suggesting the potential for further improvement in discriminative capability. Radiomics–MR fusion leverages combined models, and texture analysis can effectively monitor alterations in myocardial microstructure.

It contrasts deep learning methods with an ML algorithm incorporating manually obtained features to emulate expert operator interpretation of experiments. The deep learning approach demonstrated reasonable diagnostic accuracy, underscoring its potential to provide suitable features for training ML-based models, given its consistent depiction of cardiac architecture and deformation. The identification of amyloid molecules is a crucial step in suspected amyloidosis diagnosis, commonly accomplished through immunohistochemistry and laser capture techniques, followed by amyloid subtype determination via mass spectrometry.

Regardless of the type of amyloid protein, amyloid fibrils share a standard structure, enabling in vitro examination through x-ray diffraction, electron microscopy, and specific chemical stains. Computational fibril detection and prediction methods have emerged to cope with the vast protein sequencing data generated, albeit with inherent cost and time limitations. Protein sequencing methods enable the simultaneous execution of millions of reactions, yielding extensive datasets.

Computational investigations have deepened our understanding of amyloid formation mechanisms. The "dock-and-lock" mechanism, supported by extensive computational studies, posits rapid monomer adsorption at fibril ends, followed by slow structural rearrangement hindered by monomer–fibril interactions. Discoveries suggest additional fibril-dependent secondary mechanisms.

Further computational exploration is warranted to enhance comprehension of nucleation processes. Simulation studies have pinpointed vital factors such as appropriate peptide–fibril surface interaction strength, preferential amyloidogenic conformation of bound peptides, facile peptide movement on fibril surfaces, and simple dissociation for rapid turnover (Allegra et al., 2023).

11.5 Amyloidosis and dementia

Amyloidosis and dementia have always had an intriguing and multifaceted correlation related to neurodegenerative disorders. This section looks into this correlation, because amyloidosis shares standard pathologic features with dementia, particularly AD. This will help in understanding the underlying mechanisms and developing effective therapeutic strategies.

11.5.1 Amyloidosis and dementia: Key definitions and mechanisms

The link between cerebral amyloidosis and dementia is being investigated in contemporary neuropathology. Its preliminary origins were documented by Alois Alzheimer. In his work, plaque formations were discovered in a dementia patient's brain, which led to further investigation of the correlation between amyloidosis and dementia. It was only in 1984, when Glenner and Wong (Glenner and Wong, 1984a, 1984b) first identified the presence of amyloid-beta (Aβ) peptide in brain plaques characteristic of AD that the hypothesis was formally considered. In 1960, Neumann (1960) considered amyloidosis as a distinct pathology, and later a hypothesis later was originated that amyloid deposits in cerebral vessels and brain parenchyma could be the cause of dementia and AD (Hardy and Selkoe, 2002).

Question that arise are how this hypothesis could benefit advances in treatment and how researchers started to investigate the potential relationship between amyloidosis and dementia.

Cerebral amyloidosis relates to dementia because the Aβ peptide accumulation in the brain has a role in AD pathogenesis and other forms of dementia. Amyloidosis is the process by which amyloid proteins become abnormally deposited in organs or tissues. It may disrupt normal function. In the context of AD, amyloidosis may refer to Aβ peptide accumulation in the brain's parenchyma and blood vessels. This accumulation is characterized by the formation of amyloid plaques in the brain's extracellular space and cerebral amyloid angiopathy (CAA), in which amyloid deposits along the walls of cerebral blood vessels.

The process can be understood through some main mechanisms: plaque formation and neuronal damage (Hampel et al., 2021; Rukmangadachar and Bollu, 2023), tau pathology (Hampel et al., 2021), inflammatory response (Iliyasu et al., 2023), and CAA (Banerjee et al., 2024). The first mechanism is related to the accumulation of insoluble amyloid plaques in the brain's extracellular spaces, which is thought to interfere with neuronal communication, leading to the loss of synapses and neuronal death. Neural communication disruption and cell loss contribute to cognitive decline and dementia symptoms. In tau pathology, Aβ accumulation is also associated with the hyperphosphorylation of tau, a microtubule-associated protein. In AD, tau proteins become abnormally phosphorylated and form neurofibrillary tangles (NFTs) inside neurons, further contributing to neuronal dysfunction and death. Amyloid plaques trigger an inflammatory response in the brain. Microglia, the brain's resident immune cells, become activated to clear the plaques. However, this process can lead to chronic inflammation, causing further neuronal damage and exacerbating cognitive decline. Moreover, in CAA, amyloid deposits in the cerebral blood vessel walls weaken the vessels, increasing the risk of hemorrhage and contributing to the disruption of cerebral-related blood flow, which can lead to additional brain damage and cognitive impairment.

Fig. 11.3 shows the relationship between amyloidosis and dementia. It captures the progression of molecular alterations within the brain through the key mechanisms presented before and to their manifestation as clinical symptoms of dementia. It highlights the nature of cognitive decline and underscores the importance of each pathologic process in the overall disease path.

The relationships involved in research on AD are also perceived in Fig. 11.4, as suggested by Sims et al.(2020). This model details the risk factors associated with AD, including genetic factors and their relation to clinical and cellular models.

11.5.2 Literature survey linking amyloidosis and dementia

Understanding the link between amyloidosis and dementia has led researchers to explore therapeutic strategies aimed at reducing Aβ levels in the brain or preventing its aggregation, as mentioned at the end of this section. Mechanisms that may link amyloidosis with dementia were suggested as hypotheses and also investigated more deeply.

The amyloid hypothesis began with the work of Alois Alzheimer. In 1906, Alzheimer presented a case in which peculiar plaques and tangles were detected in the

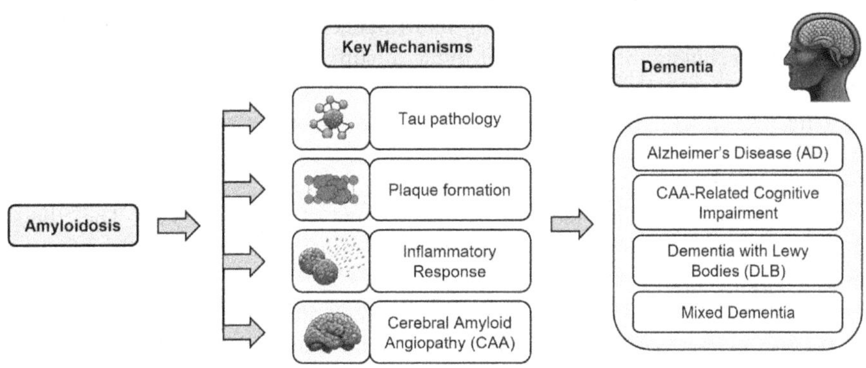

FIGURE 11.3

Key mechanisms among amyloidosis and dementia and potential diseases.

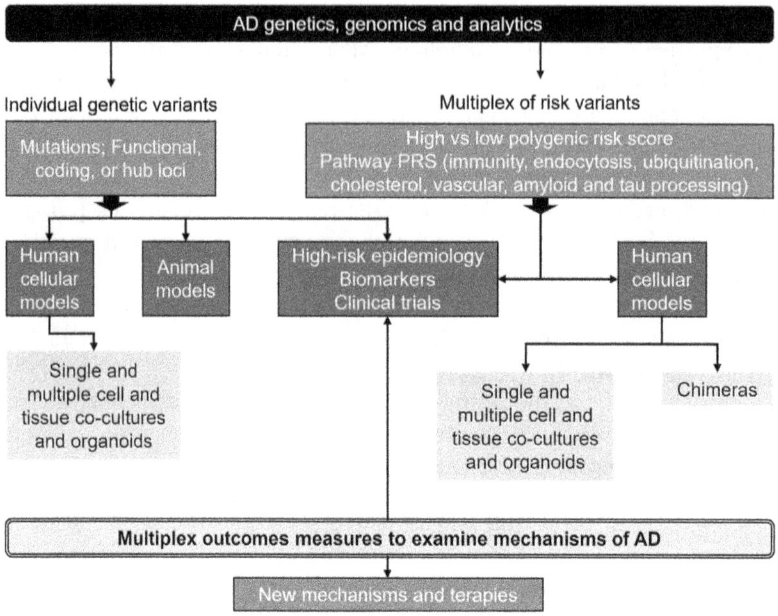

FIGURE 11.4

Alzheimer's disease (AD) mechanisms through the perspective of Sims et al. *PRS*, polygenic risk score.

Adapted from Sims, R., Hill, M., Williams, J. 2020. The multiplex model of the genetics of Alzheimer's disease. Nat. Neurosci. 23 (3), 311322.

brain of a patient who had had dementia, a report that was later published in 1907 (Alzheimer, 1910; Fischer, 1910). This case laid the groundwork for subsequent neuropathologic research into dementia.

Further histologic explorations were related to amyloid plaques. In 1927, Divry explored the use of Congo red staining for the amyloid nature of these plaques through their characteristic apple green birefringence under polarized light (Divry, 1927). This technique became crucial in the identification and study of amyloid in brain tissue.

With Alzheimer's discovery, dementia and its intricacies could be easily noted. In 1949, Benedek and McGovern (1949) reported a case of AD in which cerebral vessel amyloidosis was investigated, emphasizing the significance of vascular involvement in the disease pathology (Benedek and McGovern, 1949). Their findings suggested that amyloid deposits might contribute to the cognitive decline observed in dementia.

A crucial understanding of the systemic nature of amyloidosis was evidenced in the work of Scholz (1938), who studied the amyloid nature in senile brains (Scholz, 1938), and the comprehensive neuropathologic characterization of amyloid plaques by Morel and Wildi (Morel, 1952). Neumann founded the modern concepts of AD and cerebral amyloidosis. These early observations of amyloid vascular changes and argyrophilic plaques in the central nervous system were crucial in identifying amyloidosis as a distinct pathology (Neumann, 1960).

The focus then shifted toward a biochemical understanding of amyloid plaques, introduced by Glenner and Wong in 1984. The researchers isolated and identified Aβ, the main protein constituent of these plaques, providing a molecular basis for AD and firmly establishing the amyloid hypothesis (Glenner and Wong, 1984a). The implications of these findings are further explored in the amino acid analyses presented in Table 11.5, which lists chromatographic fractions relevant to Alzheimer's research.

Beginning in 1984, several studies and discoveries were conducted to understand phenomena linking amyloidosis with dementia. By the 1990s, it started to be more explicit that vascular contributions have a significant role in dementia.

Table 11.5 Amino acid analyses of chromatographic fractions.

	Nanomolar percentage			No. of residues	
Amino acid	**G-100 β**	**HPLC β_1**	**HPLC β_2**	**β_1 peptide**	**β_2 peptide**
Asp	11.47	11.07	11.11	3	3
Thr	3.15	1.59	1.36	1	1
Ser	5.60	6.35	5.38	2	2
Glu	11.40	12.26	11.18	4	3
Pro	T[a]	T	T	0	0
Gly	1.65	16.71	15.23	5	5
Ala	8.27	8.08	7.91	2	2
½ Cys	0.94	T	T	1	1
Val	9.47	8.69	11.74	3	4
Met	2.92	2.06	2.48	1	1
Ile	4.29	3.58	3.90	1	1
Leu	7.34	6.68	6.30	2	2
Tyr	2.71	1.89	2.07	1	1
Phe	5.76	6.13	6.95	2	2
His	4.64	5.44	5.94	2	2
Lys	5.01	6.46	5.44	2	2
Arg	3.30	2.97	2.89	1	1
Try	ND[b]	ND	ND	–	–
Totals	99.92	99.96	99.86	33	33

HPLC, *high-performance liquid chromatography;* ND, *neurodegenerative disease.*

[a] *Trace.*

[b] *Tryptophan not determined.*

Adapted from Glenner, G.G., Wong, C.W., 1984a. Alzheimer's disease: initial report of the purification and characterization of a novel cerebrovascular amyloid protein. Biochem. Biophys. Res. Commun. 120 (3), 885890; Glenner, G.G., Wong, C.W., 1984b. Alzheimer's disease and Down's syndrome: sharing of a unique cerebrovascular amyloid fibril protein. Biochem. Biophys. Res. Commun. 122 (3), 11311135.

A 1990 study (Natté et al., 2001) on hereditary cerebral hemorrhage with amyloidosis, Dutch type found a high amount of CAA in the frontal neocortex associated with dementia in patients. This independent relationship between CAA and cognitive impairment, even in the absence of Alzheimer's typical plaques and tangles, suggested that vascular health was a significant factor in the development of dementia. This realization led to a broader examination of vascular factors in cognitive decline and opened new opportunities for potential therapeutic interventions.

Petersen et al. (1997) identified a novel mutation associated with transthyretin amyloidosis, a form of amyloidosis distinct from the Aβ pathology typically observed in AD. The study traced a genetic framework associated with dementia, specifically highlighting the role of non-Alzheimer's amyloidosis in cognitive decline. The mutation's discovery in the transthyretin gene provided significant insight into the diverse genetic origins of dementia, suggesting that different forms of amyloid deposition could lead to similar clinical manifestations of dementia.

With the development of positron emission tomography (PET) techniques, advances in amyloid imaging allowed for more precise tracking of amyloid deposition in the living brain. In 2001, a study published by Jorge and Blas (Ghiso and Frangione, 2001) on cerebral amyloidosis and amyloid angiopathy indicated that not all amyloid proteins lead to deposition and subsequent dementia, suggesting a more complex relationship between the presence of amyloid and cognitive decline. These imaging studies provided a better understanding of the progression and manifestation of amyloidosis-related dementia.

In 2002, Holton et al. (2002) explored the manifestations of familial Danish dementia, a pathology characterized by extensive CAA, parenchymal protein deposits, and neurofibrillary degeneration associated with a BRI2 gene mutation. Building on such empirical evidence, the theoretical models of dementia evolved, centering on the role of amyloid plaques and tau-containing NFTs. These models gained traction with the discovery that mutations in the APP and presenilin genes contribute to nonnormal amyloid protein production, a concept reinforced by Hardy and Selkoe's cascade hypothesis (Hampel et al., 2021) presented in Fig. 11.5, which associates AD progression with Aβ-protein aggregation.

The study on senile systemic amyloidosis by Tanskanen et al. (2006) also highlighted the association between CAA and dementia, contributing to the understanding that amyloid deposition in cerebral vessels could independently influence cognitive decline separately from parenchymal amyloid plaques.

In 2010, integrating the amyloid and tau hypotheses provided a more comprehensive understanding of Alzheimer's pathogenesis. Whereas amyloid plaques could trigger a cascade of events, tau protein abnormalities were closely linked to the neurodegenerative process, as indicated by a study from Tamayev et al. (2010) that focused on memory deficits caused by the BRI2 mutation. The mutation's effects on BRI2 function rather than amyloidosis suggested an association between amyloid-independent and amyloid-related dementia.

In 2011, Klunk (2011) also evaluated the use of amyloid PET imaging in detecting amyloid plaques, which is crucial in diagnosing and researching AD. This study

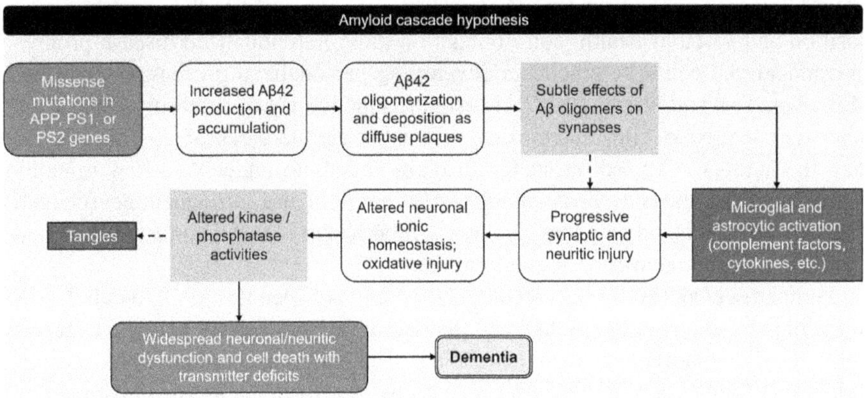

FIGURE 11.5

Hardy and Selkoe's cascade hypothesis.

Adopted from Hampel, H., Hardy, J., Blennow, K., Chen, C., Perry, G., Kim, S.H., Villemagne, V.L., Aisen, P., Vendruscolo, M., Iwatsubo, T., 2021. The amyloid-β pathway in Alzheimer's disease. Mol. Psychiatr. 26 (10), 54815503

discusses the correlation between imaging data and postmortem findings, its role in therapeutic trials, and its potential in preclinical Alzheimer's research. The importance of amyloid imaging in enhancing clinical diagnosis accuracy is highlighted with the progression of AD and the effectiveness of antiamyloid therapies.

A neuropathologic analysis from Ziskin et al. (2015) in 2015 focuses on the Tyr69His TTR variant, further clarifying the link between genetic variants and the clinical presentation of amyloidosis with dementia. It studies the existent pathophysiologic mechanisms, offering evidence that specific gene variants could predispose individuals to amyloid deposition and dementia.

Tau pathology began to emerge as a critical component of AD. Neurofibrillary tangles, composed of hyperphosphorylated tau protein, were identified. The association between tau dysfunction and amyloid deposition was highlighted in a paper by Nation et al. (2015) on pulse pressure about tau-mediated neurodegeneration, which indicated that tau pathology could affect amyloidosis on cognitive functions.

A mechanism in dementia-linked amyloidosis, specifically deamidation of the S100A9 protein, was studied by Adav et al. (2016). The study shows that this protein modification could be a key event triggering protein aggregation and subsequent amyloid formation in the brain. This discovery may open opportunities for understanding the molecular developments that lead to dementia and offer potential goals for therapeutic intervention.

In 2018, a study carried out by Lange et al. (2018) on predicting Alzheimer's dementia in patients with amnestic mild cognitive impairment suggested that when added to models that include cognitive status, biomarkers of neurodegeneration and brain amyloidosis can significantly improve the prediction of progression to

AD. This stepwise approach to understanding risk factors, including genetic predisposition and vascular health, contributes to a more individualized disease progression model and could be crucial in developing personalized prevention strategies. Moreover, Yew and Nation (2017) investigated the impact of cerebrovascular resistance on cognitive decline and cortical atrophy related to cerebral amyloidosis. For that, the authors analyzed resistance in brain vessels to understand how impaired blood flow contributes to the accumulation of amyloid and subsequent neurodegeneration. The results indicate that cerebrovascular health is a critical factor in dementia when amyloid deposition is predominant.

Fernandez et al. (2023) studied the relationship between the APOE e4 allele, one of the higher genetic risks for AD, and the risk of familial Danish dementia. Results highlighted that individuals with this allele had an earlier onset of amyloidosis and a more rapid progression to dementia. A study by de Oliveira et al. (de Oliveira et al., 2023) in 2023 addressed the challenge of differentiating dementia with Lewy bodies from Alzheimer's dementia using cerebrospinal fluid (CSF) biomarkers. The high prevalence of amyloidosis across various dementia syndromes was found to affect the performance and reliability of CSF biomarkers in the differential diagnosis. Results emphasized the need for biomarkers that can distinguish more accurately between different types of dementia, given the broad occurrence of amyloidosis. This study, like many others, contributes to refining diagnostic tools for dementia to improve early detection and tailor treatments to specific dementia subtypes.

11.5.3 Other treatments, methods, and diagnostics

A review throughout the decades on some tools for early detection and intervention has been highlighted. Here, other studies specifically for treatment and diagnosis are discussed within the scope of amyloidosis-related dementia. Antiamyloid therapies and the evolving utility of biomarkers are explored.

Aducanumab, an anti-Aβ monoclonal antibody (mAb) for treating AD, was developed in 2020 (Plascencia-Villa and Perry, 2022). It outlines the clinical efficacy of aducanumab, including its molecular characteristics and its development focused on early stage Alzheimer's patients with evidence of Aβ pathology.

Filippi et al. (2023) also outlined a plan for integrating anti-Aβ mAbs for AD into the Italian health care system. It emphasizes the need for a biomarker-based biological diagnosis of Alzheimer's considering both amyloid and tau-related biomarkers, and presents a structured approach for patient selection and management. Efficiency and limitations were evaluated through three clinical studies, with mixed results. Study 302 showed a positive effect, meeting its primary and secondary objectives with statistically significant treatment effects favoring aducanumab regarding the high-dose treatment arm. However, study 301 was negative, showing no difference from placebo in the high-dose treatment arm at the primary efficacy end point. The discrepancies between studies may lead to further analyses in the future, including considerations of dosing, rapid disease progression, and patient demographics.

Rajab et al. (2023) focused on using ML techniques to select and classify neuropathologic features associated with AD. It outlines a study conducted on a cohort from the Cognitive Function and Ageing Study to identify critical neuropathologic features that could assist in classifying dementia. Using the ML approach, various filter-based feature selection methods were employed to determine the relevance of each feature to dementia classification, highlighting the importance of features such as Braak NFT stage and CAA for dementia status classification. The ranking of neuropathology features showed high relevance for certain features such as the Braak NFT stage, which was consistently ranked in the top positions across different methods. This indicates its utility for both human and computer-aided dementia diagnosis. Other consistently highly ranked features include BrainNet tau stage, CAA type, Thal phase, subpial brain stem, and subpial tau astrogliopathy in the mesial temporal lobe.

Inoue et al. (2023) explored the impact of combining classical neuropathology features with nonstandard features (e.g., regional markers and demographic information) on classification accuracy. This approach addressed misclassifications and improved sensitivity, especially among older individuals. Including nonstandard features alongside top-ranked neuropathology, the features showed promise in reducing misclassification rates and enhancing model performance.

In 2014, Pietrzak et al. (2014) studied the impact over 3 years of anxiety symptoms on the relationship between cerebral amyloid burden and memory decline in healthy older adults without dementia. It suggests that anxiety symptoms may exacerbate the decline in verbal and episodic memory associated with higher levels of Aβ. In addition, anxiety was found to contribute to a more significant decline in executive function independent of amyloid levels and other risk factors, indicating that managing anxiety could potentially delay memory decline in this population.

Finally, Tamano et al. (2016) investigated the impact of Aβ1-42 injections into mice on inducing dementia-like behavior abnormalities under conditions of social isolation. The study reported increased aggression, anxiety-like behavior, cognitive decline, and elevated serum corticosterone levels in Aβ1-42–injected mice and study suggested that these behavior changes were associated with increased excitability of hippocampal neurons, which might be linked to elevated corticosterone levels.

Table 11.6 lists this literature, which provides a relationship between amyloidosis and dementia.

11.5.4 The amyloid clock

Advances in dementia research also led to the validation of the concept of an "amyloid clock," a theoretical model that predicts the progression of AD based on the accumulation of Aβ in the brain, leading to the progression of cognitive decline. It allows clinicians and researchers to model the course of AD more accurately. This approach also facilitates the identification of critical time intervals for intervention, potentially allowing for more effective treatment strategies to slow or impede

Table 11.6 Review linking amyloidosis and dementia.

References	Year	Key findings	Dementia cause	Amyloid type	Additional notes
Benedek and McGovern (1949)	1949	Alzheimer's disease case with cerebral cortex amyloidosis	Vascular factors	Cerebral	Case emphasizing significance of vascular involvement
Divry (1927)	1927	Histologic exploration of amyloid plaques	Plaque formation	General	Introduction of Congo red staining
Fernandez et al. (2023)	2023	APOE influence on amyloidosis and dementia	Genetic factors	N/A	Role of APOE in familial Danish dementia
de Oliveira et al. (2023)	2023	Plan for Aβ amyloid monoclonal antibodies in Alzheimer's disease	N/A	Aβ	Focus on biomarker-based biological diagnosis
Ghiso and Frangione (2001)	2001	Relationship between cerebral amyloidosis, stroke, dementia	CAA	N/A	Focus on cerebral amyloidosis and its effects
Glenner and Wong (1984a,b)	1984	Alzheimer's and Down syndrome shared amyloid protein	Plaque formation	Aβ	Discovery of shared cerebrovascular amyloid protein
Hampel et al. (2021)	2021	Overview of Aβ pathway in Alzheimer's disease	Plaque formation, tau pathology	Aβ	Comprehensive review of Alzheimer's disease mechanisms
Holton et al. (2002)	2002	Familial Danish dementia and its unique amyloidosis	Genetic factors	Danish	Novel form of cerebral amyloidosis
Iliyasu et al. (2023)	2023	Multiple pathways implicated in Alzheimer's disease by Aβ aggregation	Inflammatory response, tau pathology	Aβ	Understanding mechanisms behind Aβ aggregation
Inoue et al. (2023)	2023	Cerebral small vessel disease in vascular dementia and Alzheimer's disease	Vascular factors	N/A	Pathophysiology of cerebral small vessel disease
Klunk (2011)	2011	Amyloid imaging as biomarker for Alzheimer's disease	N/A	Aβ	Role of amyloid imaging in predicting Alzheimer's disease risk

Lange et al. (2018)	2018	Prediction of Alzheimer's dementia in patients with MCI	N/A	Aβ	Increased value of neurodegeneration and amyloidosis biomarkers
Nation et al. (2015)	2015	Pulse pressure in relation to tau-mediated neurodegeneration	Tau pathology	N/A	Link between vascular health and dementia
Natté et al. (2001)	2001	Dementia in hereditary cerebral hemorrhage with Dutch type	CAA	Dutch	CAA and its relation to dementia independent of other pathologies
Neumann (1960)	1960	Amyloid vascular changes and argyrophilic plaques	Vascular factors	General	Early observation of amyloidosis as distinct pathology
Petersen et al. (1997)	1997	New mutation in transthyretin amyloidosis linked to dementia	Genetic factors	TTR	Discovery of a mutation associated with dementia
Pietrzak et al. (2014)	2014	Impact of anxiety on memory decline and amyloid burden	N/A	Aβ	Anxiety's effect on cognitive decline and amyloid levels
Tanskanen et al. (2006)	2006	Amyloidosis, CAA, and dementia in Finnish population	CAA	Finnish	Relationship among systemic amyloidosis, CAA, and dementia
Yew and Nation (2017)	2017	Cerebrovascular resistance, amyloidosis, cognitive decline	Vascular factors	Aβ	Relationship between cerebrovascular health and dementia
Ziskin et al. (2015)	2015	Tyr69His TTR variant and dementia	Genetic factors	TTR	Link between genetic variants and clinical presentation of dementia

CAA, *cerebral amyloid angiopathy;* MCI, *mild cognitive impairment,* N/A, *not available;* TTR, *transthyretin.*

disease progression. Studies on Aβ accumulation and neurodegeneration include this concept of an amyloid clock.

Frisoni et al. (2022) revised the amyloid hypothesis, discussing dynamic biomarkers and the relationship among Aβ, tau pathology, and neurodegeneration. This supports the amyloid clock concept by understanding the mechanisms in the origin of AD progression.

Research funded by the National Institute on Aging and published in *Brain* (Betthauser et al., 2022) highlights how estimating amyloid onset can predict Alzheimer's progression. This study developed and tested algorithms using PET scans to estimate the age at which amyloid begins accumulating in the brain, demonstrating the feasibility of an amyloid clock.

A tool developed by researchers from the University of Texas at Arlington, further improved and discussed in HealthITAnalytics (McNemar, 2021), also uses a learning-based framework to predict Alzheimer's progression. This approach allows for the more precise prediction of the disease's development stages and aligns with the amyloid clock concept by using biomarkers to predict the disease's progress.

11.6 Conclusion

This chapter discussed linkage between amyloidosis and dementia by identifying fundamental mechanisms that explain such a linkage. Some main mechanisms are plaque formation, CAA, tau pathology, and inflammatory response. The historical background started with the formulation of the amyloidosis hypothesis for dementia, delving into deeper investigations on the path for better evidence and into the report of some potential diagnostics, rare cases of dementia types, and implementations of feature extractions via imaging analysis or ML analysis. Among the methods in the review, biomarker testing, either CSF analysis or blood-based, is promising for clinical practice. CSF analysis measures Aβ and tau protein levels in the CSF. Abnormal levels of these proteins can indicate AD several years before symptoms appear. Other studies focus on identifying Aβ and tau protein levels in the blood, offering a less invasive method for the early detection of AD. Moreover, PET scans can directly visualize plaques of amyloid in the brain. Amyloid PET imaging uses radiotracers to detect amyloid deposits.

Although not diagnostic, testing for the APOE η4 allele's presence, which the literature shows is a risk factor for AD development, can provide information about an individual's risk. Not frequently tested in the literature, the use of ML to select and classify neuropathologic features from imaging and biomarker data can help identify and classify dementia-related pathologies more accurately. Sometimes, a combination of genetic testing and other such analyses may help in the faster and more accurate detection of dementia.

Identified limitations may include the mixed results from clinical studies on anti-amyloid therapies, such as aducanumab. This highlights challenges in achieving consistent treatment efficacy and the need for further research to understand the variability in patient responses. There is also difficulty in differentiating among various types of dementia owing to overlapping symptoms and the presence of amyloidosis across different dementia syndromes. Therefore, more accurate biomarkers may be needed.

Future work might involve improving diagnostic tools and treatment strategies by developing novel therapeutic interventions targeting Aβ accumulation. Moreover, biomarkers for the early detection and differentiation of dementia types should be explored, as well as of the molecular mechanisms of the connection between amyloidosis and cognitive decline. Moreover, more ML-based methods should be approached to classify neuropathologic features associated with dementia and investigate nonstandard features to improve diagnosis accuracy and treatment personalization.

References

Adav, S.S., Gallart-Palau, X., Tan, K.H., Lim, S.K., Tam, J.P., Sze, S.K., 2016. Dementia-linked amyloidosis is associated with brain protein deamidation as revealed by proteomic profiling of human brain tissues. Mol. Brain 9, 1–10.

Agibetov, A., Seirer, B., Dachs, T.-M., Koschutnik, M., Dalos, D., Rettl, R., Duca, F., Schrutka, L., Agis, H., Kain, R., 2020. Machine learning enables prediction of cardiac amyloidosis by routine laboratory parameters: a proof-of-concept study. J. Clin. Med. 9 (5), 1334.

Allegra, A., Mirabile, G., Tonacci, A., Genovese, S., Pioggia, G., Gangemi, S., 2023. Machine learning approaches in diagnosis, prognosis and treatment selection of cardiac amyloidosis. Int. J. Mol. Sci. 24 (6), 5680.

Alzheimer, A., 1910. Die diagnostischen Schwierigkeiten in der Psychiatric. Zeitschrift für die gesamte Neurologie und Psychiatrie 1 (1), 1–19.

Banerjee, G., Farmer, S.F., Hyare, H., Jaunmuktane, Z., Mead, S., Ryan, N.S., Schott, J.M., Werring, D.J., Rudge, P., Collinge, J., 2024. Iatrogenic Alzheimer's disease in recipients of cadaveric pituitary-derived growth hormone. Nat. Med. 1–9.

Benedek, S., McGovern, V.J., 1949. A case of Alzheimer's disease with amyloidosis of the vessels of the cerebral cortex. Med. J. Aust. 2 (12), 429–430.

Betthauser, T.J., Bilgel, M., Koscik, R.L., Jedynak, B.M., An, Y., Kellett, K.A., Moghekar, A., Jonaitis, E.M., Stone, C.K., Engelman, C.D., 2022. Multi-method investigation of factors influencing amyloid onset and impairment in three cohorts. Brain 145 (11), 4065–4079.

Blondel, V.D., Guillaume, J.-L., Lambiotte, R., Lefebvre, E., 2008. Fast unfolding of communities in large networks. J. Stat. Mech. Theor. Exp. 2008 (10), P10008.

David, M.P.C., Concepcion, G.P., Padlan, E.A., 2010. Using simple artificial intelligence methods for predicting amyloidogenesis in antibodies. BMC Bioinf. 11, 1–13.

Desport, E., Bridoux, F., Sirac, C., Delbes, S., Bender, S., Fernandez, B., Quellard, N., Lacombe, C., Goujon, J.-M., Lavergne, D., 2012. Al amyloidosis. Orphanet J. Rare Dis. 7, 1–13.

Divry, P., 1927. Etude histochimique des plaques seniles. J. Belge Neurol. Psychiat. 27, 643–657.

Fernandez, A., Gomez, M.-T., Vidal, R., 2023. Lack of ApoE inhibits ADan amyloidosis in a mouse model of familial Danish dementia. J. Biol. Chem. 299 (1).

Filippi, M., Cecchetti, G., Cagnin, A., Marra, C., Nobili, F., Parnetti, L., Agosta, F., 2023. Redefinition of dementia care in Italy in the era of amyloid-lowering agents for the treatment of Alzheimer's disease: an expert opinion and practical guideline. J. Neurol. 270 (6), 3159–3170.

Fischer, O., 1910. Die presbyophrene Demenz, deren anatomische Grundlage und klinische Abgrenzung. Zeitschrift für die gesamte Neurologie und Psychiatrie 3 (1), 371–471.

Frisoni, G.B., Altomare, D., Thal, D.R., Ribaldi, F., van der Kant, R., Ossenkoppele, R., Blennow, K., Cummings, J., van Duijn, C., Nilsson, P.M., 2022. The probabilistic model of Alzheimer disease: the amyloid hypothesis revised. Nat. Rev. Neurosci. 23 (1), 53–66.

García-García, E., González-Romero, G.M., Martín-Pérez, E.M., Zapata Cornejo, E.de D., Escobar-Aguilar, G., Cárdenas Bonnet, M.F., 2021. Real-world data and machine learning to predict cardiac amyloidosis. Int. J. Environ. Res. Publ. Health 18 (3), 908.

Gertz, M.A., Lacy, M.Q., Dispenzieri, A., 1999. Amyloidosis: recognition, confirmation, prognosis, and therapy. Mayo Clin. Proc. 74 (5), 490–494.

Ghiso, J., Frangione, B., 2001. Cerebral amyloidosis, amyloid angiopathy, and their relationship to stroke and dementia. J. Alzheim. Dis. 3 (1), 65–73.

Ghosh, K., Halder, T., Roy, M., Biswas, C., Gayen, R.K., Chakravarty, D., 2022. A survey on medical image diagnosis systems: problems and prospects. In: Proceedings of International Conference on Computational Intelligence, Data Science and Cloud Computing: IEM-ICDC 2021. Springer, pp. 243–252.

Glenner, G.G., Wong, C.W., 1984a. Alzheimer's disease: initial report of the purification and characterization of a novel cerebrovascular amyloid protein. Biochem. Biophys. Res. Commun. 120 (3), 885–890.

Glenner, G.G., Wong, C.W., 1984b. Alzheimer's disease and Down's syndrome: sharing of a unique cerebrovascular amyloid fibril protein. Biochem. Biophys. Res. Commun. 122 (3), 1131–1135.

Hampel, H., Hardy, J., Blennow, K., Chen, C., Perry, G., Kim, S.H., Villemagne, V.L., Aisen, P., Vendruscolo, M., Iwatsubo, T., 2021. The amyloid-β pathway in Alzheimer's disease. Mol. Psychiatr. 26 (10), 5481–5503.

Hardy, J., Selkoe, D.J., 2002. The amyloid hypothesis of Alzheimer's disease: progress and problems on the road to therapeutics. Science 297 (5580), 353–356.

Holton, J.L., Lashley, T., Ghiso, J., Braendgaard, H., Vidal, R., Guerin, C.J., Gibb, G., Hanger, D.P., Rostagno, A., Anderton, B.H., 2002. Familial Danish dementia: a novel form of cerebral amyloidosis associated with deposition of both amyloid-Dan and amyloid-beta. J. Neuropathol. Exp. Neurol. 61 (3), 254–267.

Iliyasu, M.O., Musa, S.A., Oladele, S.B., Iliya, A.I., 2023. Amyloid-beta aggregation implicates multiple pathways in Alzheimer's disease: understanding the mechanisms. Front. Neurosci. 17, 1081938.

Inoue, Y., Shue, F., Bu, G., Kanekiyo, T., 2023. Pathophysiology and probable etiology of cerebral small vessel disease in vascular dementia and Alzheimer's disease. Mol. Neurodegener. 18 (1), 46.

Klunk, W.E., 2011. Amyloid imaging as a biomarker for cerebral β-amyloidosis and risk prediction for Alzheimer dementia. Neurobiol. Aging 32, S20–S36.

Kurer, M.H., Baillod, R.A., Madgwick, J.C., 1991. Musculoskeletal manifestations of amyloidosis. A review of 83 patients on haemodialysis for at least 10 years. J. Bone Jt. Surg. Br. Vol. 73 (2), 271–276.

Kyle, K.A., Bayed, K.D., 1975. Amyloidosis: review op 236 cases. Medicine 54 (4), 271–299.

Lange, C., Suppa, P., Pietrzyk, U., Makowski, M.R., Spies, L., Peters, O., Buchert, R., Alzheimer's Disease Neuroimaging Initiative, 2018. Prediction of Alzheimer's dementia in patients with amnestic mild cognitive impairment in clinical routine: incremental value of biomarkers of neurodegeneration and brain amyloidosis added stepwise to cognitive status. J. Alzheim. Dis. 61 (1), 373–388.

McCarthy, E., 2020. Amyloidosis in bone. In: Tumors and Tumor-like Lesions of Bone. Springer, pp. 901–903, 3030283143.

McNemar, E., 2021. Improving Breast Cancer Imaging with Artificial Intelligence. HealthITAnalytics.

Morel, F., 1952. General and cellular pathochemistry of senile and presenile alterations of the brain. In: Proc 1st Int Congress Neuropathol Rome, pp. 347–374.

Nation, D.A., Edmonds, E.C., Bangen, K.J., Delano-Wood, L., Scanlon, B.K., Han, S.D., Edland, S.D., Salmon, D.P., Galasko, D.R., Bondi, M.W., 2015. Pulse pressure in relation to tau-mediated neurodegeneration, cerebral amyloidosis, and progression to dementia in very old adults. JAMA Neurol. 72 (5), 546–553.

Natté, R., Maat-Schieman, M.L.C., Haan, J., Bornebroek, M., Roos, R.A.C., Van Duinen, S.G., 2001. Dementia in hereditary cerebral hemorrhage with amyloidosis-dutch type is associated with cerebral amyloid angiopathy but is independent of plaques and neurofibrillary tangles. Ann. Neurol. 50 (6), 765–772.

Neumann, M.A., 1960. Combined amyloid vascular changes and argyrophilic plaques in the central nervous system. J. Neuropathol. Exp. Neurol. 19 (3), 370–382.

de Oliveira, F.F., Miraldo, M.C., Castro-Neto, E.F., Machado, F.C., de Almeida, S.S., Matas, S.L.de A., Bertolucci, P.H.F., da Graça Naffah-Mazzacoratti, M., 2023. Differential performance of cerebrospinal fluid biomarker ratios and formulas to discriminate dementia with Lewy bodies from Alzheimer's dementia and cognitively healthy people reflects high level of amyloidosis across dementia syndromes. Alzheimer's Dementia 19, e071206.

Petersen, R.B., Goren, H., Cohen, M., Richardson, S.L., Tresser, N., Lynn, A., Gali, M., Estes, M., Gambetti, P., 1997. Transthyretin amyloidosis: a new mutation associated with dementia. Ann. Neurol. 41 (3), 307–313.

Pietrzak, R.H., Scott, J.C., Neumeister, A., Lim, Y.Y., Ames, D., Ellis, K.A., Harrington, K., Lautenschlager, N.T., Szoeke, C., Martins, R.N., 2014. Anxiety symptoms, cerebral amyloid burden and memory decline in healthy older adults without dementia: 3-year prospective cohort study. Br. J. Psychiatr. 204 (5), 400–401.

Plascencia-Villa, G., Perry, G., 2022. Neuropathologic changes provide insights into key mechanisms of Alzheimer disease and related dementia. Am. J. Pathol. 192 (10), 1340–1346.

Rajab, M.D., Jammeh, E., Taketa, T., Brayne, C., Matthews, F.E., Su, L., Ince, P.G., Wharton, S.B., Wang, D., Cognitive Function and Ageing Neuropathology Study Group, 2023. Assessment of Alzheimer-related pathologies of dementia using machine learning feature selection. Alzheimer's Res. Ther. 15 (1), 47.

Rukmangadachar, L.A., Bollu, P.C., 2023. Amyloid beta peptide. In: StatPearls. StatPearls Publishing.

Scholz, W., 1938. Studien zur Pathologie der Hirngefäße II: Die drusige Entartung der Hirnarterien und-capillaren (Eine Form seniler Gefäßerkrankung.). Zeitschrift für die gesamte Neurologie und Psychiatrie 162, 694–715.

Shenoy, S., Allon, M., Beathard, G., Brouwer-Maier, D., Dember, L.M., Glickman, M., Lee, C., Litchfield, T., Lok, C., Huber, T., 2018. Clinical trial end points for hemodialysis vascular access: background, rationale, and definitions. Clin. J. Am. Soc. Nephrol. 13 (3), 490–494.

Sims, R., Hill, M., Williams, J., 2020. The multiplex model of the genetics of Alzheimer's disease. Nat. Neurosci. 23 (3), 311–322.

Symmers, W.S.C., 1956. Primary amyloidosis: a review. J. Clin. Pathol. 9 (3), 187.

Tamano, H., Ide, K., Adlard, P.A., Bush, A.I., Takeda, A., 2016. Involvement of hippocampal excitability in amyloid β-induced behavioral and psychological symptoms of dementia. J. Toxicol. Sci. 41 (4), 449–457.

Tamayev, R., Giliberto, L., Li, W., d'Abramo, C., Arancio, O., Vidal, R., D'Adamio, L., 2010. Memory deficits due to familial British dementia BRI2 mutation are caused by loss of BRI2 function rather than amyloidosis. J. Neurosci. 30 (44), 14915–14924.

Tanskanen, M., Kiuru-Enari, S., Tienari, P., Polvikoski, T., Verkkoniemi, A., Rastas, S., Sulkava, R., Paetau, A., 2006. Senile systemic amyloidosis, cerebral amyloid angiopathy, and dementia in a very old finnish population. Amyloid 13 (3), 164–169.

Vaxman, I., Gertz, M., 2020. When to suspect a diagnosis of amyloidosis. Acta Haematol. 143 (4), 304–311.

Yew, B., Nation, D.A., 2017. Cerebrovascular resistance and cerebral amyloidosis: effects on cognitive decline, cortical atrophy and progression to dementia. Alzheimer's Dementia 13 (7), P1098.

Ziskin, J.L., Greicius, M.D., Zhu, W., Okumu, A.N., Adams, C.M., Plowey, E.D., 2015. Neuropathologic analysis of Tyr69His TTR variant meningovascular amyloidosis with dementia. Acta Neuropathologica Commun. 3 (1), 43. https://doi.org/10.1186/s40478-015-0216-0.

CHAPTER

12 Machine learning for bone deformation detection in real-world applications

Chandrakant Mahobiya[1] and Sailesh Suryanarayan Iyer[2]

[1]*Anjaneya University, Raipur, Chhattisgarh, India;* [2]*Rai School of Engineering, Rai University, Department of Computer Science and Engineering, Ahmedabad, Gujarat, India*

12.1 Introduction

The development of machine learning (ML) in health care is a necessary innovation for patients. With this technology, people will get new treatment options for many medical problems. While the diagnosis process is accelerated, patients will get more time and facilities to care for themselves. Along with all these avid services, the prominence of ML technology seems to signal about what the future holds for this sector. This chapter will involve determining health use and specifically ML potential in redesigning the current treatment options available for bone disorder diagnosis (Suzuki et al., 2019).

Bone abnormalities pose a particularly complicated problem for medical experts. Two of the body's most fundamental frameworks are bone and the human skeleton, which consists of bones, joints, and different components. Bone is one of the body's most complex structures, and it can contain specific differences from the usual structure and function that need a high level of attention to identify. Because ordinary diagnostic techniques are unable to and efficiently recognize them adequately, in most instances physicians have to rely on subjective views and grade assessments. Also, although these methods are of great utility, they are still susceptible to human error and uncertainty. Artificial intelligence (AI) algorithms combined with medical images can be used as a supplement in such locations where doctors are scarce.

Thus, faced with the numerous trifles in bone deformation identification, ML offers a way out. ML algorithms can detect even minor abnormalities and classify complex deformations according to the degree of development, which results in better descriptions and a more accurate diagnosis. For instance, scoliosis is a widespread disease characterized by a sideways curvature of the spine. Until the advent of ML, the beginning of scoliosis and further developments were diagnosed and traced by human: the hand carefully drew on x-ray images the progress of disease development, and any error could have serious consequences (Sahin, 2023). Improper diagnosis could result in endangering the life of the patient.

Diagnosing Musculoskeletal Conditions using Artificial Intelligence and Machine Learning to Aid Interpretation of Clinical Imaging
https://doi.org/10.1016/B978-0-443-32892-3.00012-9

The use of ML algorithms has already fundamentally changed the approach to the diagnosis of this pathology. Such deep learning as convolutional neural networks (CNNs) can accurately examine x-ray pictures and detect even the smallest deviations from the normal alignment of the spine. ML algorithms can also track changes in the development of curvature to provide doctors with a fundamentally new, dynamic vision of the problem. Furthermore, there is another type of ML, namely support vector machine (SVM). It is successfully used in determining the severity of scoliosis, which is important for developing therapy and intervention considering the patient's individual characteristics.

Concrete examples from real life demonstrate the significant influence of ML in therapeutic environments. Case studies show that ML algorithms can identify bone anomalies early on, enabling proactive therapies to halt the advance of disorders and enhance patient results. ML methods provide numerous opportunities in bone deformation diagnosis, such as detecting small fractures in osteoporotic patients and categorizing complex hip dysplasia patterns.

This chapter explores the relationship between ML and the complexities of osteoporosis diagnosis. ML seeks to stimulate new research, collaboration, and innovation in medical research by highlighting the transformative power of algorithms. Integrating ML into osteoporosis diagnosis offers the potential to improve patient care and support health care providers. These insights lead us into a new era of diagnostic wellness, driven by the significant potential of ML in maintaining bone health and bone integrity.

12.2 Classic issues in bone deformation analysis

Health professionals face complex issues in the identification and diagnosis of osteoporosis through orthopedic examinations, such as the identification of small structures and complex patterns. Traditional screening methods, including physical examination and radiographs, can be subjective and inconsistent owing to clinicians' homogeneous interpretations. Monitoring bone loss over time is equally difficult. Accurate diagnosis is essential for robust treatment planning and can lead to delayed prognosis and treatment. Advanced imaging techniques and mechanical mastering techniques are addressing these challenges, increasing the accuracy, efficiency, and uniformity of orthopedic diagnosis.

12.2.1 Historical context and key challenges faced

The interesting history of osteoporosis research is closely related to medical understanding and diagnostic methods. Over the centuries, medical scientists and healers observed and documented skeletal abnormalities, sometimes linking them to spiritual and biological origins. Ancient civilizations such as the Egyptians and Greeks left behind books and artifacts indicating that they had knowledge of orthopedic problems, but their treatment was mainly based on superstitious and folklore (Barhoom et al., 2022).

The Renaissance was a crucial turning point in the study of anatomy as medical science advanced. Andreas Vesalius and other pioneers advanced human anatomy through detailed dissections, establishing a structured method for studying bone formation. The techniques available at that period, such as basic x-rays in the late 19th century, restricted the level of understanding of bone deformities.

The 20th century saw significant progress in medical imaging technology, with the widespread use of x-ray imaging, computed tomography (CT) scans, and magnetic resonance imaging (MRI). The advances transformed bone deformation analysis by offering doctors comprehensive views of skeletal components. However, despite these technical advances, difficulties remained.

Finding small problems in bones has been hard. Some bone issues, especially when they start, look like just small problems. This makes it tricky for doctors to find bone problems such as weak bones or curved spines early on.

Bones can change in many ways. They may bend strangely or grow in an odd way. To find the right treatment, doctors must understand these problems well. They need to group the bone changes by how serious they are. This takes a lot of knowledge and care. Studying how bones change shape has gone from old myths to modern science. Even with new imaging tools, doctors still struggle to spot small issues, label complex deformities, and make a correct diagnosis. Using ML is one way to analyze bone shape changes more precisely, usefully, and fairly today. The doctors see problems. Bone shapes differ. ML helps spot the changes. Science helps cure diseases.

Two doctors may see one x-ray but they might have different views on it. This could mean a wrong diagnosis for the patient. Looking for bone problems by hand is hard work. It takes a lot of time, too. Mistakes can happen easily. Better mechanisms can be developed to check patients.

12.3 Importance of precision and efficiency in diagnosis

Finding bone problems correctly is important. It helps choose the best way to treat the problem. It also makes results better for the patient. The human body has many bones and joints. So, we need to check it carefully to find even small differences from what is normal.

12.3.1 Precision diagnosis

Getting the right diagnosis is important. It helps doctors understand what kind of bone problem you have. They can figure out if it is a broken bone, arthritis, curved spine, or weak bones. Knowing the exact problem lets them plan the best treatment for you. If you have weak bones, the doctor can give you medicine. They might also suggest diet changes or physical therapy. The right treatment can stop bones from getting weaker. It can prevent broken bones, too (Barhoom et al., 2022).

12.3.2 Impact on treatment planning

A correct diagnosis is important for good treatment results. Identifying the type and level of bone problems helps pick the best treatment options (Lippi and Plebani, 2016). The treatment plan must match the details of the problem. It may involve surgery, braces, medicine, or exercise therapy. Accuracy ensures treatments target the right issues. Tailored therapies work better and reduce risks or side effects.

12.3.3 Early detection and prevention

Finding bone problems early is important. It helps with treatment and long-term results. An accurate diagnosis lets doctors find issues before symptoms start. Early detection allows actions to stop problems from getting worse. It also helps improve the patient's quality of life. Finding osteoporosis early means treatments can start quickly. These treatments make bones stronger. They also lower the chances of bone breaks.

12.3.4 Diagnostic efficiency

Getting the test results quickly is important in hospitals. Slow tests can make patients' health more critical or cause new problems. Fast tests show doctors what bone issues a person has right away. Getting treatment faster feels good. Patients feel happy quickly when all goes well. Quick treatment starts without a long wait. This is good for sick people. Efficiency also saves money by avoiding extra tests or visits.

12.3.5 Technological advancements and precision

Developments such as specialized imaging machines and computer programs help doctors do better work. High-tech imaging such as CT and MRI scans let doctors see bones clearly to find problems more easily. Computers with smart programs can look at imaging data carefully to spot patterns and types of bone issues, which helps doctors make the right diagnosis.

12.4 Machine learning solutions for bone deformation

Closely studying how machine learning works in bone deformation entails reliance on large datasets due to the numerous factors such as angle, orientation, internal structures among others. This section presents ways in which mastering the machine such as deep learning, support vector machines, and ensemble methods can help in the aforementioned issue. In medical practices, convolutional neural networks (CNNs) are the most preferred deep learning models to evaluate advanced complex bone diseases present in X-ray and CT images. Bone deformation classification is an efficient scope for support vector machines, while ensemble methods manages and improves predictive accuracy and reliability of these models. Development in this area will allow soon diagnosis of bone problems, resulting in precise treatments of these issues and overall better patient recovery (Litjens et al., 2017).

12.4.1 Leveraging large datasets for pattern recognition

Many pictures are needed to spot bone problems. Doctors take x-rays, CT scans, and MRI pictures to look at bones. These acquired images are fed into a computer program. The program can find tiny details that doctors might miss. Having lots of different images helps the program learn better (Shahzad et al., 2021).

12.4.1.1 Comprehensive data insights

Large sets of data help us learn a lot about bone shape changes. We can look at different ages, body parts, and situations. This variety lets computers study many cases. It sees simple issues and hard changes. By learning from a wide range, computers can spot patterns people may miss. This leads to the faster and better detection of bone problems.

12.4.1.2 Improved early detection

During extensive data analysis, clinicians can disclose the initial phases of bone deformities. Changes in bone structure or density, such as scoliosis or osteoporosis are distinguished in a better fashion. Early detection will be highly valuable in terms of starting prompt therapies to stop the advance of diseases and improve outcomes for the patients.

12.4.1.3 Customized treatment strategies

By using more understanding data, the healthcare organizations are able to develop targeted treatment plans. ML helps classify abnormalities by the type, severity, and development, helping specialists in selecting appropriate treatments. Personalized alternatives such as medications, surgical procedures, or adjustments to the way of life play a critical function in improving affected well-being using strategic care.

Large data can reveal complex patterns and connections not obvious to the human eye. Algorithms can recognize similarities in patients with similar bone abnormalities and thus set up prognostic models, estimate the risk, and so on. Deeper understanding of the data environment would help better clinical decisions in bone health and deformity.

12.4.1.4 Challenges and opportunities

The use of large datasets for pattern recognition comes with constraints such as data quality assurance, privacy, and computational resource constraints. Interdisciplinary collaboration among clinicians, data scientists, and ethicists can be used to overcome these challenges and develop a strong ethical framework for data use.

Using multiple datasets to identify patterns in osteoporosis research could transform diagnostic power. This database provides valuable information for early diagnosis, personalized treatment strategies, and a deeper understanding of bone health. The development of ML algorithms and expanded datasets has the potential to increase patient outcomes and advance bone health research.

12.4.2 Deep dive into machine learning techniques: Deep learning, support vector machine, and ensemble methods

Deep learning, SVM, and ensemble techniques have been the recent revolution in bone deformation analysis. In this respect, three key ML techniques-deep learning, support vector machine, and ensemble methods-are discussed, and their relative strengths in applications are highlighted.

12.4.2.1 Deep learning: Revealing intricate patterns

CNNs have revolutionized medical imaging analysis through deep learning. CNNs are proficient at recognizing patterns in images, enabling the identification of intricate anomalies in bone architecture (Murphy, 2012).

To find out if some patterns can show up before time or below the surface, CNNs utilize many tiers of neurons for feature extraction from images. Such intricate pattern detection enhances accuracy in diagnosing bone deformities, giving doctors deeper insights into skeletal health.

12.4.2.2 Support vector machine

Bone deformation analysis (James et al., 2021) has found SVM algorithms to be useful in classification because they are robust (James et al., 2021). These programs are good at dividing data points into groups, so we can correctly recognize faults based on their characteristics. A bone disease may be rightly determined with the help of SVM that analyzes signs taken from medical images. Doing so means the doctor can identify the kind of deformity and its extent, which contributes when choosing treatment tactics and considerations (Abe, 2005).

12.4.2.3 Ensemble methods: Improving precision and dependability

Ensemble methods such as random forest and gradient boosting demonstrated high modeling accuracy and reliability in the analysis of bone deformation; workers in the field can have more confidence in the machine-generated results (Mahobiya et al., 2022, 2023a). Different algorithms are somehow composed and bundled together to make the final forecast more resistant and accurate (Abe, 2005). Conjoined approaches are characterized by a combination of several ML systems, which are used to complement each other's flaws (Mahobiya et al., 2023b). It means that the diagnoses have the larger degree of reliability and it is less likely that they will be erroneous. Additionally, that opens the way to the correct and efficient planning of further treatment.

12.4.2.4 Integration of machine learning techniques

Using such ML algorithms could prove to be a systematic and accurate approach to studying the deformation of bone. Deep learning allows doctors to see hidden things that cannot be spotted with the naked eye with pictures, SVM is such an impressive classifier, and ensemble models raise the accuracy and reliability in a diagnostic test. Hence, an accurate diagnosis is possible if these methods are used simultaneously and can be of great help in preparing a treatment that is specifically made for the patient (Asimit et al., 2022).

12.4.2.5 Clinical uses and future prospects

These capabilities of ML algorithms in medical treatments have revealed its impact, which is positive. ML-guided bone deformation analysis is the most important factor that transforms the patient's course of treatment by detecting the initiation of osteoporosis and classification of scoliosis precisely. In the future, we can expect advancements in machine learning techniques, multimodal imaging, and predictive models, which will allow for more thorough examinations. The involvement of bone deformation in stress analysis is now being changed by the latest technologies. Prompt diagnosis and individualized intervention are believed to be significant factors contributing to the prognosis, and thus advance medical and bone structure research.

12.5 Deep learning for bone deformations

CNNs are relevant in medical analysis using an advanced deep learning technique for bone deformations. CNNs are accurate because they are able to extract complex features from x-rays, CT scans, and MRI images, detecting even the most minute abnormalities with great accuracy. RNNs come useful whenever time series data is under consideration in the structure deforming process. Recurrent neural networks, or RNNs, look at the sequence of data from medical imaging of a patient in order to help physicians to have a better understanding of the disease progression. With support, this leads to better treatment planning and monitoring.

12.5.1 Understanding convolutional neural networks for image analysis

Convolutional neural networks (CNNs) are an extremely effective tool in the field of medical image processing, particularly when it comes to analyzing bone deformities. A model of deep learning is created, which is designed to extract important information from images automatically so that clinical signs of minor abuses of bones can be found quickly and effectively (Sharma et al., 2018).

Convolution operations act as the backbone of the CNNs because they are the feature extractor blocks (Chen et al., 2016). The scale uses filters that move across the input image while leaving nearby areas untouched, concentrating on edges, corners, or textures. Bone deformities can be identified by CNNs with the aid of the rich detail learned through multilayer convolution, owing to neural networks that enable hierarchical learning. Pooling layers, positioned after the convolutional layer, aim to decrease the spatial dimensions of feature maps while maintaining essential information (Yadahalli et al., 2021). The most common approaches to feature map down-sampling are max pooling and average pooling. Here, the highest value is selected or averaged from certain areas. Subsampling selects the most crucial details by making it easier to correspond to systems computation.

For instance, output activation functions such as rectified linear unit (ReLU) add nonlinearity to the CNN. They make the network able to estimate and recognize complicated patterns. The ReLU activation function is usually applied to it because it is not difficult and also is helpful in solving the vanishing gradient problems. The fact that pixels images of bones can be discerned with CNNs by means of the ReLU method in the output of convolution and pooling layers is one successful aspect of bone investigation with CNNs.

12.5.1.1 Fully connected layers

The next part is to use convolutional and pooling layers to extract the features. Then, through the fully connected layers, the final predictions or output are generated. For this to take place, the layers have a working link between neurons in the adjacent levels, allowing the CNN to decipher interconnected correlations in the extracted data (Shen et al., 2021). The output layer is usually driven by the SoftMax activation function for the multiclass classification. This function is responsible for providing probabilities for different anomalies that might be involved in the bone deformation.

The CNN training is about uploading the images with labels and using backpropagation to compute how network weights and biases must be modified to reduce prediction errors. Algorithms such as stochastic gradient descent (SGD) and Adam are used to make the process easier by changing the network parameters at each step with an aim of making the model better. After the corporate with many input photos of the pattern, then another, it trains for periods of time and CNN as an expert does correct identification and recognition of bone deformities.

12.5.1.2 Utilization in bone deformation analysis

CNNs have turned the buzz word in identifying bone deformities with high accuracy helping clinician to make clinical decisions as easy as never before by registering correct observations. They use this to analyze the x-rays, CT scans and MRI images in order to detect small abnormalities such as osteoporosis, fractures or scoliosis. Investigations of huge sets of data are required before the early identification of deformities is possible and proper categorization. This helps to implement corrective surgery or a customized treatment strategy on time, which will improve the outcome.

CNNs have made images processing in bone deformation diagnosis effective, which is enabling a more efficient and accurate diagnosis by doctors because they are discovering those small lesions. CNNs are a tool through which the pictures to be analyzed and their properties can be automatically learned, leading to developments in precision medicine for skeletal health.

12.5.1.3 Recurrent neural networks for temporal data in deformation tracking

RNNs are considered the most suitable neural networks to be used for processing the sequenced data of bone deformation every time. RNNs are the most important type of neural networks used for processing a sequence of data related to bone deformations (Yu et al., 2020). Long short-term memory (LSTM) cells, a key innovation based on RNNs, are crucial for time-series data processing, implementing LSTM alongside other RNN architectures will further enhance model performance. LSTMs address the vanishing-gradient problem to which the traditional RNNs are exposed, thereby enabling the networks to acquire long-term consistencies and interconnections in the sequential data (Agarwal et al., 2022). This becomes more of a plus when it comes to deformation tracking, which plays a major role in knowing a certain illness because of the changes in bone structures over time.

12.5.1.4 Managing time-series data

The process of reading the tracking of deformity in bone involves scanning a set of medical imaging scans at intervals of time log. RNNs with LSTM cells are designed to handle sequential numbers in the series of time events. Each of the scans provides consequent feedback, which enables the network to adapt to different lengths of time and intervals by which bone densities are changed or produced (Agarwal et al., 2022).

12.5.1.5 Understanding temporal dependencies

By analyzing recurrent input sequences over the layers of LSTMs, RNNs are able to deduce dependence and relations by successively scanning the data. This model facilitates the recognition of a change in bone shapes or developments of small pattern variations over time, which can signal deteriorating disease or healing process reactions (Agarwal et al., 2022).

12.5.1.6 Predictive modeling

RNNs can be employed in predictive modeling for bone health by supervising deformations along with conformation (Wu et al., 2016). It is possible to predict the future growth of bone organs through looking at historical images of patients and comparing them to outcomes of the research. For clinicians and management at the level of patients, predicting which complications or adverse events may occur is important. By making it possible for them to evade ill results and make informed decisions, they get the tools they need in time without changing the patient.

12.5.1.7 Treatment monitoring and decision support

RNNs in deformation tracking, which bring data with both a historical view of the ethnic trajectory of a disease and real-time monitoring capabilities, are a great example of this. The system is able to pinpoint anomalies from regular images and notifies clinicians when a prompt review may be necessary owing to a potential issue (Agarwal et al., 2022). This ongoing tracking becomes a decision support tool that provides doctors with an opportunity to make the related changes quickly once required. Such synergic usage of RNNs along with imaging devices as MRI and CT scans increases the accuracy of bone deformity analysis.

These networks may be able to rebuild 3D bone models using photo sequences and therefore provide detailed information on the rate of disease progression. Combining RNNs with imaging systems has the latent potential to provide more insight into changes in the skeletal system.

12.6 Support vector machine in bone deformation

SVM is reliable and helpful because is enables bone abnormalities to be identified precisely and accurately, and medical practitioners may possess more exact diagnostic abilities. SVM has done well to prove its efficacy in several ways of analyzing bone deformation; the evidence is convincing. X-rays and MRI images can be used in SVM to obtain an accurate diagnosis of osteoporosis. Moreover, the potential of the technique can be seen with its ability to grade the status of scoliosis (Niculescu et al., 2020). Using SVM in basic of bone tissue diseases diagnosis to make more accurate and easy diagnoses to give the exact treatment as soon as poosible.

12.6.1 Support vector machine for classification of bone irregularities

SVM, a strong, versatile ML technique of classification for bone anomalies, is also used in many applications. It is the clinicians' insider instrument for a true diagnosis of skeletal diseases and their conditions.

12.6.1.1 Support vector machine principles

SVM performs these tasks by determining the largest possible hyperplane in a space consisting of several high-dimensional dimensions that separates the distinguished classes of data points. The classification of bone deformation is tackled through SVM after examining examples from a dataset in which bone properties and their corresponding classes are present. SVM can properly classify bone abnormalities by extracting input features from diagnostic imaging data (such as x-ray images or MRI scans) and then clustering them into different groups (Wu et al., 2016).

12.6.1.2 Kernel trick for nonlinear classification

SVM assists nonlinear data through the help of a kernel method called the kernel trick. SVM converts lower dimensional data points into lifting dimensions whereas they are separable by a linear boundary (Gholamalinezhad and Khosravi, 2020). Deep learning algorithms employ geometric image processing and SVM can clearly detect delicate features of bone formations indicating hidden diseases. SVM is successful in multiclass classification, particularly with varieties of deformities and categories in factors of bone aberration.

SVM for bone abnormalities may involve multiclass classification using one-vs-all or one-vs-one strategies, incorporating all available information about a patient's bone health. This approach provides comprehensive insights into skeletal health.

12.6.1.3 Feature selection and dimensionality reduction

SVM is indispensable in medical imaging to pick the best features and reduce dimensionality, causing the model to become simpler (Yu et al., 2020).

SVM assists in the diagnostic process by highlighting specific features of bone structures in imaging data. It provides clarity and reduces the computational complexity for researchers studying large datasets.

12.6.1.4 Medical uses and advantages

The use of SVM has a direct influence on bone deformation analysis and has led to the clinic's benefits. In studies, the application of SVM proved helpful at rightly diagnosing pathologies such as osteoporosis, scoliosis, lesser arm/greater arm fractures, and other skeletal defects.

SVM's precision would help the physicians to design unique courses of treatment suitable for different stages, suggest timely disease detection, and track illness from advances. SVM has become an indispensable tool for clinicians for both the diagnosis and characterization of bone anomalies, provide clinicians with precise and fast classification techniques.

SVM is a promising tool in medical imaging because of its ability to process nonlinear data, carry multiclass classification, and research and extract major features from images. It therefore has great potential for increasing patients' health assessment and improving the care outcome.

12.7 Ensemble methods for improved accuracy

Ensemble methods comprised using multiple ML models for predictions are efficient in increasing the accuracy of bone deformation recognition. They use the individual strengths of each type of model to overcome their weaknesses (Sainath et al., 2015). In these algorithms the robustness and reliability of bone deformation classification are enhanced and the capability of the model of differentiating and classifying new data conclusively is established. The complexity of a model and memory resources are the most crucial problems that should be solved when dealing with an ensemble approach in medical image analysis. Such challenges presented by the group are effective in identifying skeletal abnormalities, making it a valuable tool for evaluating bone health.

12.7.1 Ensemble learning strategies in bone deformation identification

Using ensemble learning algorithm is a strong way to enhance the certainty of bone deformation categorization, which makes the exact diagnosis and bone classification effortless for doctors. Random forest is a groundbreaking ensemble learning method used in medical imaging data assessment, which achieves high accuracy in the classification of bone irregularities. Random forest can thoroughly check a lot of complex patterns in photos, producing many decision trees and the results of their predictions (Mahobiya et al., 2023a).

The method provides the opportunity for small changes that are primary symptoms of illnesses such as osteoporosis or fractures, assisting the diagnosis of clinicians. Gradient boosting prepares an ensemble of classifiers focused on increasing the accuracy of abnormal bone deformation detection. Such an approach consists of developing a chain of weak learners successively making up for the mistakes of their predecessors (Haji and Abdulazeez, 2021). In those skeletal health evaluations, gradient boosting algorithms enable the complexity of bones to be used to classify unusual structures with their individual features (Malviya et al., 2024).

12.7.1.1 AdaBoost for improved classification

The most widely used boosting algorithm, AdaBoost, is referred to as the adaptive boosting algorithm. It applies a number of weak learners that result in improving the accuracy of classifiers (Pandey and Ram Janghel, 2019). The groups of AdaBoost algorithms are characterized by high precision when it comes to grouping skeletal abnormalities into specific classes; therefore, they work best with computer-assisted detection. AdaBoost produces high precision rates and plausible results in skeletal state assignment by recursively sorting inadequately labeled data.

12.7.2 Advantages and challenges of ensemble techniques

Merits of Ensemble Methods: Even though one strategy is applied in the study of bone deformity detection in this case, it overrelies on the knowledge that combining different systems proposes definite advantages. A balanced ensemble learning paradigm offers several advantages in the detection of bone deformities. They impact this by counterbalancing different models and prediction errors from individual model learning, improving the prediction accuracy of the overall model (Salehinejad et al., 2017). The ensemble techniques increase the model's ability to generalize to unknown data points, which give the system robustness to provide consistent results in different clinical scenarios. By analyzing x-rays or CT scans, doctors gain detailed information about deviations in bone structure, leading to more precise diagnoses.

12.7.3 Challenges in ensemble techniques

Ensemble learning algorithms are important assets, but they consume significant amount of resource energy and computer capacity. Using multiple models and creating your own ensemble training method can make the training process more complex and demanding, requiring sufficient processing power and resources. However, the need to incorporate a batch of datasets from different group of models to be trained collectively with the required quality data creates a challenge in medical image analytics, especially when there is a shortage of labeled data.

12.8 Ethical, practical, and technical considerations

Critical points for integrating ML in medicine include ethics, the workflow, and technical questions. Advanced encryption and anonymization techniques were used to protect the data and to comply with legislative requirements. Enhancing the explainability of AI models for doctors boosts their confidence in

AI-supported diagnoses and improves their ability to make timely and accurate decisions. The interdisciplinary collaboration across data scientists, physicians and policymakers can be used in creating ethically responsible and effective ML solutions that facilitate the health status of patients and solve ethical dilemmas.

12.8.1 Privacy protection measures in health care data handling

Preserving privacy remains a key aspect when handling health care data, especially when this sector utilizes ML algorithms. Advanced cryptographic techniques, robust access controls, and pseudonymization are used to protect private data classified as patient data. Health insurance portability and accountability act (HIPAA) and general data protection regulation (GDPR) (Kwak, 2013) state that the patient's data have to be managed in a secured and ethical way. These measures are significant in giving credibility to patients as part of health care delivery and virtuously proving the safety of sensitive records.

12.8.2 Interpretability of machine learning models for clinical decision-making

As ML models are being adopted into health care, models' understanding becomes even more essential for the right decisions to be made. Not only is it crucial for ML algorithms to provide precise forecasts, but understanding the reasons behind these predictions and aiding in decision-making is also paramount for healthcare professionals. Feature-related approaches such as feature importance analysis and model visualization will help in interpreting medical decision process models (Mladenić, 2005). The algorithm's transparency increases credibility and gives clinical specialists a basis for making quality treatment decisions on the use of ML systems. Improvement in the interpretability of ML models by health care professionals helps design health care technologies as well as patient care outcomes.

12.8.3 Importance of interdisciplinary collaboration in medical artificial intelligence

The cooperation of interdisciplinary cooperation is an important aspect in the area of medical AI application. It is highly important to come up with a policy-making approach based on data science, medicine, ethics, as well as policy-making considerations in developing and applying AI technology in health care. Data scientists in partnership with physicians implement any ML model to see whether the model is applicable in clinical settings or satisfies patients' demands. Ethicists are of great help in this sense by dealing with ethical problems such a disregarding bias, assuring fairness, and fostering transparency within AI systems. Policy -makers thus become active players in setting up laws and regulations governing the real-world implementation of AI in medical environments. This cooperative approach ensures that AI-themed technologies are focused on ethical guidelines and are constantly validated and incorporated in the health care process. The participation of diverse specialists produces synergies to promote common development, resolve ethical dilemmas, and maximize benefits associated with AI as a tool to improve the quality of patients and the entire performance of clinics.

12.9 Case study

12.9.1 Background

Now, let's discuss Mr. Rakshit Sisodia, an Assistant Professor in the Department of Aeronautical Engineering at MATS University. Unfortunately, he suffered a traumatic incident on December 18, 2023, which resulted in serious injuries. He experienced a multifragmented complex tibia plateau fracture. Above-knee amputation was suggested for him by the orthopedic consultant. Additionally, there was a substantial wound (10 × 2 cm) at the site of the fracture, which was also infected.

12.9.2 Introduction

Thus, it is crucial to say the case study is about the high importance of both new technologies, especially ML algorithms, which assist experts in diagnosing the case and selecting the best methods for the injury to Mr. Rakshit.

12.9.3 Diagnosis

At that time, Mr. Rakshit underwent a set of medical procedures, including scans and investigations, which were necessary to examine the level of the injuries. Despite this, the diagnosis in his case was complicated because he had a complex fracture and noninfected wound, and therefore an accurate and early diagnosis was crucial and could have influenced treatment.

12.9.4 Application of machine learning

By the help of ML, the algorithm is being built into the hospital's imaging system to identify and help clinicians to know accurately the type and severity of the fracture that Mr. Rakshit has, using x-rays and CT scans. These methods utilize personalization pattern recognition and statistical approaches to discover distinctive marks that enable the identification of bone fractures and other bone defects.

12.9.5 Result

Using the ML algorithms, this time clinician established an accurate diagnosis for the right leg's compound comminuted fracture of the proximal tibia and dirty fracture site almost simultaneously. The medical team was able to make an accurate and initial examination from which a suitable treatment plan perfectly tailored to Mr. Rakshit's unique injury could be readily instituted.

12.9.6 Conclusion

Introducing ML technology into diagnostics influenced the process positively and allowed us to make a diagnosis far earlier and focus on treatment and prevention rather than a diagnosis. The physicians became adept at making wise decisions, and patients received the crucial attention they sought through the use of highly sophisticated images for analysis, which ultimately led to Mr. Rakshit's recovery from the condition.

This study demonstrates one of the bright sides of ML, more broadly speaking, especially with regard to diagnostic imaging and treatment planning for individual doctors. Because technology is constantly developing, we can surely predict even more refined diagnoses and treatment methods that will be effective and lead to good results for patients (Fig. 12.1 and Table 12.1).

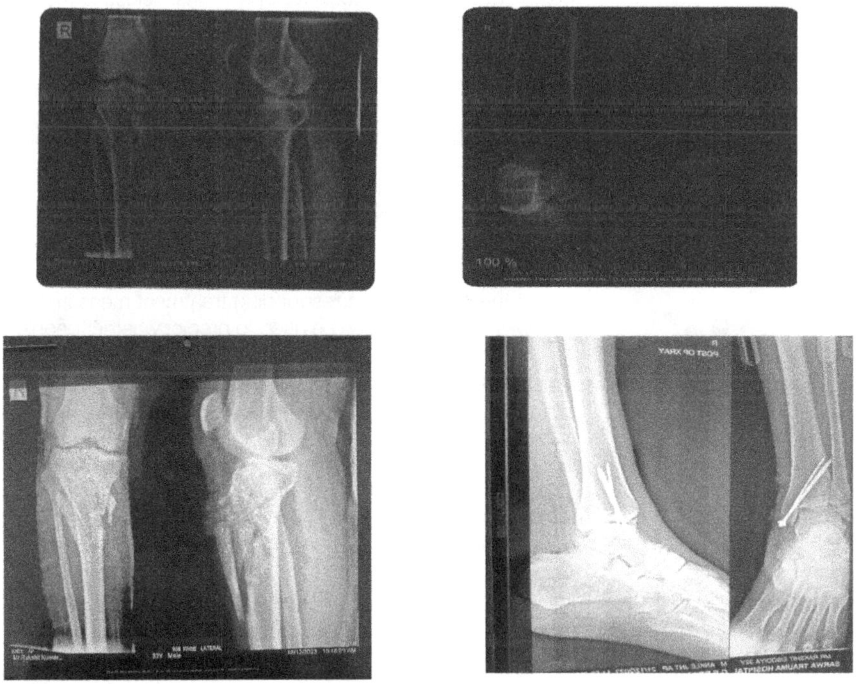

FIGURE 12.1

X-ray images of a compound comminuted fracture of the proximal tibia on the right side, with a 10 × 2 cm wound.

12.10 Conclusion

Table 12.1 Potential applications of machine learning in bone deformation care.

Area bone deformation disease care	Details
Diagnosis and recognition	Machine learning will take the streaming of medical imaging data such as magnetic resonance imaging or computed tomography scans a step further by analyzing them and automatically classifying certain spinal diseases such as spinal stenosis. Artificial intelligence (AI) algorithms may cater to this segment for precise and exact diagnoses, which are visible to the health care providers.
Treatment planning	Machine learning can help the medical team through designing personalized treatment procedures for spine diseases. AI models with the help of patient data such as x-rays, clinical notes, and results can explore different treatment options and tools for operations, selecting most suitable treatment plan and surgical technique.
Surgical Guidance	Machine learning has proven capable of delivering assistance when necessary in spine surgeries. The AI technology makes it possible to perform a surgery at a different level by complementing preoperative imaging data with intraoperative data, permitting the exact positioning of complicated spinal structures and allowing surgeons to make proper judgments instead.
Predictive modeling	Machine learning has the potential to design a predictive model that can assess the progress of spinal conditions and detect the efficacy of corresponding treatment measures. This way of modeling can be used to predict whether people will get better, if they need new treatment ideas, and if they and their care team will find working together easier.
Rehabilitation support	Machine learning has the capabilities of intelligent rehabilitation programs that can address spinal diseases, which vary depending on individual characteristics. AI systems can accomplish personalized recommendations and consult patients during a rehabilitation process through data collection, such as movement patterns and sensor data.
Remote monitoring	ML-enabled remote surveillance systems will enhance the monitoring of patients suffering from spinal diseases outside health care facilities. Such systems allow the constant surveillance of the patients for any modifications in symptoms or movements in a given time interval which might be beneficial. Such changes are immediately communicated to physicians or caregivers so that the patients can be treated when required even if they are not in a hospital.

12.11 Transforming bone deformation diagnosis

The introduction of the ML has rendered the field of bone deformation diagnosis to be totally dissimilar to how it had been before, because it has almost changed the way clinical evaluations are made and patient treatment is given. Today ML algorithms have helped with the precision of diagnostics and type of malignancy determinations. Besides, these greatly refine treatment strategies that are individually tailored to patients. The prospects of AI in the health sector are promising as long as we accomplish advances in deep learning, multimodal data fusion, and real-time monitoring. This progress leads to more precise diagnostics and proactive adjustments in the understanding and treatment of the underlying causes (etiology). We are witnessing an unprecedented period of scientific discovery and technological transformation. Adding ML to detecting bone misalignment is worth investing and calls for further research and collaboration, becoming the picture of a data-driven, patient-focused medical field. Such methods increase the likelihood of positive effects and better patient outcomes so that the life standards of people with skeletal disorders improve.

12.11.1 Recap of machine learning's role in improving clinical assessments

Medical image analysis, particularly for the identification and evaluation of bone deformities, has proven to be a boon in the clinical area owing to ML. As the power of ML algorithms has increased, there has also been high accuracy and precision in diagnosing skeletal diseases on time, leading to the ability to devise proper plans of treatment for individual patients. AI has allowed for the diagnostic accuracy of health care providers to increase. This higher accuracy in the process of diagnosis has been achieved using advanced methods such as deep learning, ensemble techniques, and data fusion from different imaging modalities. They facilitate both the process of the evaluation and the process of treatment for better patient care results by canvassing skeletal health. The implementation of smart ML in the clinical workflow offers a promising opportunity to bring revolutionary changes in the diagnosis of deformation of bones and quality of health care.

12.11.2 Future prospects and emerging trends in medical artificial intelligence

Medical AI is a field that has markedly huge potential for growth in the diagnosis of bone deformities. Breakthroughs and advances in network deep learning methods such as CNNs and RNNs could lead to enhanced accuracy and effectiveness in detecting minor bone irregularities. Integrating data derived from various imaging techniques such as x-rays, CT scans, and MRI ML models creates a more in-depth and versatile assessment of the patient owing to the availability of wider datasets. The emergence of real-time monitoring systems and wearable health devices opens the prospect for preventative monitoring of skeletal health, making early intervention possible with individually tailored interventions. Medical AI is an emerging area of research able to lead to a great revolution in the field of diagnosing bone deformities, throwing more doors open for the good outcomes of patients and improved health care products.

12.11.3 Inspiring further research and collaboration opportunities

Finally, integrating ML into the diagnosis of bone deformation within clinical medicine (CM) not only sparks new ideas but also encourages research collaboration. The dynamic capabilities of ML algorithms enhance accuracy optimization, enable more personalized treatment options, and support precise predictive models for bone health. Decision-making in the field of AI falls on data scientists, clinicians, researchers, ethicists, and lawmakers. Collaboration among them is compulsory for driving innovation and solving all ethical, performance, and technical issues. Through facilitating transdisciplinary teamwork and guiding research work, we have a real chance to reveal the boundaries of ML in enhancing the clinical examination of patients' skeletal systems. This risk-sharing and collaborative attitude drives new cancer therapies in medical AI and guarantees that patients will be provided with the highest quality care through personalized and data-driven intervention methods. While we explore the development of skeletal conditioning tools and personalized healthcare, the future holds great promise for transformative advancements.

References

Abe, S., 2005. Support Vector Machines for Pattern Classification, vol. 2. Springer, London.

Agarwal, R., Singh, J., Gupta, V., 2022. Predicting the compressive strength of additively manufactured PLA-based orthopedic bone screws: a machine learning framework. Polym. Compos. 43 (8), 5663–5674.

Asimit, A.V., Kyriakou, I., Santoni, S., Scognamiglio, S., Zhu, R., 2022. Robust classification via support vector machines. Risks 10, 154. https://doi.org/10.3390/risks10080154.

Barhoom, A.M., Al-Hiealy, M.R.J., Abu-Naser, S.S., 2022. Bone abnormalities detection and classification using deep learning-Vgg16 algorithm. J. Theor. Appl. Inf. Technol. 100 (20), 6173–6184.

Chen, Y., et al., 2016. Deep feature extraction and classification of hyperspectral images based on convolutional neural networks. IEEE Trans. Geosci. Rem. Sens. 54 (10), 6232–6251.

Gholamalinezhad, H., Khosravi, H., 2020. Pooling Methods in Deep Neural Networks, A Review arXiv preprint arXiv:2009.07485.

Haji, S.H., Abdulazeez, A.M., 2021. Comparison of optimization techniques based on gradient descent algorithm: a review. PalArch's J. Archaeol. Egypt/Egyptology 18 (4), 2715–2743.

James, G., et al., 2021. Support vector machines. In: An Introduction to Statistical Learning: With Applications in R, pp. 367–402.

Kwak, N., 2013. Nonlinear projection trick in kernel methods: an alternative to the kernel trick. IEEE Transact. Neural Net. Learn. Syst. 24 (12), 2113–2119.

Lippi, G., Plebani, M., 2016. Improving accuracy of diagnostic studies in a world with limited resources: a road ahead. Ann. Transl. Med. 4 (3), 43. https://doi.org/10.3978/j.issn.2305-5839.2015.11.12.

Litjens, G., et al., 2017. A survey on deep learning in medical image analysis. Med. Image Anal. 42, 60–88.

Mahobiya, C., Sonar, S., Kolhe, S., 2022. Prediction of important insect of Soybean according to weather parameters using genetic algorithm based optimized ensemble classifier. NeuroQuantology 20 (12), 501–512.

Mahobiya, C., Iyer, S., Kolhe, S., 2023a. Hybrid features based classification of insect and leaf disease of Soybean plants using random forest classifier. Int. J. Eng. Trends Technol. 71 (2), 408–420.

Mahobiya, C., Iyer, S., Kolhe, S., 2023b. A novel approach for detection of Soybean leaf disease using Bayesian optimized-KNN classifier with NCA-based feature selection. J. Agric. Sci. Technol. 25 (1), 119–133.

Malviya, R., Rajput, S., Vaidya, M., 2024. Artificial Intelligence for Bone Disorder: Diagnosis and Treatment. John Wiley & Sons.

Mladenić, D., 2005. Feature selection for dimensionality reduction. In: International Statistical and Optimization Perspectives Workshop. Subspace, Latent Structure and Feature Selection. Springer Berlin Heidelberg, Berlin, Heidelberg.

Murphy, K.P., 2012. Machine Learning: A Probabilistic Perspective. MIT press.

Niculescu, B., Faur, C.I., Tataru, T., Diaconu, B.M., Cruceru, M., 2020. Investigation of biomechanical characteristics of orthopedic implants for tibial plateau fractures by means of deep learning and support vector machine classification. Appl. Sci. 10 (14), 4697.

Pandey, S.K., Ram Janghel, R., 2019. Recent deep learning techniques, challenges and its applications for medical healthcare system: a review. Neural Process. Lett. 50, 1907–1935.

Sahin, M.E., 2023. Image processing and machine learning-based bone fracture detection and classification using X-ray images. Int. J. Imag. Syst. Technol. 33 (3), 853–865.

Sainath, T.N., et al., 2015. Convolutional, long short-term memory, fully connected deep neural networks. In: 2015 IEEE International Conference on Acoustics, Speech and Signal Processing (ICASSP). Ieee.

Salehinejad, H., et al., 2017. Recent Advances in Recurrent Neural Networks arXiv preprint arXiv:1801.01078.

Shahzad, D., et al., 2021. GaN-based high-power-density AC–DC–AC converter for single-phase transformerless online uninterruptible power supply. IEEE Trans. Power Electron. 36 (12), 13968–13984.

Sharma, N., Jain, V., Mishra, A., 2018. An analysis of convolutional neural networks for image classification. Proc. Comput. Sci. 132, 377–384. https://doi.org/10.1016/j.procs.2018.05.198.

Shen, S.C.Y., Fernández, M.P., Tozzi, G., Buehler, M.J., 2021. Deep learning approach to assess damage mechanics of bone tissue. J. Mech. Behav. Biomed. Mater. 123, 104761.

Suzuki, K., et al., 2019. Machine learning in medical imaging. IEEE J. Biomed. Health Inform. 23, 1361–1362.

Wu, H.Y., Gong, C.S.A., Lin, S.P., Chang, K.Y., Tsou, M.Y., Ting, C.K., 2016. Predicting postoperative vomiting among orthopedic patients receiving patient-controlled epidural analgesia using SVM and LR. Sci. Rep. 6 (1), 27041.

Yadahalli, S., Parmar, A., Zambare, P., Sawant, R., 2021. Bone deformity identification using machine learning. In: 2021 5th International Conference on Intelligent Computing and Control Systems (ICICCS). IEEE, Madurai, India, pp. 1355–1358. https://doi.org/10.1109/ICICCS51141.2021.9432272.

Yu, Y., et al., 2020. Rmaf: Relu-memristor-like activation function for deep learning. IEEE Access 8, 72727–72741.

CHAPTER

Artificial intelligence and machine learning for foot and ankle disorders

13

Shubham Gupta[1], Harashleen Kour[1], Paul Antony[2]

[1]*Model Institute of Engineering and Technology, Jammu, Jammu and Kashmir, India;* [2]*Department of Orthopedics, Apollo Super Speciality Hospital, Muscat, Oman*

13.1 Introduction

Artificial intelligence (AI) and machine learning (ML) are revolutionizing the global health care sector, leading to improved median patient outcomes, personalized treatment programs, and better understanding when it comes to diagnosis. The adoption of AI and ML in health care represents an astounding achievement as well as the culmination of medical science, in which patients are guaranteed a more accurate diagnoses and doctors have the freedom to create customized solutions for them. Regardless, the effective application of these technologies will be hampered because of initiatives related to data privacy, prejudice, incorporation, and containment. As medical technology continues to progress, AI and ML will increasingly steer the future of patient care, ensuring that these initiatives are more effective, more specific, and more open to everyone. In many ways, the conscious application of these technologies will help us to unleash their full capacity today: the future of health care depends on them.

AI and ML tap into large volumes of medical data, including electronic health records (EHRs), medical imaging data, and genetic information. An AI and ML algorithm may analyze the data, identify hidden patterns, forecast potential disease premises, and recommend ways to personalize treatment. AI and ML could fundamentally alter the way health care is delivered by improving diagnostic accuracy and widening the spectrum of therapies and patient outcomes (Hida et al., 2023). The problem has been investigated to a greater degree in a variety of fields, including spinal care, allowing for the inclusion of imaging, learning from experiences to realize patient results, and providing personalized treatment. AI models, which are connectionist, like deep learning, and ML, are a useful system for detecting, diagnosing, treating, monitoring, and controlling a wide range of health issues, including infectious disease and mental issues. Despite these benefits, the use of AI in the health sector still has difficulties (Choo et al., 2021). For AI to be implemented responsibly, factors such as the condition of the data, safety, and semiotics must be addressed. AI in the health area might be a game-changing concept that automates duties, assesses complete data collections, and enhances the quality of services in general.

Fig. 13.1 shows the steps involved in developing and deploying an ML model to diagnose foot and ankle disorders.

https://doi.org/10.1016/B978-0-443-32892-3.00013-0

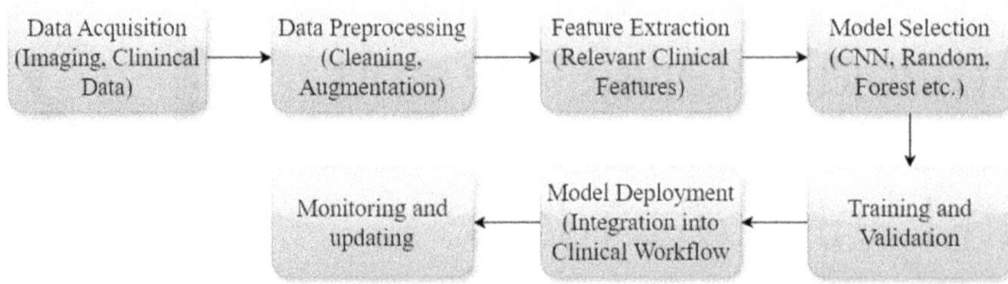

FIGURE 13.1

Machine learning algorithm pipeline.
CNN, convolutional neural network.

13.1.1 Health care monitoring

The health care business is facing a significant problem in ensuring compatibility among apps offered by various manufacturers. Multiple applications can be used by each department of hospital management or medical clinic to share organizational and clinical data. Such applications could be technologically advanced. Regrettably, each program may support several communication interfaces, necessitating ongoing development and maintenance. The development of interoperability among heterogeneous health care data systems is critical considering the decrease in health care costs, because it will lead to better and more effective treatment of patients, management, and care. Health monitoring systems have had an important role in lowering hospitalization costs, medical staff load, time, and so forth.

Smart health tracking systems are an innovative technology and a novel method for monitoring health. Heart rate, blood pressure, ECG, and temperature are all measured as part of an overall health monitoring system.

The health-care system has advanced significantly. Smart health monitoring is essential in hospitals, residential settings, and outdoor settings because it is enabled by various communication techniques such as GPS and radio-frequency identification (Fig. 13.2).

13.2 Machine learning in health care

ML is a subset of AI that teaches systems how to use data to behave or anticipate without explicitly being programmed. Fig. 13.3 demonstrates how the use of ML for medical purposes may completely alter how disease is recognized, treated, and prevented. Potential uses of ML for health care include:

- Diagnosis and treatment: ML approaches are taught to examine medical pictures such as computed tomography (CT) scans to assist in a patient's diagnosis or choose the best course of action.
- Personalized medicine: Using a patient's unique traits such as genetics and medical history, ML can be used to forecast which medicines are most likely to be beneficial for that patient.

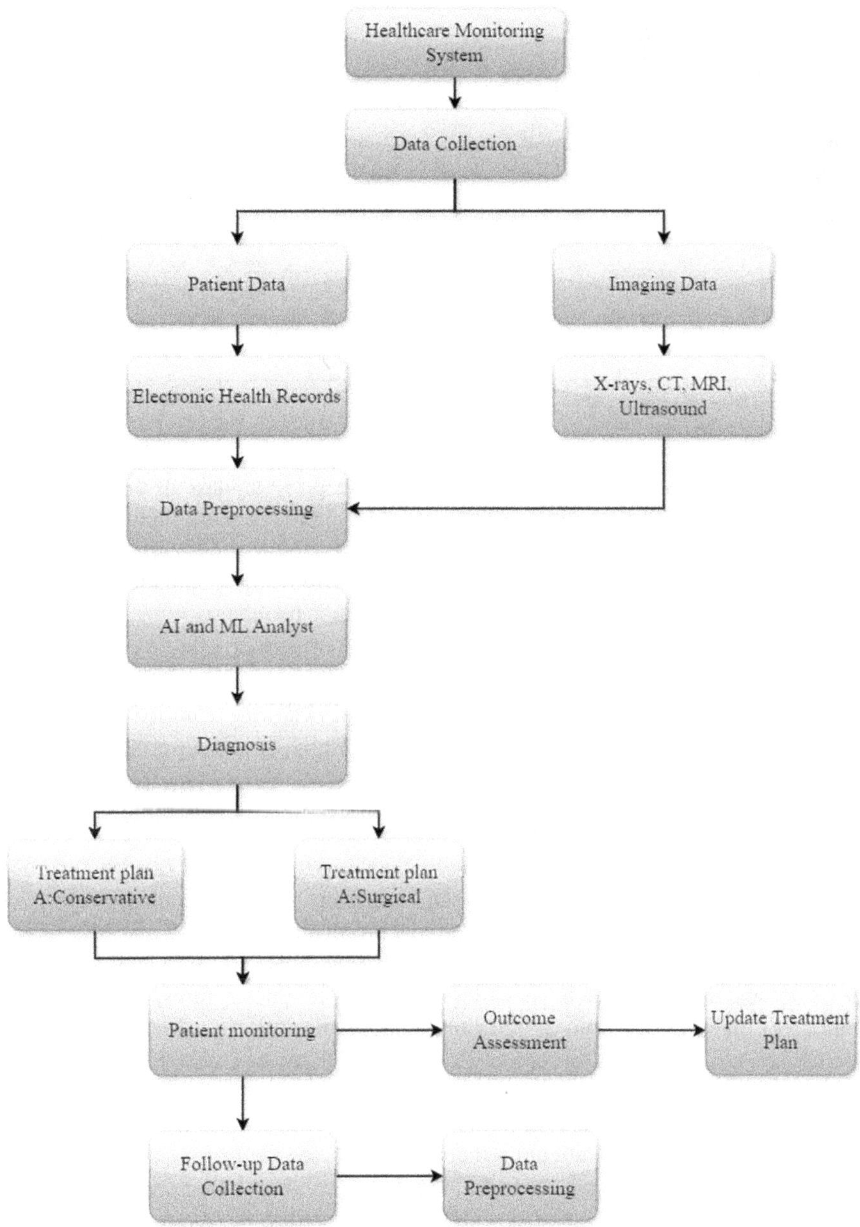

FIGURE 13.2

Classification of health care monitoring system.

AI, artificial intelligence; *CT*, computed tomography; *ML*, machine learning; *MRI*, magnetic resonance imaging.

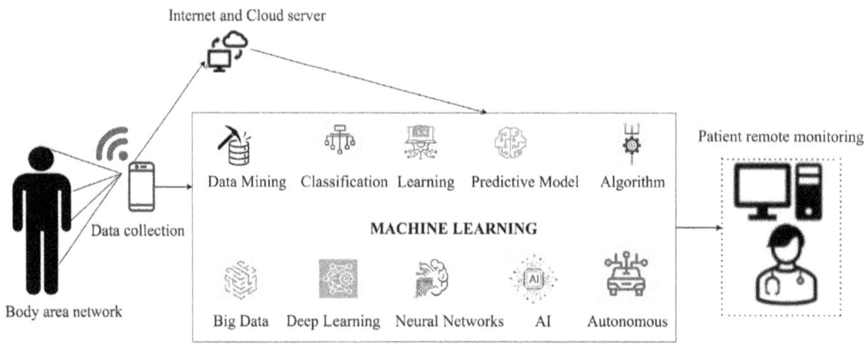

FIGURE 13.3

Machine learning in health care.
AI, artificial intelligence.

- Clinical decision support: To assist health care professionals in making more informed decisions regarding patient care, ML approaches are included in clinical decision support systems.
- Population health management: By analyzing data from huge populations, ML may be used to spot sequences that help in creating public health programs. In general, applying ML to health care has the potential to boost system effectiveness, lower costs, and improve patient outcomes. Fig. 13.3 shows ML in health care.

13.3 Artificial intelligence and machine learning for foot and ankle disorders

The development of AI and ML is rapidly changing the health care sector, facilitating new ways for the diagnosis, treatment, and management of numerous medical conditions. Foot and ankle disease is the most promising field of AI and ML application that can generally be diagnosed effectively and treated with individual strategies. Foot and ankle disease is versatile, ranging from minor sprains, fractures, and bunions to major issues such as arthritis, diabetic ulceration, and Charcot-Marie-Tooth disease. It manifests through a wide range of illnesses and conditions and requires special identification and treatment to maintain mobility and quality of life. AI and ML are used (Merrill et al., 2019) to process images to diagnose ankle fractures and clinically assess patient predictions. ML-based algorithms are involved in the automatic analysis of gait as part of the foot mechanical examination, helping with planning surgery. ML is also widely used to predict the need for ankle-foot orthosis (AFO) in patients with stroke. Deep neural networks have great potential and can accurately predict the need for AFO, boosting AI technology methods. AI-classified foot gestures, on the other hand, are used in robotic ankle systems in low-limb prostheses, demonstrating a good classification rate. These achievements illustrate the many uses of AI and ML in improving diagnosis accuracy, treatment planning, and functional results for foot and ankle problems.

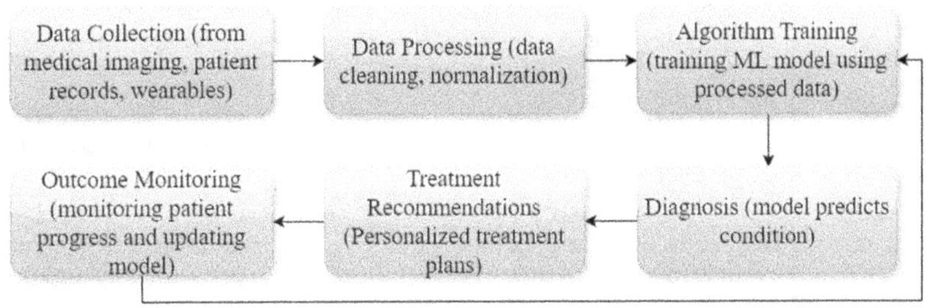

FIGURE 13.4

Overview of artificial intelligence and machine learning (ML) in foot and ankle disorders.

Fig. 13.4 outlines the general workflow of how AI and ML are integrated into the diagnosis and treatment of foot and ankle disorders.

13.4 Enhancing the accuracy of diagnosing foot disorders using artificial intelligence

AI improves foot problem diagnostics via various approaches. ML techniques such as SVM (Support Vector Machine), RF (Random Forest), KNN (k-nearest neighbors), LR (Logistic Regression), and MV (Multivariate) may aid in feature selection and classification. Transfer learning using convolutional neural network (CNN) models, which employ pressure sensors for heat map extraction, has displayed remarkable accuracy in identifying gait problems. Additionally, data synthesis systems using statistical approaches and NeuralProphet models might generate synthetic time series data for training diabetic foot diagnostics models, assisting in overcoming data scarcity difficulties. Furthermore, the use of decision trees and neural networks for data mining might reveal critical correlations among types of foot disease and biomechanical characteristics, enhancing diagnostic accuracy (Kim and Choi, 2023). These AI driven technologies offer rapid and precise tools for recognizing foot disease, which is advantageous for both patients and health care providers.

Fig. 13.5 illustrates the specific process of how AI assists in diagnosing a foot or ankle disorder, from the initial patient visit to the diagnosis (Tables 13.1–13.3).

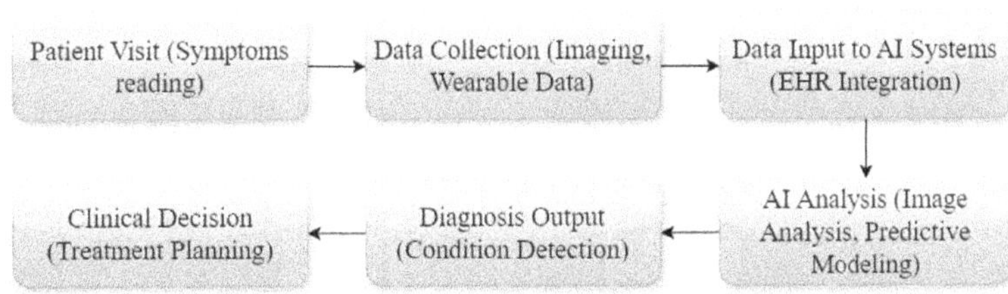

FIGURE 13.5

Artificial intelligence (AI)-driven diagnosis workflow.
EHR, electronic health record.

Table 13.1 Diagnosing foot disorders using artificial intelligence (AI).

Method	Description	Key findings
Deep learning algorithms	Use of convolutional neural networks to analyze imaging data (x-rays, magnetic resonance imaging, computed tomography scans)	Significantly improves accuracy in detecting fractures and soft tissue injuries, often exceeding 90% accuracy
Integration with electronic health record	Combining AI analysis with electronic health records to provide comprehensive patient insights	Enhances diagnostic accuracy by considering historical data and clinical notes alongside imaging
Predictive modeling	Machine learning models predicting surgical outcomes and rehabilitation needs based on patient data	Optimizes treatment plans and predicts complications, improving overall patient management
Mobile AI diagnostics	Using AI tools on mobile devices to analyze images and patient data for remote diagnostics	Increases accessibility and early detection of foot disorders, particularly in resource-limited settings
Automated x-ray analysis	AI systems specifically designed to interpret foot and ankle x-rays	Reduces the rate of missed diagnoses and supports less experienced clinicians in making accurate decisions

Table 13.2 Role of artificial intelligence (AI) in foot disorder diagnostics.

Role of AI	Description	Impact/benefits
Image analysis	AI algorithms, particularly convolutional neural networks, analyze medical images (such as x-rays, magnetic resonance imaging and computed tomography [CT] scans) for abnormalities.	Increases accuracy and speed in diagnosing fractures, deformities, and soft tissue injuries. Reduces human error.
Data integration	AI integrates diverse data sources, including electronic health records, patient history, and imaging.	Provides a comprehensive view of patient health, improving diagnostic accuracy and personalized treatment plans.
Predictive analytics	Machine learning models predict outcomes such as treatment success, risk of complications, and recovery time lines.	Optimizes treatment strategies, enhances patient management, and reduces complications.
Automated diagnostics	AI systems perform the automated analysis of diagnostic images, identifying conditions such as fractures and joint issues.	Supports clinicians by providing second opinions, reducing diagnostic time, and improving diagnostic confidence.
Remote monitoring and diagnosis	AI tools on mobile platforms analyze patient data and images remotely.	Enhances accessibility of diagnostic services, particularly in remote or underserved areas, and allows for continuous monitoring.
Weight-Bearing imaging analysis	AI analyzes weight-bearing x-rays and CT scans to diagnose instabilities and deformities under functional load conditions.	Improves diagnosis of conditions that are not apparent in non–weight bearing images, such as syndesmosis and Lisfranc injuries.

Table 13.3 Applications of artificial intelligence (AI) and machine learning (ML) in foot and ankle disorders.

Application	Description	AI/ML techniques used	Benefits	Challenges
Diagnostic imaging analysis	Automated analysis of x-rays and magnetic resonance imaging and computed tomography scans to identify foot and ankle disorders.	Convolutional neural networks (CNNs), deep learning	Faster and more accurate diagnosis, reduced human error	High-quality annotated datasets needed for training
Predictive modeling	Predicting the progression of conditions such as osteoarthritis in the foot and ankle.	Regression models, random forest, neural networks	Early intervention, personalized treatment plans	Complexity in modeling owing to individual variability
Biomechanical analysis	Analyzing gait patterns and foot mechanics for custom orthotic design.	Motion capture analysis, neural networks	Improved patient outcomes, customized treatments	Requires sophisticated equipment and data processing
Surgical planning	Assisting surgeons in planning complex foot and ankle surgeries.	Three-dimensional (3D) modeling, simulation, reinforcement learning	Enhanced precision in surgery, better recovery outcomes	High computational resources and validation required
Rehabilitation monitoring	Tracking and optimizing rehabilitation exercises after surgery or injury.	Wearable sensors, time series analysis, ML algorithms	Improved adherence to rehabilitation protocols, personalized adjustments	Privacy concerns, need for continuous monitoring
Early detection of diabetic foot ulcers	Identifying signs of diabetic foot ulcers before they become severe.	Image recognition, CNNs, ML	Prevent severe complications, cost-effective management	Need for regular monitoring, data privacy issues
Custom orthotics and prosthetics design	Designing patient-specific orthotics and prosthetics using biomechanical data.	3D scanning, ML algorithms	Better fit and comfort, improved mobility	Integration with current manufacturing processes
Telemedicine and remote Consultation	Providing remote diagnosis and treatment recommendations.	Natural language processing, chatbots, ML algorithms	Access to care in remote areas, reduced need for travel	Ensuring accuracy and reliability of AI recommendations
Pain management	Predicting and managing chronic pain related to foot and ankle disorders.	Predictive analytics, ML algorithms	Personalized pain management plans, better patient outcomes	Complexity in pain assessment, subjective nature of pain
Outcome prediction	Predicting surgical and treatment outcomes to guide decision-making.	Predictive analytics, ML algorithms	Improved surgical success rates, tailored patient care	Data integration from various sources, validation needed

13.5 Role of artificial intelligence in foot disorder diagnostics

AI has a critical role in foot issue diagnosis by automating the procedure using ML techniques and unreal neural structures (Seeliger et al., 2019). Progressive ML techniques are employed to automate feature selection, foot condition categorization, and result interpretation. AI's ability to understand causal linkages from data, even without previous structure-based information, upgrades diagnostic accuracy (Gupta et al., 2023). In the diabetic foot care industry, AI research aids to resolve problems such as delayed diagnosis and treatment compliance by giving feasible resolutions using artificial neural structures. Furthermore, a sensor-supported medical time series data synthesis system was showcased to train diabetic foot diagnostic models, highlighting the application of AI in generating artificial data for training (Gupta et al., 2023). Overall, AI hastens the diagnosis of foot problems, elevates accuracy, and allows individualized health care solutions.

13.6 Applications of artificial intelligence and machine learning in foot and ankle disorders

AI and ML have revolutionized the field of foot and ankle disorders by significantly improving diagnosis and treatment. In diagnostic imaging, AI-driven tools, including CNN and deep learning algorithms, have enabled the more accurate and quick interpretation of x-rays and magnetic resonance imaging (MRI), and CT scans. AI has helped to diagnose fractures, arthritis, and soft tissue injuries more accurately than can human experts, avoid human errors, and facilitate early intervention (Cooperman and Brandão, 2024). In pavements such as regression modeling, random forests, and neural networks predict the occurrence of degenerative diseases such as osteoarthritis. It is beneficial because physicians can initiate timely and proactive treatment when they know the likely trajectory of such diseases. Current systems analyze gait patterns and foot mechanics through biomechanical analysis, a tool that uses motion capture systems and neural networks. They design custom orthotics that enable the provision of individual-specific care to increase the treatment success rate. AI also has a crucial role in complex foot and ankle surgery planning. Three-dimensional (3D) modeling, risk assessment, simulation, reinforcement learning tools, and systems help surgeons mark out complex surgeries (Paleyes et al., 2022). It is beneficial in case where the presurgical context, and surgeons can triangulate ensuring the postsurgical context is healthy and thus minimal recovery consequences, thus making the results a success. These systems are used to monitor rehabilitation through wearable sensors and time-series analysis to determine the patient's progress. The patient's rehabilitation pattern therapy is then adjusted for the desired outcome. AI also uses image identification tool kits and learning CNN to initiate early detection and initiate early treatment to avoid severe consequences.

AI has also revolutionized the design of custom orthotics and prosthetics, with 3D scans and ML algorithms ensuring that such devices are perfectly designed to match patients' unique biomechanical profiles (Anastasio et al., 2023). Through this approach, the level of comfort and mobility in patients with foot and ankle disorders has significantly improved, leading to an enhanced quality of life. Other applications in which AI has improved foot and ankle care include telemedicine and remote consultations. Natural language processing (NLP) and chatbots, among other ML algorithms, have enabled the provision of services to patients in remote areas. Patients can reach specialized care without the need for travel. AI is also being used in pain management, enabling the creation of predictive models for personalized pain management plans. Thus the outcome in patients has relatively improved, especially for parameters relevant to pain management. AI also generates predictive tools that can predict outcomes after surgery or other therapies (Figs. 13.6 and 13.7).

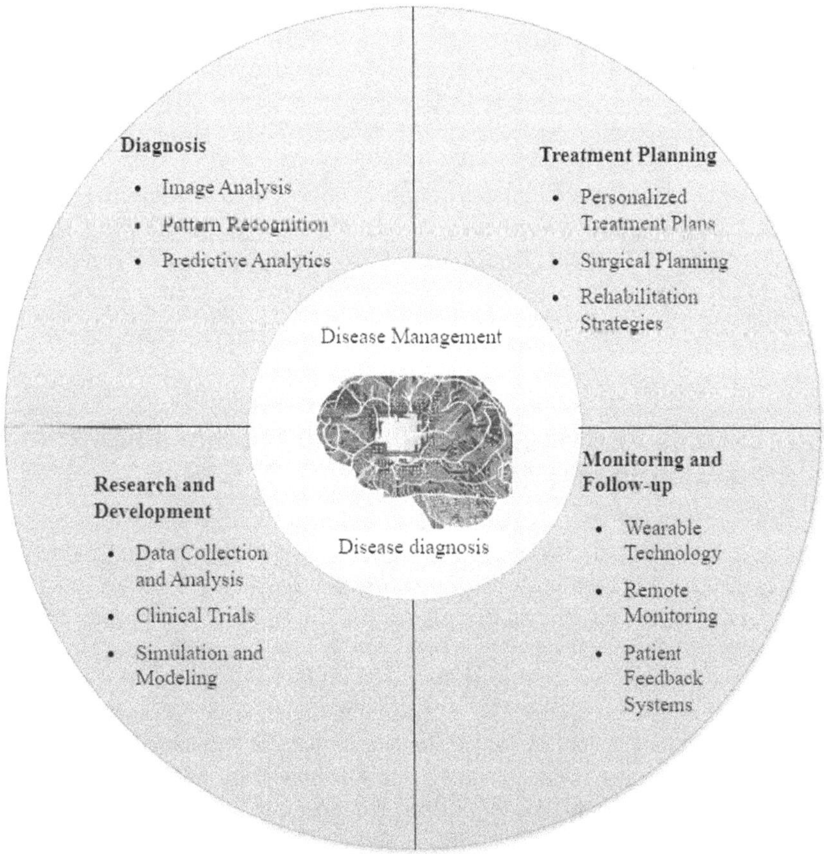

FIGURE 13.6

Applications of artificial intelligence and machine learning in foot and ankle disorders.

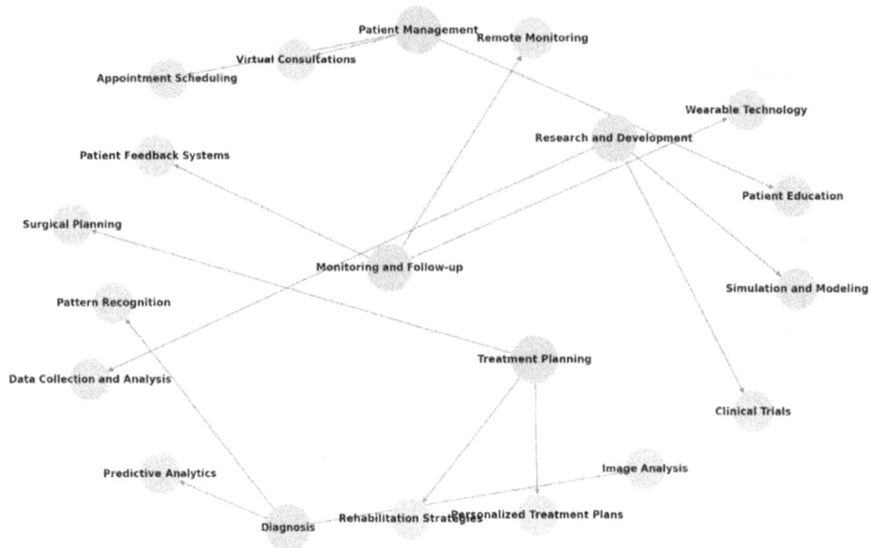

FIGURE 13.7

Representation of artificial intelligence and machine learning in foot and ankle disorders.

The diagram shows the main areas of application and their respective subcategories, including diagnosis, treatment planning, monitoring and follow-up, research and development, and patient management. This visual representation demonstrates how these advanced technologies are being integrated into different aspects of managing foot and ankle disorders.

13.7 Benefits of artificial intelligence and machine learning in foot and ankle care

Rapid advances in AI and ML have led to wide applications in various medical fields, including foot and ankle care. These technologies have several merits; for instance, they have a significant role in improving diagnostic accuracy, treatment planning, patient monitoring, and overall health care delivery. This chapter has discussed some benefits of introducing AI and ML in foot and ankle care. Diagnostic accuracy is among the benefits associated with AI and ML in foot and ankle care (Lisacek-Kiosoglous et al., 2023). Conventional diagnostic approaches highly depend on clinicians' expertise; thus, diagnoses may be subjective or associated with errors. For example, the analysis of diagnostic images by radiologists may yield inconsistent results. AI and ML algorithms that specifically rely on deep learning demonstrated better performance in the analysis of x-rays, MRI, and CT images. They can identify subtle patterns or abnormalities in these images that may be ignored by human radiologists. As a result, these algorithms aid in the early and correct identification of conditions such as fractures, ligament tears, and arthritic changes. For example, CNNs have demonstrated high accuracy in fracture detection within x-ray images.

Another benefit of AI and ML in foot and ankle disorders is the development of personalized treatment plans. AI systems can analyze the patient's medical history, genetic predisposition, lifestyle, and previous treatment responses, to develop an algorithm that predicts the most effective treatment plan for a specific patient (Kitamura et al., 2019). Such an approach is particularly favorable for chronic disorders such as arthritis or diabetic foot ulcers (DFUs), because the effectiveness of treatment can differ immensely. ML algorithms can predict the likely outcome of treatment for a patient; therefore, the clinician can choose the most suitable therapeutic options, which results in better patient outcomes and diminishes the risk-based planning of treatment. AI and ML also improve the monitoring and handling of patient conditions. For instance, most foot and ankle clinics work with patients for several weeks, months, or even years. Therefore, it is paramount to ensure continuous monitoring and intervention. AI algorithms powered by data collected from embedded sensors in wearable devices can monitor a patient's gait, physiological movements, and pressure. Through such data analysis, the AI system can find the behavioral patterns of patients, analyze deviations, and predict likely complications to develop algorithms to detect deviation and predict complications This method reduces the risk of serious complications, because early interventions can be made. For example, in an observation of a DFU, early stage infection might be detected even with no change entered, precluding the result of a patient with ulcers who needs amputation. (Traditionally, this sort of operation is done not until the severe stage.) AI technology and ML hold immense value for foot and ankle care, the biggest investment that can be made in AI and ML technology is improving workflow and resource management (Rao et al., 2012). By automatically analyzing medical images, AI lightens the workload of radiologists and other professionals in this field and frees the capacity for difficult cases and patient care. Furthermore, AI-based predictive analytics can improve scheduling and resource planning by telling us how many patients to expect, and hence when we should start each route.

AI technology can boost resource use and system performance, reducing patient wait times. It can help empower patients and improve their attitudes toward health knowledge, shifting their mindset from a sick state to a healthier one. Another area for AI investment is rehabilitation and recovery. Rehabilitation is a crucial part of managing most foot and ankle conditions. AI may help in these domains by creating and monitoring rehabilitation plans (Choo et al., 2021; Roscher et al., 2020). By combining data collected from wearable devices, mobile health apps, and AI, a patient's course of recovery can be monitored and recommendations can be made in real time to adjust exercise routines. Thus patients might recover faster from illness and have a much higher level of function.

13.8 Rehabilitation and monitoring of recovery in foot and ankle disorders

Rehabilitation and monitoring of recovery from disorders of the foot and ankle is a complex and multidimensional process that requires an array of tools and techniques. Advanced technologies such as AI and wearables bring continuous surveillance data based on real-time information to adjusted personalized regimens based on individual patient care. Using wearable devices and mobile apps for rehabilitation efforts increases their efficiency and effectiveness, ultimately resulting in higher-quality patient care with respect to foot and ankle disorders (Chae et al., 2020; Saraswat et al., 2022). The rehabilitation process for body recovery begins with an initial assessment. Here, details are collected, such as the medical history, symptoms, and baseline data from imaging techniques such as x-rays and MRI and CT scans that will be important to retrieval. Functional testing, gait analysis, walking tests, and mobility tests are used to determine areas of the patient's baseline condition in need of further study. This assessment is vital because it provides the nurse with the necessary data to form a personalized plan (López-Martínez et al., 2020; Pawar et al., 2020). A personalized plan is developed based on the patient assessment and the goals that the patient hopes to achieve. Typically, this plan involves physical therapy, exercise, and, when necessary, surgery. There should be short-term and long-term goals to motivate the patient and determine the success of the treatment plan. Continuous monitoring is a critical step in the healing process. Wearable devices such as smart insoles and activity trackers are critical to monitor the patient's gait parameters and general activity levels. The use of mobile apps to log in exercises, symptom tracking, and access to educational and instructional information is also essential.

AI and Data Analysis: Another critical aspect is the use of AI algorithms to process massive amounts of data generated by wearable gadgets and other monitoring devices. Anomalies in the data, complications likely to occur that should be predicted, and the effectiveness of the rehabilitation must be determined from time to time. An AI system can help the clinician regularly monitor data and act promptly (Stauffer et al., 2023). For example, if the gait analysis indicates an irregularity, AI alerts the physician to investigate further and may suggest adjusting the rehabilitation as necessary.

Feedback and Adjustment: Additionally, regular feedback and adjustment are required to help track the progress and plan for a suitable path to achieve the set goal. Information generated by AI and data analysis is useful in regulating exercise regimens, changing clinician able therapy procedures or introducing new interventions where that has been identified. Thus the rehabilitation plan is evolving as the patient recovers and rapidly addresses the problem affecting him or her.

Patient Engagement: The patient's engagement in the plan is crucial to determining whether the rehabilitation program will be successful. It is essential to educate the patient on the causes, mechanism, prevention, need for adherence, and self-care practices. Some devices ported through virtual apps can remind patients when it is time to perform an exercise, and assess the patient (Li et al., 2021). The analysis assesses whether the patient is following the nurse's instructions or if device logs indicate a lack of performance. This contributes to outcome evaluation.

However, long-term monitoring is crucial to avoid reinjury and control chronic conditions even after the main rehabilitation phase (Vaish et al., 2023). Checkups, constant monitoring of wearable devices powered with predictive analytics to determine the most significant risks easily, and the prevention of additional problems with the help of exercises and lifestyle changes that fit the specific needs of the client can prevent similar problems in future and enable sufficient healing.

13.9 Diagnostic tools and techniques in foot and ankle disorders

For foot and ankle disorders, the diagnosis is vital and requires a variety of approaches. A diagnosis must be accurate to provide the most appropriate treatment plan and improve patients' clinical outcomes. Technological advances within the field of medicine have promoted an evolution in the diagnostic tools available. Within this field is the important domain of technological developments (Kitamura et al., 2019; Kasula, 2022). There is a wide variety of diagnostic tools and equipment for doctors to use. A new piece comes out every day. Meanwhile traditional techniques and modern equipment are used for the diagnostic armory. There are several diagnostic tools available for assessing foot and ankle disorders, such as X-rays (Khedkar et al., 2020). Wherever reference is made to examination of the foot or ankle, x-ray radiography is the first technique for a detailed look at the bony framework. Fractures of the foot, arthritis, hallux valgus, and heel spur can be diagnosed by x-rays and magnetic resonance imaging (MRI). With the support of MRI, the soft tissues of the feet and ankles are revealed rather than the hard tissue. Moreover, MRI is used to examine the foot and ankle joints. The diagnostic standard suggests that MRI is the favorable method of choice for suspected medical conditions of ligaments and tendon, as well as bone marrow disease. CT scans are valuable for examining subcapital fractures that would be difficult to diagnose with ordinary radiographs.

An all-encompassing physical examination is essential for the diagnosis of foot and ankle disorders. On clinical examination, the foot is observed carefully for deformity, swelling, or changes of the skin. It is also necessary to measure range of motion, power, and sensation. Apart from a general examination, specific tests are available to diagnose ligament and tendon injuries, such as the Thompson test for Achilles tendon rupture and the anterior drawer test.

Gait analysis is essential for assessing biomechanical anomalies and functional deficits in patients with foot and ankle disorders. Various technologies produce gait parameters, such as the length of stride, forefoot distribution, and gait symmetry, using pressure-sensitive mats, motion capture systems, and wearable trackers (Jung et al., 2023). Gait analysis is especially beneficial in identifying gait anomalies such as those associated with flat feet, ankle instability, and neuropathies. Emerging diagnostic technologies, digital health solutions, and digital health technologies including telehealth services, smartphone apps, and wearable devices ensure the diagnosis of foot and ankle disorders. Telemedicine allows patients to consult with foot and ankle experts without visiting the hospital, ensuring prompt diagnosis and care recommendations. Smartphone applications and wearable technology let persons monitor symptoms and therapy adherence, and communicate in real time with medical professionals. Artificial intelligence (AI) and machine learning (ML) can inspect large amounts of medical information to enhance diagnostic accuracy and timeliness, providing an alternative to traditional diagnostic procedures such as imaging investigations and electromyography (Kim et al., 2021). ML-based computer-assisted diagnostics can detect minor aberrations in imaging reports, classify their anatomical findings, and get updated advice for treating patients with individualized genetics based on their demographics.

The diagnostic outlook for foot and ankle disorders has experienced profound changes to innovations in imaging studies, clinical evaluation methods, and novel digital health applications. An increased variety of diagnostic equipment and modalities currently enable the precise evaluation and management of multiple foot and ankle pathologies. Implementing the use of novel techniques in clinics may help enhance diagnostic quality, treatment possibility, and patient outcomes in foot and ankle disorders (Okay et al., 2021).

13.10 Case studies

13.10.1 Case studies of artificial intelligence in diagnosis of foot and ankle disorders

The applications of AI have revolutionized the diagnosis and treatment of several medical diseases, including foot and ankle disorders. AI technology, especially ML techniques and deep learning models, have significantly improved diagnostic accuracy, patient outcomes, and clinical practices. The subsequent section provides multiple case studies of AI applications in identifying foot and ankle problems.

These case studies exhibit how AI may modify the diagnosis of foot and ankle tribulations. More progressive algorithms and hefty datasets may be employed by AI schemes to refine the diagnostic accuracy, predict issues, and customize treatment tactics. Assimilating AI in clinical practice not only ameliorates patient results, it optimizes health care reserves and aids physicians in making more informed choices (Cheng et al., 2023). As AI technology progresses, its applications in foot and ankle care are foreseen to grow, granting even more benefits to patients and health care practitioners.

Case Study 1: Artificial intelligence–Assisted Diagnosis of Ankle Fractures.

Background: Ankle fracture is a frequent injury requiring a thorough diagnosis for proper treatment. Traditional techniques need radiologists to examine radiographs manually, which is time-consuming and prone to human error.

AI Application: Researchers made a deep learning model using CNNs to identify ankle fractures automatically in radiographs (Ahmed et al., 2020). The model was trained using a huge collection of annotated ankle x-rays.

Outcome: The AI model achieved a diagnosis accuracy equivalent to that of expert radiologists. It greatly lowered the diagnostic time and enhanced the identification of small fractures that humans might miss. Thus, the case study demonstrates the promise of AI for improving the diagnosis of accuracy in health care ventures.

Case Study 2: Machine Learning in Predicting Diabetic Foot Ulcers.

DFUs are one of the most catastrophic consequences of diabetes because they often result in an infection or cut. Early prediction and its outcomes are vital to prevent damage.

AI application: The machine learning model was created to predict patients at risk of developing diabetic foot ulcers (DFUs) based on data, including demographics, medical history, clinical findings, and tests, using Random Forest and Support Vector Machine algorithms. The predictive approach effectively identified patients at high risk of developing DFUs, allowing early intervention and individual care plans. The predictive approach was associated with a decrease in the number of DFUs and overall improved patient care.

Case Study 3: Automated Detection of Achilles Tendon Tears.

This type of accident and similar occurrences frequently result in Achilles tendon breaks, a consequence that can have dire consequences if not immediately and correctly diagnosed and treated.

AI Application: Researchers have created a deep learning–based framework for diagnosing Achilles tendon tears using a compilation of MRIs (Antunes et al., 2022). To increase treatment performance, the framework combined CNNs and transfer educational practices.

Results: The AI tool accurately diagnosed Achilles tendon tears with strong sensitivity and specificity. The automated detection system serves as a valuable quantitative tool for confirming the diagnosis of Achilles tendon tears in patients and reduces the workload for radiologists. These developments clearly demonstrate the utility of AI in improving diagnostic accuracy and quality clinical decision-making to deliver improved patient outcomes.

Case Study 4: Artificial intelligence–Driven Identification of Plantar Fasciitis.

Plantar fasciitis is a common problem in the heel, usually established by a clinical examination and imaging. An AI model was developed to evaluate ultrasound imagery and clinical data of diagnosed plantar fasciitis. The study model included logistic regression and neural networks. The accuracy and early diagnosis of plantar fasciitis were determined for the AI model. Intervention and therapy to treat the disorder were swiftly performed owing to the diagnosis. The use of AI in this case has an impact on diagnostic measurement through more consistent and subjective evaluations of diagnoses to be conducting.

Case Study 5: Predictive Analytics for Postoperative Complications.

Background: Postoperative complications arising from foot and ankle surgery can greatly affect a patient's recovery and hospital bills.

AI Application: An AI mechanism using predictive analytics was built to scrutinize preoperative and perioperative data and forecast which patients would most likely have complications after surgery (Bharadwaj et al., 2021). In this the algorithms used are gradient boosting and decision tree algorithms.

Result: The AI mechanism was able to predict problems well, such as infections and slow recovery. This helped hospital staff to channel treatments and closely monitor patients with the potential to have issues, reducing such occurrences and resulting in better surgery.

13.10.2 Case studies and Clinical Trials on successful implementations of artificial intelligence in diagnosing foot and ankle disorders

The use of AI in the identification and therapy of foot and ankle complications has garnered favorable reviews. Several in-depth case studies and research are provided that exemplify the effective use of AI systems. Such instances display the capacity of AI to enhance diagnostic accuracy, provide quicker health care output, and benefit patients.

Case Study 1: Artificial intelligence–Assisted Diagnosis of Ankle Fractures (Deep Learning for Fracture Detection: A Case Study in Ankle Radiographs)

This study aims to evaluate the effectiveness of deep learning models in identifying ankle fractures from radiographs. The researchers trained a CNN using a large dataset of ankle radiographs. The dataset included pictures of both fractured and nonfractured ankles. These pictures were annotated by qualified radiologists. The model was designed to identify fractures automatically and was tested on a different set of pictures not used in the training process. The AI model achieved an accuracy of 94%, a sensitivity of 91%, and a specificity of 96%, which was equal to, and in some respects better than those of human radiologists in terms of diagnosis. Implementation significantly reduced the time spent diagnosing fractures and allowed for the identification of small fractures (Caruana et al., 2020). This AI system will be beneficial to radiologists in that it will provide a second opinion, ensure fractures do not go undetected, reduce the time required to diagnose, and may save a patient's life.

Case Study 2: Predicting Diabetic Foot Ulcer Risk with Artificial Intelligence (AI) (AI-Based Gait Analysis for Diabetic Foot Ulcer Prediction)

The objective is to develop an AI model predicting DFU risk from gait analysis data. The study used an ML model created from gait analysis data collected from people with diabetes. The model had three features: foot pressure distribution, step length, and walking speed. Using the random forest and neural network algorithms, the model analyzed people's data to predict DFUs. The AI model predicted the occurrence of DFUs by 92% whereas the sensitivity and specificity for high-risk individuals were 88% and 94%, respectively. The predictions may help the health care sector intervene early and take preventative measures. Early identification of individuals at high risk of DFU can greatly reduce ulcer occurrence, decrease health care prices, and improve the quality of life of people with diabetes. In addition, an AI model allows for individualized monitoring and proactively decides which high-risk individuals need treatment.

Case Study 3: Artificial intelligence in the Diagnosis of Achilles Tendon Tears (Automated Detection of Achilles Tendon Tears Using Deep Learning Models)

An AI model, particularly a CNN, was trained with an annotated dataset of MRI images created by radiologists focused on musculoskeletal imaging. AI model performance was assessed against expert musculoskeletal radiologists' performance grading Achilles tendon tears. The AI model recognized Achilles tendon tears with a sensitivity of 95% and a specificity of 93% for average performance. It was well-matched in performance to radiologists as an automated evaluation for trustful examination, reducing the diagnostic load. AI for the diagnostic support of Achilles tendon injury can improve the speed of the diagnosis, promote accuracy, and allow radiologists to focus on patients with potential complications. It enhances the clinic's processing capability while ensuring the proper management and care of patients.

Clinical Trial 1: Artificial intelligence for Personalized Treatment Plans in Foot and Ankle Surgery (Predicting Outcomes of Foot and Ankle Surgery Using Machine Learning)

The goal was to predict surgery outcomes using ML methods in patients who have had foot and ankle surgery, followed by surgery-based advice to patients. The research conducted a retrospective study based on patients' records, which include demographic data, the medical background of the patient, and surgery results. Learning algorithms such as random forest and support vector machine were used to develop predictive indicators. Learning algorithms forecast surgery resulted in 89% precision. The ability to identify high-risk individuals enabled distinct treatment techniques and preoperative interventions to be tailored to various predicted issues based on individual patient profiles (Chen et al., 2021). When assessments of potential surgery results are performed before the operation, surgeons can prepare better for expected problems and provide individual treatment methods. This guarantees patient safety, achieves high success rates of surgical intervention, and shortens the recovery time.

Clinical Trial 2: Artificial intelligence in Postoperative Complication Prediction (Machine Learning in Predicting the Healing Time of Diabetic Foot Ulcers)

In this work, an AI model on clinical data from people with diabetes who underwent foot surgery. AI used gradient boosting, an ML method, and decision tree to project complications and healing times. The AI model was correct in 87% of projections and detected danger for 90% of patients. As a result, there were therapeutic material improvements and preventative measures has been taken. By tracking the number of complications during surgery and healing, health care professionals can adopt an individualized approach that reduces the number of complications and shortens recovery time. The advantage of this approach is that patient satisfaction is improved and resources are used wisely.

13.11 Real-world examples of artificial intelligence implementations in diagnosing foot and ankle disorders

The researchers have now described several fully integrated medical diagnostics with a large role that are obvious candidates for the AI reconstruction of foot and ankle identification. The following examples are not only possibilities for enhancing diagnostic accuracy, they may also result in quantifiable patient and procedural outcomes in practice and real-life.

Example 1: Zebra Medical Vision's artificial intelligence for Fracture Detection.

Overview: Zebra Medical Vision, a medical imaging corporation that uses AI, has developed an AI that can recognize fractures in x-ray images, particularly the foot and ankle. Many institutions around the world have implemented the Zebra AI into their x-ray workflows. The AI software analyzes x-ray images for conceivable fractures and signals radiologists when it finds suspected fractures that need further examination. The AI improved the percentage of fractures identified and the rate of fracture recognition, especially when the fractures were subtle and many times missed by radiologists. The AI decreased diagnostic time by assisting radiologists, providing excellent and quick initial reviews, and eliminating the time radiologists spent examining each image, which enabled them to come to a faster diagnosis and potential treatment, which was excellent. This AI software app was especially effective in emergency rooms, where the quick identification of fractures is critical for compassionate starts.

Example 2: Aidoc's artificial intelligence Solution for Radiology.

Aidoc, the AI startup that specializes in point solutions for radiology, developed AI software that could diagnose thousands of conditions using imaging, including foot and ankle problems. Several hospitals and clinics installed Aidoc's AI into their existing PACS (Picture Archiving and Communication System). The AI goes through the imaging reports and flags unusual cases and then ranks which conditions need to be treated right away. The use of point solutions increased the effectiveness with which the process currently worked. It sorted cases based on their level of need, so that urgent cases would be picked up as soon as possible. In terms of accuracy, radiologists reported the AI was a reliable second opinion for identifying fractures and other issues in most foot and ankle images. From a clinical point of view, clinicians stated that using AI diminished their load and increased their confidence in diagnoses.

Example 3: PathAI for Diabetic Foot Ulcer Assessment.

PathAI, a leading provider of AI-enabled pathology services, developed an AI system specifically designed to assist in the diagnosis and management of DFUs. The AI system was fully integrated into the workflow of diabetic clinics and wound care centers where clinical photographs of foot ulcers are taken, and the system provided information about the degree of severity and healing prediction. Achieving fully accurate measurements and comprehensive assessment of the size and depth of the ulcer and the tissue type, achieving superior precision and elimination of interobserver variability. Additionally, it recorded feature modifications with time to help physicians track healing of the ulcer and modify the management plan accordingly. Using AI to make more informed clinical decisions and individualized therapies helped heal patients' DFUs faster.

Example 4: IBM Watson Health for Personalized Treatment Plans.

IBM Watson Health used AI to create personalized therapeutic protocols for patients with different illnesses, including those of the foot and ankle. Based on IBM Watson's AI analysis of EHR data of patient histories, therapy outcomes, and guidelines, it provided individualized treatment plans for patients with foot and ankle disease. The therapy AI included specific treatment plans in its recommendations according to the individual patient as well as treatments to be effective, rapidly increasing the likelihood the patient would undergo them. Using AI suggestions, medical workers made good decisions regarding surgery, physiotherapy, and preventive measures, which helped to reduce the number of patients with foot and ankle problems admitted to hospitals Patient care gave information resources and advice to aid patients in understanding the disease and follow up therapy regimens better.

Example 5: Google Health's artificial intelligence for Diabetic Retinopathy and Foot Ulcers.

With AI models in place for identifying diabetic retinopathy, Google Health went further in the field and expanded its study to include DFUs based on prior experience in medical picture processing. The AI model created for DFUs was tested with wound care professionals and later applied in clinical practice. It analyzed pictures of foot ulcers, looking for signs of infection, necrosis, and other effects. The AI demonstrated excellent accuracy in identifying issues concerning DFUs, allowing early management. Timely identification of initial signs of decay enabled prompt treatment, reducing the risk of severe complications and amputation. The AI model's efficacy in actual clinical practice served as an opportunity to introduce it into general use to treat diabetes, and this translated into improved outcomes in a vast number of patients.

These real-life use cases showcase the transformative power of AI in foot and ankle issue detection and treatment. Implementation of AI software in clinical practice enables health care professionals to achieve higher diagnostic precision, conduct operations faster, and provide more personalized treatments. These examples demonstrate that AI is not of the future; it is a tangible tool with a significant impact on the medical field that exists and is rapidly developing, particularly in niche areas such as foot care and ankle treatment. The role of AI in health care is bound to grow with advances in technology, and the range of benefits professionals and patients will be able to get from it will likely increase further.

13.12 Future directions and emerging trends in artificial intelligence and machine learning for foot and ankle disorders

AI and ML applications for the diagnosis and treatment of foot and ankle ailments are projected to progress dramatically because of numerous trends and advances. One of the most exciting advances is the integration of AI with wearable technologies (Chua et al., 2023). Widespread access to wearable systems with advanced sensors and AI algorithms could allow for density monitoring of patients' biomechanics and gait, enabling accident forecasting and avoidance and improving the lives of patients with neurological illnesses, such as those with DFUs. Furthermore, the use of AI to design therapy may rapidly become feasible.

AI may also create highly customized treatment regimens that achieve better therapeutic results using large quantities of patient information, such as genetic data, personal data, and the patient's reactions to therapy. Another important trend is the use of AI in image analysis (Kumar et al., 2024). AI will become more effective as 3D imaging and augmented reality technologies advance. These advances will result in more exact diagnostic capabilities and surgical treatment planning, which will decrease the margin of error and ultimately more successful patient outcomes. In addition, AI-powered predictive analytics will forecast disease progress and its future effects, making enhanced preventive therapy familiar (Crespo-Gonzalez et al., 2022). Such predictive activity will be extremely useful in treating major conditions such as arthritis and repetitive tendonitis, for which early detection might have a major impact on disease progression.

Furthermore, the use of AI in telemedicine is expected to narrow the gap in access to specialist treatment, particularly in distant and underserved locations. AI-powered diagnostic tools may help general practitioners make more accurate recommendations to specialists, resulting in timely and appropriate treatment (Hasas et al., 2024). The incorporation of NLP into EHR systems will expedite documentation, enabling physicians to focus more on patient care than administrative responsibilities. Additionally, NLP may aid in extracting relevant insights from enormous volumes of unstructured health care data, increasing decision-making.

13.13 Conclusion

The combined capabilities of AI and ML have revolutionized orthopedics to a new level. AI's potential to analyze large volumes of medical data, recognize patterns, identify deviations, and influence to facilitate accurate diagnosis and treatment planner creation. This innovation also helps to improve the efficiency of health care delivery and patient outcomes by enabling providers to offer truly patient-centered care tailored to individual needs. In addition, in some cases, AI-powered image-based diagnostic technologies can detect fractures, arthritis, and ligament tears more precisely during their early stages, offering patients more effective, less costly, and thus less traumatic treatment (Jayatilake and Ganegoda, 2021). Finally, AI and ML technologies help automate mundane operations and streamline workflow management systems, which may lead to increased patient flow at health facilities, shortening wait times and lowering medical services costs. Integrating AI and ML into foot and ankle care, however, has problems. In addition to data privacy issues, these sources can have algorithm bias, making them dubious for actual use without moral standards that should be set by physicians and lawmakers (Jin et al., 2022). In future, AI will be used to investigate and develop the cure in this field. Moreover, care will become much more personalized and also more financially available to everyone. The important thing is not to abuse the new freedoms.

References

Ahmed, Z., Mohamed, K., Zeeshan, S., Dong, X., 2020. Artificial intelligence with multi-functional machine learning platform development for better healthcare and precision medicine. Database 2020, baaa010.

Anastasio, A.T., Mills, F.B., Karavan, M.P., Adams, S.B., 2023. Evaluating the quality and usability of artificial intelligence–generated responses to common patient questions in foot and ankle surgery. Foot Ankle Orthop. 8 (4). https://doi.org/10.1177/24730114231209919.

Antunes, R.S., André da Costa, C., Küderle, A., Yari, I.A., Eskofier, B., 2022. Federated learning for healthcare: systematic review and architecture proposal. ACM Transact. Intell. Syst. Technol. (TIST) 13 (4), 1–23.

Bharadwaj, H.K., Agarwal, A., Chamola, V., Lakkaniga, N.R., Hassija, V., Guizani, M., Sikdar, B., 2021. A review on the role of machine learning in enabling IoT based healthcare applications. IEEE Access 9, 38859–38890.

Caruana, R., Lundberg, S., Ribeiro, M.T., Nori, H., Jenkins, S., 2020. Intelligible and explainable machine learning: best practices and practical challenges. In: Proceedings Of The 26th Acm Sigkdd International Conference on Knowledge Discovery & Data Mining, pp. 3511–3512.

Chae, J., Kang, Y.J., Noh, Y., 2020. A deep-learning approach for foot-type classification using heterogeneous pressure data. Sensors 20 (16), 1–19. https://doi.org/10.3390/s20164481.

Chen, I.Y., Pierson, E., Rose, S., Joshi, S., Ferryman, K., Ghassemi, M., 2021. Ethical machine learning in healthcare. Annu. Rev. Biomedic. Data Sci. 4, 123–144.

Cheng, C.T., et al., 2023. Evaluation of ensemble strategy on the development of multiple view ankle fracture detection algorithm. Br. J. Radiol. 96 (1145), 15–20. https://doi.org/10.1259/bjr.20220924.

Choo, Y.J., Kim, J.K., Kim, J.H., Chang, M.C., Park, D., 2021. Machine learning analysis to predict the need for ankle foot orthosis in patients with stroke. Sci. Rep. 11 (1), 1–8. https://doi.org/10.1038/s41598-021-87826-3.

Chua, M., Kim, D., Choi, J., Lee, N.G., Deshpande, V., Schwab, J., Lev, M.H., Gonzalez, R.G., Gee, M.S., Do, S., 2023. Tackling prediction uncertainty in machine learning for healthcare. Nat. Biomed. Eng. 7 (6), 711–718.

Cooperman, S.R., Brandão, R.A., 2024. Investigating the proficiency of an AI tool in summarizing foot and ankle literature: a quantitative, qualitative and accuracy analysis. Foot Ankle Surg.: Tech. Rep. Cases 4 (March), 100384. https://doi.org/10.1016/j.fastrc.2024.100384.

Crespo-Gonzalez, C., Dineen-Griffin, S., Rae, J., Hill, R.A., 2022. A qualitative exploration of mental health services provided in community pharmacies. PLoS One 17 (5), e0268259.

Gupta, P., Kingston, K.A., O'Malley, M., Williams, R.J., Ramkumar, P.N., 2023. Advancements in artificial intelligence for foot and ankle surgery: a systematic review. Foot Ankle Orthop. 8 (1). https://doi.org/10.1177/24730114221151079.

Hasas, A., Hakimi, M., Shahidzay, A.K., Fazil, A.W., 2024. AI for social good: leveraging artificial intelligence for community development. J. Commun. Serv. Soc. Empower. 2 (02), 196–210.

Hida, M., et al., 2023. Development of hallux valgus classification using digital foot images with machine learning. Life 13 (5). https://doi.org/10.3390/life13051146.

Jayatilake, S.M.D.A.C., Ganegoda, G.U., 2021. Involvement of machine learning tools in healthcare decision making. J. Healthcare Eng. 2021.

Jin, D., Sergeeva, E., Weng, W.H., Chauhan, G., Szolovits, P., 2022. Explainable deep learning in healthcare: a methodological survey from an attribution view. WIREs Mech. Dis. 14 (3), e1548.

Jung, S., Bong, J.H., Kim, K., Park, S., 2023. Machine-learning-based coordination of powered ankle–foot orthosis and functional electrical stimulation for gait control. Front. Bioeng. Biotechnol. 11 (January), 1–15. https://doi.org/10.3389/fbioe.2023.1272693.

Kasula, B.Y., 2022. Machine learning applications for early detection and intervention in chronic diseases. Internat. Transact. Artif. Intell. 6 (6), 1–7.

Khedkar, S., Gandhi, P., Shinde, G., Subramanian, V., 2020. Deep learning and explainable AI in healthcare using EHR. In: Deep Learning Techniques for Biomedical and Health Informatics, pp. 129–148.

Kim, J.H., Mo, Y.C., Choi, S.M., Hyun, Y., Lee, J.W., 2021. Detecting ankle fractures in plain radiographs using deep learning with accurately labeled datasets aided by computed tomography: a retrospective observational study. Appl. Sci. 11 (19). https://doi.org/10.3390/app11198791.

Kim, Y.-C., Choi, Y.-H., 2023. AI-based foot X-ray reading in real-world: evaluating the accuracy of assistive decisions for diagnosing foot & ankle disorders. Foot Ankle Orthop. 8 (4), 2023. https://doi.org/10.1177/2473011423s00022.

Kitamura, G., Chung, C.Y., Moore, B.E., 2019. Ankle fracture detection utilizing a convolutional neural network ensemble implemented with a small sample, de novo training, and multiview incorporation. J. Digit. Imag. 32 (4), 672–677. https://doi.org/10.1007/s10278-018-0167-7.

Kumar, R., Gupta, M., Abraham, A., 2024. A critical analysis on vertebra identification and cobb angle estimation using deep learning for scoliosis detection. IEEE Access 12, 11170–11184. https://doi.org/10.1109/ACCESS.2024.3353794.

Li, W., Chai, Y., Khan, F., Jan, S.R.U., Verma, S., Menon, V.G., Kavita, F., Li, X., 2021. A comprehensive survey on machine learning-based big data analytics for IoT-enabled smart healthcare system. Mobile Network. Appl. 26, 234–252.

Lisacek-Kiosoglous, A.B., Powling, A.S., Fontalis, A., Gabr, A., Mazomenos, E., Haddad, F.S., 2023. Artificial intelligence in orthopaedic surgery. Bone Joint Res. 12 (7), 447–454. https://doi.org/10.1302/2046-3758.127.bjr-2023-0111.r1.

López-Martínez, F., Núñez-Valdez, E.R., García-Díaz, V., Bursac, Z., 2020. A case study for a big data and machine learning platform to improve medical decision support in population health management. Algorithms 13 (4), 102.

Merrill, R.K., Ferrandino, R.M., Hoffman, R., Shaffer, G.W., Ndu, A., 2019. Machine learning accurately predicts short-term outcomes following open reduction and internal fixation of ankle fractures. J. Foot Ankle Surg. 58 (3), 410–416. https://doi.org/10.1053/j.jfas.2018.09.004.

Okay, F.Y., Yıldırım, M., Özdemir, S., 2021. Interpretable machine learning: a case study of healthcare. In: 2021 International Symposium on Networks, Computers and Communications (ISNCC). IEEE, pp. 1–6.

Paleyes, A., Urma, R.G., Lawrence, N.D., 2022. Challenges in deploying machine learning: a survey of case studies. ACM Comput. Surv. 55 (6), 1–29.

Pawar, U., O'Shea, D., Rea, S., O'Reilly, R., 2020. Incorporating explainable artificial intelligence (XAI) to aid the understanding of machine learning in the healthcare domain. In: Aics, pp. 169–180.

Rao, S., Riskowski, J.L., Hannan, M.T., 2012. Musculoskeletal conditions of the foot and ankle: assessments and treatment options. Best Pract. Res. Clin. Rheumatol. 26 (3), 345–368. https://doi.org/10.1016/j.berh.2012.05.009.

Roscher, R., Bohn, B., Duarte, M.F., Garcke, J., 2020. Explainable machine learning for scientific insights and discoveries. IEEE Access 8, 42200–42216.

Saraswat, D., Bhattacharya, P., Verma, A., Prasad, V.K., Tanwar, S., Sharma, G., Bokoro, P.N., Sharma, R., 2022. Explainable AI for healthcare 5.0: opportunities and challenges. IEEE Access 10, 84486–84517.

Seeliger, A., Pfaff, M., Krcmar, H., 2019. Semantic web technologies for explainable machine learning models: a literature review. PROFILES/SEMEX@ ISWC 2465, 1–16.

Stauffer, T.P., Kim, B.I., Grant, C., Adams, S.B., Anastasio, A.T., 2023. Robotic technology in foot and ankle surgery: a comprehensive review. Sensors 23 (2). https://doi.org/10.3390/s23020686.

Vaish, A., Migliorini, F., Vaishya, R., 2023. Artificial intelligence in foot and ankle surgery: current concepts. Orthopadie 52 (12), 1011–1016. https://doi.org/10.1007/s00132-023-04426-x.

CHAPTER

14 Conclusion: A future perspective on diagnosing musculoskeletal conditions using artificial intelligence and machine learning

Swetza Singh, Vamakshi Thaker and Shivam Verma
School of Pharmacy, BBD University, Lucknow, Uttar Pradesh, India

14.1 Introduction

A considerable financial burden is placed on society by the large number of patients claiming to attend hospitals for emergencies and seeking urgent care owing to musculoskeletal trauma. Medical imaging is necessary to ensure that those with trauma receive the best possible care and evaluation. However, because of this system's intricacy and numerous layers, there is a potential for human error to impair several features of the efficacy of medical imaging. Artificial intelligence (AI) and machine learning (ML), two technological developments, have the potential to change our health care systems fundamentally.

This overview represents the descriptive statement of ML and AI in different aspects of trauma imaging as well as a prediction of how these developments may be used to improve patient outcomes and diagnostic imaging systems. Orthopedics and traumatology are two areas in which AI and digitization are being applied (Shukla Shubhendu and Vijay, 2013).

Concerning clinical guidelines, diagnostic accuracy is increased using AI tools for osteoarthritis (OA) in the knee, and different bone and joint disorders are also easily traceable through radiological surveillance. AI enables judgments about the course of the disease and prognosis and the detection of early radiological symptoms. The workload of radiologists may be reduced and patient throughput can be raised by automating the process of creating reports. AI also decreases the influence of interrater variability in the assessment of radiographic x-ray morphologies and generates more uniform values in results.

Worldwide, musculoskeletal disorders (MSDs) account for most disabilities, with low back pain constituting the main cause of disability in the majority of nations. However, because of disparate case definitions that are divided based on body region-specific

Diagnosing Musculoskeletal Conditions using Artificial Intelligence and Machine Learning to Aid Interpretation of Clinical Imaging
https://doi.org/10.1016/B978-0-443-32892-3.00014-2

disease, there is still a dearth of comparative data. Although MSDs have historically been associated with physical stress, research indicates that inherited, congenital, or acquired pathologic processes may be to blame. When there is a documented reasonable relationship between physical and psychological circumstances and MSDs, the term work-related MSDs is employed (Luo and El Naqa, 2019).

MSDs are thought to affect 1.71 billion individuals globally, accounting for the majority of disability. Of 204 nations examined, low back pain is the leading cause of disability in 160. MSDs are an umbrella term encompassing numerous illnesses and injuries affecting muscles, tendons, ligaments, joints, and bones. MSDs can cause discomfort and dysfunction in the upper and lower limbs, the back, and the cervical area. In 2017, the prevalence of MSDs was greater in women than in men worldwide, and it increased gradually in older age groups with considerable intercountry variance (Russell and Bohannon, 2015).

MSDs have several underlying causes, including personal characteristics such as age, sex, anthropometry, and genetic background. Most noteworthy, though, is that several modifiable risk variables have been linked to the occurrence of MSDs. These include biomechanical overuse, exposure to environments and organizations, psychosocial risk factors, and other lifestyle diseases including smoking and obesity (Erickson et al., 2017).

John McCarthy coined the term artificial intelligence in 1956, whereas the term machine learning was given by Arthur Samuel in 1959. AI pertains to a domain within computer science dedicated to enabling computers to replicate human cognitive abilities. ML is a subset of AI that empowers computers to learn from data and recognize patterns without requiring explicit programming.

AI is increasingly being used in various domains including musculoskeletal radiology, skeletal trauma, orthopedic surgery, physical and rehabilitation medicine, and sports medicine (Román-Belmonte et al., 2023)

AI replicate human intelligence in tasks and is quickly growing within diagnostic imaging. This technology holds the promise of enhancing workflow efficiency in this field (D'Angelo et al., 2022).

14.2 Various musculoskeletal disorders

14.2.1 Neck and shoulders

An optimum cashier posture and workspace ergonomically are guided by a number of different factors that affect shoulder and neck postures, muscles, and ligament alignment topics that will be covered later in this chapter. When the cervical spine is in a straight and vertical position, the shoulder and neck muscular activity levels drop. However, when the trunk is slightly inclined backward and the neck is vertical, levels increase (Schüldt et al., 1986).

Furthermore, working with flexed and/or abducted shoulders increases activity in the cervical and thoracic erector spinal as well as the descending trapezius.

Most back and neck discomfort is brought on by problems with the spine's structure. A common problem, muscular pain is usually brought on by stimulation of the dorsal rami of the spinal nerve, which affects the deeper muscles, and a local reaction to the injury to the spine, which affects the superficial muscles. In the lumbar and cervical spines, strain is extremely rare. Although it is less common to produce back or neck discomfort on its own, fibromyalgia can coexist with these conditions. Sometimes extraspinal disease (especially gastrointestinal, genitourinary, or vascular) or herpes zoster refers to discomfort (Schüldt et al., 1986). Both spinal and extraspinal rare causes are dangerous. It is debatable whether any anatomic anomalies such as disk bulging or degeneration, osteophytes, spondylolysis, and facet abnormalities are the cause of pain because they are frequently observed in individuals who do not have neck or back discomfort. Back pain, especially mechanical back pain, has a complex etiology, with underlying disorders that can be made worse by physical deconditioning, fatigue, muscle pain, poor posture, weakness in stabilizing muscles, decreased flexibility, and occasionally psychosocial stress or psychiatric abnormalities. As a result, pinpointing a single cause is frequently challenging or impossible (Schüldt et al., 1987).

The symptoms of a widespread myofascial pain condition such as fibromyalgia often involves back and/or neck pain (Fig. 14.1).

14.2.2 Osteoarthritis

The most prevalent joint condition, OA, mostly affects the diarthrodial joints and is linked to a growing socioeconomic effect owing to the aging population. Many risk factors contribute to primary OA, the two most common of which are obesity and advancing age. Pathophysiology is still being understood; it has evolved from being thought of as a cartilage-limited disease to a multifactorial condition affecting the

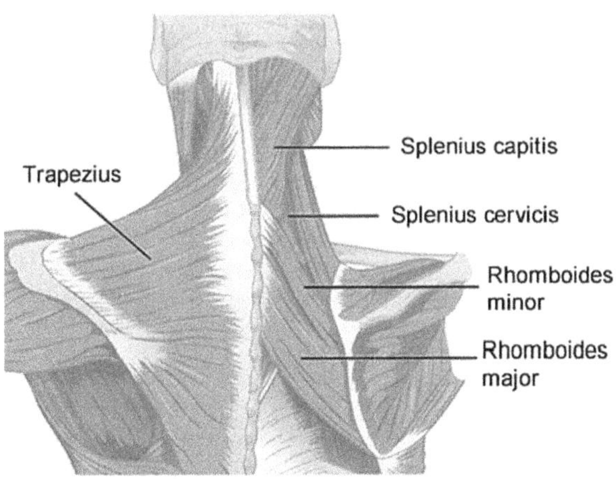

FIGURE 14.1

Neck and shoulder disorder (Lim et al., 2023).

entire joint. Convolutional neural network (CNN) model by training it with 420 Gy unit hip radiographs to identify hip OA through a binary classification system refined by Xue et al. (2017). They found that the model achieved a performance level comparable to that of a radiologist with a decade of experience (Lim et al., 2023).

Tiulpin and colleagues employed comprehensive datasets from the Osteoarthritis Initiative and Multicenter Osteoarthritis Study to construct and assess a CNN model aimed at automatically gauging the severity of knee OA using the Kellgren–Lawrence grading scale. Their results were promising, yielding an area under the curve (AUC) of 0.80 (Tiulpin et al., 2018).

14.2.3 Osteoporosis

Consider the term porous with osteoporosis. The definition of this term is porous bone. Under a microscope, the typical appearance of your bone is that of a hexagon. The spaces in the hexagon structure become abnormally large when osteoporosis develops. Your bones no longer have a typical density and your bones are more prone to breaking because they are weaker.

You may have experienced height loss and excruciating back pain if you have osteoporosis in the spinal cord. Osteoporosis disorders have been observed in over 50 million Americans owing to poor bone mass, because to the menopausal loss of estrogen. A shortage of estrogen leads to a decrease in osteoblasts, which produce new bone to replace old bones, and an increase in bone degradation (Zhang et al., 2024) (Fig. 14.2).

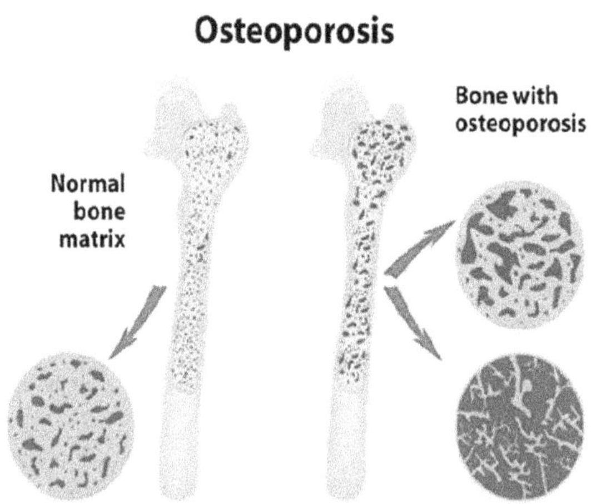

FIGURE 14.2

Condition of osteoporosis (Prevent broken bones).

14.2.4 Fragility of bones

The term fragility describes the sensitivity of bones to fractures or breaks resulting from a variety of causes, mostly linked to the density, strength, and structure of the bone. Bone fragility is caused by several circumstances.

Low bone density and bone tissue degeneration are the hallmarks of osteoporosis, a disorder that increases the risk of fractures, especially those in the wrist, hip, and backbone. Osteoporosis is more common in older persons, especially in women after menopause, although it can affect men as well.

- **Age:** Bone density tends to decline with age, and bones may become more brittle, raising the possibility of fractures.
- **Nutritional deficiencies:** Inadequate consumption of important nutrients, such as protein, calcium, and vitamin D, can deteriorate bone health and cause brittleness.
- **Hormonal factors:** Hormones are essential for preserving bone strength and density. Bone loss and fracturing can be brought on by hormonal fluctuations such as those caused by endocrine disorders or menopause.
- **Lifestyle factors:** A person's choice of lifestyle can affect bone health, including smoking, drinking too much alcohol, not exercising, and having a low body weight.
- **Genetic factors:** Bone fragility and disorders such as osteoporosis can be influenced by genetic predispositions that affect bone density and structure.
- **Medication:** A number of drugs, including corticosteroids and several anti-convulsants, have the potential to weaken bones and raise the risk of fractures.
- **Medical conditions:** Some illnesses, such as hyperparathyroidism, celiac disease, and rheumatoid arthritis, might have an impact on bone health and raise the possibility of fragility fractures (Yang et al., 2014).

Deep learning (DL) algorithms are developed to integrate bone mineral density (BMD) and get information combining multidetector computed tomography (CT) scan data on bone microarchitecture with dual x-ray absorptiometry (DXA). An analysis demonstrated that by incorporating trabecular bone characterization and ML techniques alongside conventional DXA BMD data, there was a significant improvement in predicting biomechanical strength (Roth et al., 2016).

14.2.5 Fractures

ML has been used to detect posterior element fractures and vertebral body compression fractures on CT scans. Roth et al. demonstrated the viability of employing a deep CNN to automate the detection of posterior element fractures, even with a modest dataset size. Their study yielded an AUC of 0.857, with sensitivities of 71% and 81% (Thodberg et al., 2008) (Fig. 14.3).

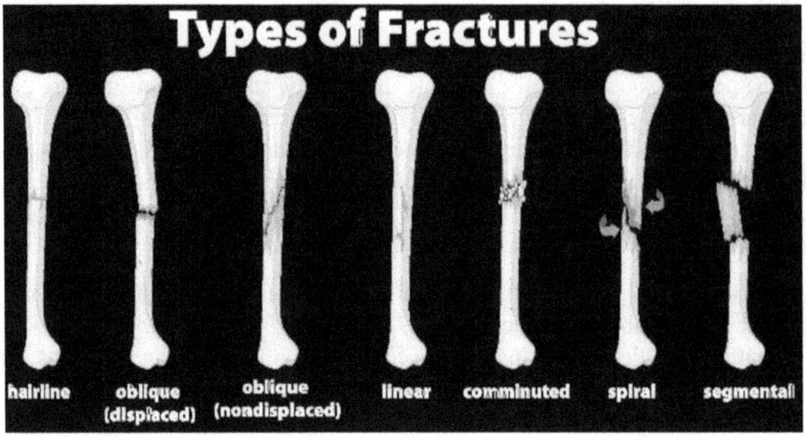

FIGURE 14.3

Types of fractures (Dev, 2023).

14.2.6 Determination of bone age

Accurate evaluation of bone age is important for various clinical domains including orthodontics and orthopedics. With progress in AI, numerous studies have explored AI-driven solutions for assessing bone age. A prominent example is BoneXpert, an automatic AI system introduced in 2008, which has gained widespread adoption, particularly in Europe. Using feature extraction, this system assesses bone age by analyzing radiographs of the left hand, focusing on 13 specific bones. This approach enhances time efficiency in routine clinical settings (Cui et al., 2022).

The integration of AI into assessing bone age proves to be a valuable resource for radiologists, lessening the workload and minimizing variability among observers. Furthermore, AI-supported interpretation of bone age can enhance accuracy, particularly benefiting less experienced readers (Jimenez-Pastor et al., 2020).

14.3 Spine imaging

When it comes to the diagnosis of different spinal abnormalities and disorders, spine magnetic resonance imaging (MRI) is essential. AI tools are being developed to help with interpreting and analyzing spine imaging owing to improvements in ML and AI. The following list of typical spine imaging methods and their applications for AI tools.

14.3.1 X-rays

X-rays are commonly used to visualize the bones in the spinal column and detect degenerative changes, fractures, and alignment issues. AI algorithms can automate the diagnosis of anomalies such as alignment problems, vertebral fractures, and deformities of the spine.

14.3.2 Computed tomography

CT scans are an important tool in evaluating bones and the spine, and diagnosing many conditions such as tumors, disc herniations, and spinal cord injuries. They give detailed information about the shape of the spine. Radiologists can use AI to detect bone density, detect abnormalities in CT scans, and map the anatomy.

14.3.3 Magnetic resonance imaging

MRI is a valuable method for evaluating soft tissues such as spinal discs, nerves, and the spinal cord. AI methods can assist in segmenting and examining spinal structures, detecting issues such as spinal tumors, disc herniations, and spinal cord compression. They offer quantitative evaluations of features such as spinal curvature and disc degeneration.

14.3.4 Functional magnetic resonance imaging

Disorders such as multiple sclerosis and spinal cord damage can be diagnosed by functional MRI (fMRI) because it can measure blood flow and function. fMRI data can be easily analyzed using AI to identify and anticipate changes in the spinal cord.

14.3.5 Positron emission tomography

Positron emission tomography (PET) scans are useful for detecting spinal tumors and metastases. AI technology can distinguish between benign and malignant spinal lesions by examining metabolic activity patterns and can analyze PET images to identify potentially dangerous lesions.

AI technologies for interpreting spine images can help radiologists identify subtle abnormalities and make better therapeutic judgments. They can also expedite the workflow and increase diagnostic accuracy. Before they are used in clinical settings, these AI algorithms must be thoroughly tested and shown through clinical studies to be safe, effective, and reliable. Several studies have showcased AI tools able to identify and label spinal vertebrae and intervertebral discs accurately on MRI and CT scans, eliminating the need for manual labeling by humans and simplifying image review processes. These models have potential for automating additional tasks (Jimenez-Pastor et al., 2020; Lessmann et al., 2019; Wimmer et al., 2018; Stinner and Edwards, 2017).

14.4 Musculoskeletal diagnosis

Trauma is among the primary reasons why people attend the emergency room. It has a substantial financial impact on society. Half of the overall costs associated with nonfatal injuries are attributed to injuries to the extremities. It places a heavy financial strain on health care systems and can lead to increased rates of morbidity and death if a diagnosis is made slowly, or not at all. For the diagnosis of various illnesses, fractures, musculoskeletal injuries, bone tumors, and infections, radiology is an essential diagnostic technique (Stinner and Edwards, 2017; Laur and Wang, 2022).

AI has been applied to radiography, CT, and MRI, among other imaging modalities. Deidentified radiography datasets such as musculoskeletal radiographs (MURA) have been created to improve case compilation and algorithm development. About 41,000 upper extremity-focused radiographs make up MURA. Professional radiologists classify each as normal or abnormal (Pankhania, 2020). Many AI techniques have been used to identify and categorize fractures. A deep network of neural networks was developed by Lindsey and colleagues to locate and identify fractures on x-ray images. AI-assisted emergency medicine professionals in identifying fractures with more accuracy as a result of this innovation (Rajpurkar et al., 2017).

Gale et al. (2017) created a DenseNet-based structure to identify hip fractures in frontal pelvic x-rays. This architecture achieved a level of performance comparable to that of a human radiologist.

14.5 Musculoskeletal oncology

According to early articles on the topic by Lodwick et al., AI has been assisting in identifying primary bone tumors since the 1960s (Lindsey et al., 2018). He et al. used an Inception V3 CNN to forecast the local recurrence of a giant cell tumor of the bone using magnetic resonance and clinical data, demonstrating that AI is also able to predict tumor recurrence (Gale et al., 2017).

In a similar vein, AI might aid in identifying soft tissue tumors. Malinauskaite et al., for instance, published an ML classifier that distinguished between lipoma and liposarcoma on MRI scans using radiomic features. This classifier performed better than three musculoskeletal radiologists, with an AUC of 0.926 (Lodwick et al., 1963; He et al., 2019; Chen et al., 2009).

Furthermore, with an AUC of up to 0.92, it is feasible to predict the histological grade of soft tissue sarcomas using radiometric characteristics and ML classification techniques. MRI scans used to track posttreatment changes in soft tissue sarcoma can also benefit from ML approaches (Malinauskaite et al., 2020; Zhang et al., 2019a).

14.6 Detection of anterior cruciate ligament and meniscal structures

Using a dataset of 1130 MRI images for training and 120 for validation, Bien and colleagues created a CNN model targeted at identifying meniscus and anterior cruciate ligament (ACL) injuries. The model's AUC for meniscal tears was 0.847 in internal validation, 0.937 for ACL tears, and 0.824 in the external validation collection (Blackledge et al., 2019).

Using endoscopy as a benchmark, researchers led by Liu trained a CNN to detect ACL lacerations, achieving an AUC of 0.98 and a sensitivity of 96%. Similarly, Ma et al. developed a CNN for meniscal injury diagnosis that performed as well as a chief physician, identifying anterior horn injuries with an accuracy of 85.6% and posterior horn injuries at 92% (Bien et al., 2018; Liu et al., 2019; Ma et al., 2023).

Chang et al. showed that by isolating particular joint structures, the model performed better. They achieved an AUC of 0.97 and a sensitivity of 100% using a CNN U-Net to isolate the ACL on coronal proton density two-dimensional (2D) MRI, followed by a CNN classifier to assess isolated ACLs for pathology. Using fat-eliminated MRI sequences, Pedoia et al. examined a CNN model for meniscal segmentation and found that it could achieve up to 90% sensitivity, 82% specificity, and an AUC of 0.89 (Couteaux et al., 2019).

14.7 Spinal structure

Numerous studies were undertaken to assess the extent of scoliosis and various other calibrations related to the spine. Researchers evaluated the capacity of AI to forecast scoliosis progression, evaluate the higher stage, identify indications of scoliosis treatment using radiography, and generate 3D spine reconstructions from biplanar images (Pedoia et al., 2019; Garcia-Cano et al., 2018; Kaddioui et al., 2020; Vergari et al., 2020).

Studies explored additional future deployments of AI, including joint space assessment in rheumatoid arthritis (Aubert et al., 2019; Zheng et al., 2020), detection of epidural masses on CT scans, identification of rotator cuff pathology, detection of femoroacetabular impingement, and hip dysplasia (Bhalodia et al., 2020).

Research efforts are ongoing; in coming years there may be further attempts to automate repetitive and time-consuming tasks to expedite the process of imaging interpretation (Ha et al., 2020; Zhang et al., 2019b).

14.8 Other potential pediatric musculoskeletal applications

14.8.1 Hip dysplasia during development

Several studies are related to the diagnosis of hip dysplasia in children, assessed by AI using neural networks (Yune et al., 2019; Offiah, 2022).

Li et al. used 11,473 anteroposterior pelvic radiographs for training and 101 images for testing to diagnose dysplastic transformation of the hip using the Sharp angle, also known as the acetabular index. Their study revealed an accuracy comparable to that of orthopedic surgeons but with reduced time requirements. Consequently, they suggested that the AI tool had the potential to substitute for orthopedic surgeons in this context (Pfeil et al., 2013).

Zhang et al. (2019a,b) employed 9081 anteroposterior pelvic radiographs for training and an additional 1138 anteroposterior pelvic radiographs for testing to diagnose developmental hip dysplasia using the acetabular index. The researchers decided that the DL system enhanced uniformity, ease, and efficacy compared with the diagnosis led by clinicians, and proposed that it could streamline existing screening processes (Liu et al., 2014; Stotter et al., 2023) (Tables 14.1 and 14.2).

Table 14.1 Artificial intelligence tools used in musculoskeletal imaging.

S. No.	Application in musculoskeletal diagnosis	Algorithm type	References
1.	Identification of hip fractures.	Convolutional neural network (CNN)	Yu et al. (2020)
2.	Detection of hip osteoarthritis	CNN	Xue et al. (2017)
3.	Identification of bone age	Google Inception V3 network	Halabi et al. (2019)
4.	Screening for osteoporosis through the utilization of low-dose chest computed tomography scans	Three-dimensional U-net CNN	Pan et al. (2020)
5.	Prediction of bone strength	Scaling index method	Yang et al. (2014)
6.	Segmentation of spinal structure	Generative adversarial network	Han et al. (2018)
7.	Segmentation knee cartilage and bone	CNN	Liu et al. (2019)

Table 14.2 Clinical responsibilities accomplished via artificial intelligence systems in musculoskeletal imaging.

S. No.	Medical term	Clinical problem	Diagnostic technique	References
1.	Trauma	Detection of fractures around hip and spine	Radiography/ computed tomography	Chen et al. (2009), Gale et al. (2017), Jones et al. (2020)
2.	Osteoarthritis	Grading of knee osteoarthritis	Radiography	Tiulpin et al. (2018)
3.	Bone tumor	Osteosarcoma and soft tissue sarcoma	Computed tomography/ magnetic resonance imaging	Fritz et al. (2018), Gitto et al. (2021), Peeken et al. (2019)
4.	Orthopedic implant	Knee or hip arthroplasty loosening	Radiography/ magnetic resonance imaging	Shah et al. (2020)

14.9 Technical aspects

14.9.1 Artificial intelligence/machine learning and deep learning

John McCarthy coined the term artificial intelligence in 1956, and Arthur Samuel introduced the term machine learning in 1959 (Russell and Bohannon, 2015). AI replicates human cognitive abilities whereas ML enables computers to learn from data without explicit programming. AI is used in various fields such as musculoskeletal radiology and orthopedic surgery, enhancing workflow efficiency in diagnostic imaging (Etli et al., 2019; Kapiński et al., 2019). ML includes supervised and unsupervised learning methods (Debs and Fayad, 2023; Li et al., 2019). DL, a branch of ML, trains deep neural networks to solve tasks without explicit programming. It excels in tasks such as image recognition and natural language processing (Yu et al., 2020).

Feedforward neural networks process data unidirectionally from input to output through hidden layers using weighted connections and activation functions for node output. They are used for classification and regression tasks (Halabi et al., 2019).

CNNs specialize in grid-like data such as images using convolutional layers to learn hierarchical features automatically. They excel in tasks such as picture classification, object detection, and image segmentation owing to their ability to capture spatial hierarchies (Chen et al., 2021).

Recurrent neural networks are suited for sequential data, maintaining information persistence through directed loops. They find applications in natural language processing, speech recognition, and time series analysis. Advanced designs such as long short-term memory (LSTM) and gated recurrent unit (GRU) address issues with capturing long-term dependencies (Liu et al., 2018).

Generative adversarial networks (GANs) employ adversarial training to train a discriminator and a generator simultaneously. The discriminator distinguishes real from synthetic samples, whereas the generator creates artificial data resembling real data. GANs are used for tasks such as image generation, style transfer, and data augmentation.

Autoencoders, designed for unsupervised learning, compress input data into a lower-dimensional encoding using an encoder and reconstruct the original data using a decoder. They perform functions such as data denoising, dimensionality reduction, and anomaly detection (Burns et al., 2017).

Deep belief networks are probabilistic models with multiple layers of stochastic and latent variables, blending neural networks and probabilistic graphical model principles for flexible feature learning and representation.

Different neural network topologies, including artificial neural networks (ANNs) and CNNs, are crucial for addressing complex problems. ANNs simulate the structure and function of the human brain whereas CNNs excel at analyzing grid-like data such as images, enabling tasks such as image recognition and natural language processing (Jones et al., 2020) (Fig. 14.4).

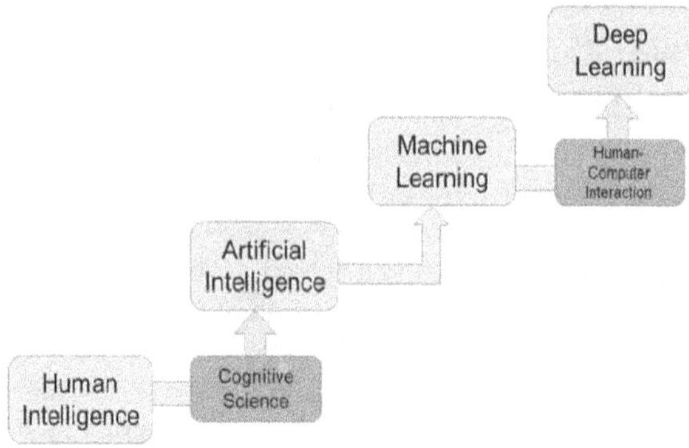

FIGURE 14.4

Singularity of human intelligence to deep learning (Kasthuri, 2023).

14.9.2 Other artificial intelligence–based tools for musculoskeletal diagnosis

1. To detect vertebral fractures in abdomen radiographs, Chen et al. (2009) applied the Residual Neural Network (ResNet) architecture, with an average accuracy of 73%. This performance lagged behind the 76.8% accuracy of radiologists and orthopedic surgeons (Chen et al., 2009). As an example, their research emphasized using the ResNeXt architecture in radiographic image-based fracture detection for musculoskeletal diagnostics (Kasthuri, 2023).
2. A DL model designed specifically for MRI analysis and able to recognize meniscal tears and ACL injuries was presented by Bien et al. (2018). Despite this, the model performed worse than radiologists' interpretations in terms of sensitivity for ACL tears and specificity for meniscal tears (Bien et al., 2018). This highlights the difficulties of reaching parity in subtle musculoskeletal imaging interpretation tasks with human experts (Gitto et al., 2021).
3. To improve sensitivity in identifying ACL rupture, Liu et al. (2019) developed a DL model that performed on par with radiologists. They created a cartilage lesion-focused model that showed strong diagnostic abilities for acute injuries and cartilage degradation (Liu et al., 2019). Their findings highlight the potential of DL to solve particular problems in diagnosing MSD, including the identification of tiny cartilage lesions and tears in ligaments (Peeken et al., 2019).
4. Using large datasets, Tiulpin et al. (2018) created a CNN model to determine the degree of OA in the knee automatically using the Kellgren–Lawrence grading system. With an AUC of 0.80, their findings showed promise and demonstrated CNNs' potential for objectively assessing the severity of OA (Shah et al., 2020).

5. Using support vector machine regression, Burns et al. (2017) designed an automated ML system that could recognize and classify lumbar and thoracic spinal compression fractures. For fracture identification, they attained a high sensitivity of 95.7% and a low error rate of 0.29 per individuals (Burns et al., 2017). This work is an excellent example of how ML may be used to identify spinal fractures accurately and quickly (Langs et al., 2007).
6. Yang et al. (2014) combined BMD values from dual-energy bone density measurements with bone microarchitecture information from CT images to forecast proximal femur failure loads using DL algorithms. According to Yang et al. (2014), their work showed a notable improvement in biomechanical strength prediction, highlighting the potential of ML to advance musculoskeletal biomechanics research (Pan et al., 2020).
7. Using MRI data, Huber et al. (2010) forecasted trabecular bone strength in the proximal tibia using ML techniques. According to their findings, MRI bone quality evaluation can be improved by combining ML approaches with data on the bone structure (Huber et al., 2010). This demonstrates how important ML is for deriving insights from intricate musculoskeletal imaging modalities (Shah et al., 2020; Han et al., 2018).
8. Using a CNN, Jamaludin et al. (2017) successfully identified vertebrae bodies and diffuse vertebrae bodies on MRI scans with 95.6% accuracy. Their model performed comparably to work by radiologists in evaluating different spinal findings (Jamaludin et al., 2017). This highlights how CNNs can be used to automate and standardize operations related to the interpretation of musculoskeletal images (Huber et al., 2010).
9. Li et al. used radiography and the Sharp angle, also known as the acetabular index, to identify hip developmental dysplasia with an accuracy on par with that of orthopedic surgeons. According to the work of Li et al from the previous year, AI technology may be able to speed and standardize the assessment of hip dysplasia (Jamaludin et al., 2017; Li and Ganz, 2003).
10. Using acetabular morphology on radiographs, Zhang et al. (2019a,b) used a DL technique to identify developing hip dysplasia, showing advantages in consistency, convenience, and efficacy over clinician-led diagnosis (Zhang et al., 2019a,b). This demonstrates how DL may improve MSD screening protocols (Krizhevsky et al., 2012).

14.10 Discussion and conclusion

Emerging AI shows the capability to change the paradigm in MSD care from understanding biological pathways to forecasting disease risk. The shift toward predictive medicine signifies a fundamental change in health care approaches. The various algorithms may also make it easier to obtain and analyze radiological images, offer details on surgical procedures, support physicians in making decisions, and improve patient education. AI has gradually advanced musculoskeletal imaging technology to improve the visualization of anatomical structures and automate quantitative assessments. Research on AI-based musculoskeletal ultrasound (US) imaging has raised the possibility that DL methods may be used as next-generation diagnostic instruments to track the health of ligaments, muscles, joints, and cartilage. Real-time diagnostics and help in making decisions for sonographers might be possible using DL's picture recognition capacity. AI-based musculoskeletal imaging may enable better patient treatment by delivering high-quality grayscale pictures, evaluating the suitability of US images, and guaranteeing consistency.

Clinicians are essential to patient care because they provide knowledge, experiences, competence, and humanity. An algorithm cannot replace their job. When integrating AI technologies into care operations, caution and moderation are required. Every link in the imaging value chain might be significantly improved by applying AI. AI can enhance the value that musculoskeletal imagers provide to their patients and to referring clinicians by enhancing image quality, patient centricity, imaging efficiency, and diagnostic accuracy. This includes evaluating the suitability of imaging orders and assisting in the prediction of patients at risk for fracture.

Several AI techniques used in musculoskeletal radiology, such as algorithms for OA quantification, fracture diagnosis, and bone age estimation, are commercially accessible. It will be essential for radiologists to be involved in selecting and implementing AI solutions as the number of these grows. Therefore, rather than taking the role of radiologists, AI may be used to improve workflow, advance diagnostic performance, and assist radiologists in keeping up with an ever-growing workload. This also means that a human authority, the radiologist, will eventually bear legal responsibility, and they ought to accept it.

References

Aubert, B., Vazquez, C., Cresson, T., Parent, S., de Guise, J.A., 2019. Toward automated 3D spine reconstruction from biplanar radiographs using CNN for statistical spine model fitting. IEEE Trans. Med. Imag. 38 (12), 2796–2806.

Bhalodia, R., Dvoracek, L.A., Ayyash, A.M., Kavan, L., Whitaker, R., Goldstein, J.A., 2020. Quantifying the severity of metopic craniosynostosis: a pilot study application of machine learning in craniofacial surgery. J. Craniofac. Surg. 31 (3), 697.

Bien, N., et al., 2018. Deep-learning-assisted diagnosis for knee magnetic resonance imaging: development and retrospective validation of MRNet. PLoS Med. 15 (11), e1002699.

Blackledge, M.D., et al., 2019. Supervised machine-learning enables segmentation and evaluation of heterogeneous post-treatment changes in multi-parametric MRI of soft-tissue sarcoma. Front. Oncol. 9, 941.

Burns, J.E., Yao, J., Summers, R.M., 2017. Vertebral body compression fractures and bone density: automated detection and classification on CT images. Radiology 284 (3), 788–797.

Chen, C.-Y., et al., 2009. Computer-aided diagnosis of soft-tissue tumors using sonographic morphologic and texture features. Acad. Radiol. 16 (12), 1531–1538.

Chen, H.-Y., et al., 2021. Application of deep learning algorithm to detect and visualize vertebral fractures on plain frontal radiographs. PLoS One 16 (1), e0245992.

Couteaux, V., et al., 2019. Automatic knee meniscus tear detection and orientation classification with Mask-RCNN. Diagnostic Interventional Imaging 100 (4), 235–242.

Cui, Y., Zhu, J., Duan, Z., Liao, Z., Wang, S., Liu, W., 2022. Artificial intelligence in spinal imaging: current status and future directions. Int. J. Environ. Res. Publ. Health 19 (18), 11708.

D'Angelo, T., et al., 2022. Artificial intelligence, machine learning and deep learning in musculoskeletal imaging: current applications. J. Clin. Ultrasound 50 (9), 1414–1431.

Debs, P., Fayad, L.M., 2023. The promise and limitations of artificial intelligence in musculoskeletal imaging. Front. Radiol. 3.

Dev, D., October 26, 2023. Types of fractures - the things you need to know | OPA. OPA Ortho. https://www.opaortho.com/types-of-bone-fractures/.

Erickson, B.J., Korfiatis, P., Akkus, Z., Kline, T.L., 2017. Machine learning for medical imaging. Radiographics 37 (2), 505–515.

Etli, Y., Asirdizer, M., Hekimoglu, Y., Keskin, S., Yavuz, A., 2019. Sex estimation from sacrum and coccyx with discriminant analyses and neural networks in an equally distributed population by age and sex. Forensic Sci. Int. 303, 109955.

Gale, W., Oakden-Rayner, L., Carneiro, G., Bradley, A.P., Palmer, L.J., 2017. Detecting hip fractures with radiologist-level performance using deep neural networks. arXiv preprint arXiv:1711.06504.

Garcia-Cano, E., et al., 2018. Prediction of spinal curve progression in adolescent idiopathic scoliosis using random forest regression. Comput. Biol. Med. 103, 34–43.

Gitto, S., et al., 2021. CT radiomics-based machine learning classification of atypical cartilaginous tumours and appendicular chondrosarcomas. EBioMedicine 68.

Ha, C.W., Kim, S.H., Lee, D.H., Kim, H., Park, Y.B., 2020. Predictive validity of radiographic signs of complete discoid lateral meniscus in children using machine learning techniques. J. Orthop. Res. 38 (6), 1279–1288.

Halabi, S.S., et al., 2019. The RSNA pediatric bone age machine learning challenge. Radiology 290 (2), 498–503.

Han, Z., Wei, B., Mercado, A., Leung, S., Li, S., 2018. Spine-GAN: semantic segmentation of multiple spinal structures. Med. Image Anal. 50, 23–35.

He, Y., et al., 2019. Convolutional neural network to predict the local recurrence of giant cell tumor of bone after curettage based on pre-surgery magnetic resonance images. Eur. Radiol. 29, 5441–5451.

Huber, M.B., Lancianese, S.L., Ikpot, I., Nagarajan, M.B., Lerner, A.L., Wismüller, A., 2010. Prediction of biomechanical trabecular bone properties with geometric features using MR imaging. Medical Imaging 2010: Biomedical Applications in Molecular, Structural, and Functional Imaging 7626: SPIE, 501–508.

Jamaludin, A., et al., 2017. ISSLS prize in bioengineering science 2017: automation of reading of radiological features from magnetic resonance images (MRIs) of the lumbar spine without human intervention is comparable with an expert radiologist. Eur. Spine J. 26, 1374–1383.

Jimenez-Pastor, A., et al., 2020. Automated vertebrae localization and identification by decision forests and image-based refinement on real-world CT data. La Radiologia Medica 125, 48–56.

Jones, R.M., et al., 2020. Assessment of a deep-learning system for fracture detection in musculoskeletal radiographs. NPJ Digit. Med. 3 (1), 144.

Kaddioui, H., et al., 2020. Convolutional neural networks for automatic risser stage assessment. Radiol. Artif. Intell. 2 (3), e180063.

Kapiński, N., et al., 2019. Monitoring of the Achilles tendon healing process: can artificial intelligence be helpful? Acta Bioeng. Biomech. 21 (1), 103–111.

Kasthuri, M., December 5, 2023. A quick look at new age trends in AI. Open Source for You. https://www.opensourceforu.com/2023/12/a-quick-look-at-new-age-trends-in-ai/.

Krizhevsky, I.S., Hinton, G.E., 2012. Imagenet classification with deep convolutional neural networks. Adv. Neural Inf. Process. Syst. 25.

Langs, G., Peloschek, P., Bischof, H., Kainberger, F., 2007. Model-based erosion spotting and visualization in rheumatoid arthritis. Acad. Radiol. 14 (10), 1179–1188.

Laur, O., Wang, B., 2022. Musculoskeletal trauma and artifcial intelligence: current trends and projections. Skeletal Radiol. 51, 257–269. https://doi.org/10.1007/s00256-021-03824-6.

Lessmann, N., Van Ginneken, B., De Jong, P.A., Išgum, I., 2019. Iterative fully convolutional neural networks for automatic vertebra segmentation and identification. Med. Image Anal. 53, 142–155.

Li, P.L., Ganz, R., 2003. Morphologic features of congenital acetabular dysplasia: one in six is retroverted. Clin. Orthop. Relat. Res. 416, 245–253.

Li, Q., et al., 2019. Auxiliary diagnosis of developmental dysplasia of the hip by automated detection of Sharp's angle on standardized anteroposterior pelvic radiographs. Medicine 98 (52).

Lim, J., Lee, D., Kim, S., Lee, S., Ryu, J.S., August 22, 2023. Analysis of abnormal muscle activities in patients with loss of cervical lordosis: a cross-sectional study. BMC Muscoskel. Disord. 24 (1). https://doi.org/10.1186/s12891-023-06782-3.

Lindsey, R., et al., 2018. Deep neural network improves fracture detection by clinicians. Proc. Natl. Acad. Sci. U. S. A. 115 (45), 11591–11596.

Liu, J., et al., 2014. Computer aided detection of epidural masses on computed tomography scans. Comput. Med. Imag. Graph. 38 (7), 606–612.

Liu, F., et al., 2018. Deep learning approach for evaluating knee MR images: achieving high diagnostic performance for cartilage lesion detection. Radiology 289 (1), 160–169.

Liu, F., et al., 2019. Fully automated diagnosis of anterior cruciate ligament tears on knee MR images by using deep learning. Radiol. Artif. Intell. 1 (3), 180091.

Lodwick, G.S., Haun, C.L., Smith, W.E., Keller, R.F., Robertson, E.D., 1963. Computer diagnosis of primary bone tumors: a preliminary report. Radiology 80 (2), 273–275.

Luo, Y., El Naqa, I., 2019. Machine learning for radiation oncology. Big Data in Radiation Oncology 41–59.

Ma, Y., et al., 2023. Visual cascaded-progressive convolutional neural network (C-PCNN) for diagnosis of meniscus injury. Diagnostics 13 (12), 2049.

Malinauskaite, I., et al., 2020. Radiomics and machine learning differentiate soft-tissue lipoma and liposarcoma better than musculoskeletal radiologists. Sarcoma 2020.

Offiah, A.C., 2022. Current and emerging artificial intelligence applications for pediatric musculoskeletal radiology. Pediatr. Radiol. 52 (11), 2149–2158.

Osteoarthritis | Symptoms, Complications & Treatment. (January 26, 2024). Drugwatch.com. https://www.drugwatch.com/health/arthritis/osteoarthritis/.

Pan, Y., et al., 2020. Automatic opportunistic osteoporosis screening using low-dose chest computed tomography scans obtained for lung cancer screening. Eur. Radiol. 30, 4107–4116.

Pankhania, M., 2020. Artificial intelligence in musculoskeletal radiology: past, present, and future. Ind. J. Musculoskeletal Radiol. 2 (89), 89–96.

Pedoia, V., Norman, B., Mehany, S.N., Bucknor, M.D., Link, T.M., Majumdar, S., 2019. 3D convolutional neural networks for detection and severity staging of meniscus and PFJ cartilage morphological degenerative changes in osteoarthritis and anterior cruciate ligament subjects. J. Magn. Reson. Imag. 49 (2), 400–410.

Peeken, J.C., et al., 2019. Tumor grading of soft tissue sarcomas using MRI-based radiomics. EBioMedicine 48, 332–340.

Pfeil, A., et al., 2013. The usefulness of computer-aided joint space analysis in the assessment of rheumatoid arthritis. Joint Bone Spine 80 (4), 380–385.

Prevent Broken Bones. (n.d.). https://www.osmifw.com/sports-medicine/prevent-broken-bones/.

Rajpurkar, P., et.al., 2017. Mura: Large Dataset for Abnormality Detection in Musculoskeletal Radiographs. arXiv preprint arXiv:1712.06957.

Román-Belmonte, J.M., De la Corte-Rodríguez, H., Rodríguez-Damiani, B.A., Rodríguez-Merchán, E.C., 2023. Artificial Intelligence in Musculoskeletal Conditions.

Roth, H.R., Wang, Y., Yao, J., Lu, L., Burns, J.E., Summers, R.M., 2016. Deep convolutional networks for automated detection of posterior-element fractures on spine CT. Medical Imaging 2016: Computer-Aided Diagnosis 9785: SPIE, 165–171.

Russell, S., Bohannon, J., 2015. Artificial intelligence. Fears of an AI pioneer. Science (New York, NY) 349 (6245), 252.

Schüldt, K., Ekholm, J., Harms-Ringdahl, K., Németh, G., Arborelius, U.P., 1986. Effects of changes in sitting work posture on static neck and shoulder muscle activity. Ergonomics 29 (12), 1525–1537.

Schüldt, K., Ekholm, J., Harms-Ringdahl, K., Arborelius, U.P., Németh, G., 1987. Influence of sitting postures on neck and shoulder emg during arm-hand work movements. Clin. Bio-Mech. 2 (3), 126–139.

Shah, R.F., Bini, S.A., Martinez, A.M., Pedoia, V., Vail, T.P., 2020. Incremental inputs improve the automated detection of implant loosening using machine-learning algorithms. Bone Joint J. 102 (6 Suppl. A), 101–106.

Should You Exercise with Osteoarthritis? (October 21, 2019). News-Medical. https://www.news-medical.net/health/Should-You-Exercise-With-Osteoarthritis.aspx.

Shukla Shubhendu, S., Vijay, J., 2013. Applicability of artificial intelligence in different fields of life. Int. J. Sci. Eng. Res. 1 (1), 28–35.

Stinner, D.J., Edwards, D., 2017. Surgical management of musculoskeletal trauma. Surgical Clinics 97 (5), 1119–1131.

Stotter, C., et al., 2023. Deep learning for fully automated radiographic measurements of the pelvis and hip. Diagnostics 13 (3), 497.

Thodberg, H.H., Kreiborg, S., Juul, A., Pedersen, K.D., 2008. The BoneXpert method for automated determination of skeletal maturity. IEEE Trans. Med. Imag. 28 (1), 52–66.

Tiulpin, A., Thevenot, J., Rahtu, E., Lehenkari, P., Saarakkala, S., 2018. Automatic knee osteoarthritis diagnosis from plain radiographs: a deep learning-based approach. Sci. Rep. 8 (1), 1727.

Vergari, C., Skalli, W., Gajny, L., 2020. A convolutional neural network to detect scoliosis treatment in radiographs. Int. J. Comput. Assist. Radiol. Surg. 15, 1069–1074.

Wimmer, M., Major, D., Novikov, A.A., Bühler, K., 2018. Fully automatic cross-modality localization and labeling of vertebral bodies and intervertebral discs in 3D spinal images. Int. J. Comput. Assist. Radiol. Surg. 13, 1591–1603.

Yang, C.-C., et al., 2014. Improving bone strength prediction in human proximal femur specimens through geometrical characterization of trabecular bone microarchitecture and support vector regression. J. Electron. Imag. 23 (1), 013013.

Yu, J., et al., 2020. Detection and localisation of hip fractures on anteroposterior radiographs with artificial intelligence: proof of concept. Clin. Radiol. 75 (3), 237.e1–237.e9.

Yune, S., Lee, H., Kim, M., Tajmir, S.H., Gee, M.S., Do, S., 2019. Beyond human perception: sexual dimorphism in hand and wrist radiographs is discernible by a deep learning model. J. Digit. Imag. 32, 665–671.

Zhang, Y., et al., 2019a. Soft tissue sarcomas: preoperative predictive histopathological grading based on radiomics of MRI. Acad. Radiol. 26 (9), 1262–1268.

Zhang, C., et al., 2019b. Statistical models to assess leg muscle mass in ambulatory children with spastic cerebral palsy using dual-energy X-ray absorptiometry. J. Clin. Densitom. 22 (3), 391–400.

Zhang, M., et al., 2024. Research advances of nanomaterials for the acceleration of fracture healing. Bioact. Mater. 31, 368–394.

Zheng, Q., Shellikeri, S., Huang, H., Hwang, M., Sze, R.W., 2020. Deep learning measurement of leg length discrepancy in children based on radiographs. Radiology 296 (1), 152–158.

Index

'*Note:* Page numbers followed by "f" indicate figures and "t" indicate tables.'

A

B

C

D

E

H

I

J

K

L

N

O

P

www.ingramcontent.com/pod-product-compliance
Lightning Source LLC
Chambersburg PA
CBHW080030150125
20411CB00010B/493